AF245544

Stem Cells in Clinical Applications

Series Editor
Phuc Van Pham

More information about this series at http://www.springer.com/series/14002

Phuc Van Pham • Achim Rosemann

Editors

Safety, Ethics and Regulations

 Springer

Editors
Phuc Van Pham
Stem Cell Institute
University of Science
Vietnam National University
Ho Chi Minh City, Vietnam

Achim Rosemann
Centre for Education Studies
Faculty of Social Sciences
University of Warwick
Coventry, UK

ISSN 2365-4198   ISSN 2365-4201   (electronic)
Stem Cells in Clinical Applications
ISBN 978-3-319-59164-3   ISBN 978-3-319-59165-0   (eBook)
DOI 10.1007/978-3-319-59165-0

Library of Congress Control Number: 2017945217

Printed on acid-free paper

This Springer imprint is published by Springer Nature
The registered company is Springer International Publishing AG
The registered company address is: Gewerbestrasse 11, 6330 Cham, Switzerland

Preface

Stem cell research and therapy herald a new era of medicine, called regenerative medicine. Stem cell transplantation brings some benefits for age-related degenerative diseases as well as genetic diseases. Although stem cell transplantation has a long history which is more than 50 years old, it still faces some safety, ethical, and regulatory issues. This volume, *Stem Cells in Clinical Applications: Safety, Ethics, and Regulations*, provides safety evaluations of stem cell treatments for some diseases and the ethical and regulatory dimensions of stem cell-based clinical applications in different countries.

The four chapters in Part I provide an introduction to the safety of stem cell transplantation. In chapter one, Gero Hütter discusses the safety of allogenic stem cell transplantation. Chapter two, by Erden Eren and colleagues introduces safety issues when using induced pluripotent stem cells in the treatment of neurodegenerative disease. Chapter three, written by Carlo S. Jackson, Marco Alessandrini and Michael S. Pepper provides an introduction to safety concerns of stem cell gene therapy. Finally, in chapter four Dimitrios Kouroupis and colleagues record the safety of non-expanded stem cells in clinical applications.

The 12 chapters in Part II address the ethical and regulatory dimensions of stem cell-based clinical applications. In chapter five, Fikile M. Mnisi explores the ethical controversies on the patenting of human embryonic stem cells in South Africa. Chapter six, by John D. Banja provides an overview of the ethical considerations in clinical stem cell research for neurological and orthopedic conditions. Chapter seven, by Barbara von Tigerstrom assesses current and emerging regulatory models for clinical stem cell research in the USA, the EU, Japan, and Australia. In chapter eight, Christine Hauskeller and Nicole Baur study the regulatory conditions for clinical stem cell research in the European Union and comment on the practical challenges for multi-country stem cell trials in this global region. In chapter nine, Tamra Lysaght examines differences in the framing of ethical concerns in professional guidelines by the International Society for Stem Cell Research (ISSCR) and the International Cellular Medicine Society (ICMS). Chapter ten, by Li Jiang focuses on the regulatory and legal situation for stem cell research in China. Jiang shows how a booming stem cell industry in China is, through an ongoing process of regulatory reform, slowly brought under the control of the state. In chapter eleven, Shashank S. Tiwari, Paul Martin, and Sujatha Raman provide insights into regulatory developments in the governance of stem cell therapies in India, which are discussed in the light of the country's social and health-care context. In chapter twelve, Iñigo de Miguel Beriain offers a detailed analysis of the ethical and legal conflicts and positions surrounding the patenting of hESC in the context of the European Union. In chapter thirteen Achim Rosemann and colleagues provide an overview of the regulatory conditions for basic, preclinical, and clinical research in China that have emerged since the early 2000s. Chapter fourteen, by Achim Rosemann addresses the ethical aspects of the donation of human embryos and oocytes for hESC research by focusing on the critical role of clinicians and researchers. Chapter fifteen, also by Achim Rosemann, illustrates some of the key challenges for international stem cell trials in the light of the ongoing process of global regulatory diversification in the stem cell field.

We are indebted to our authors who graciously accepted their assignments and who have infused the text with their energetic contributions. We are incredibly thankful to the staff of Springer Science+Business Media that published this book.

Ho Chi Minh City, Vietnam   Phuc Van Pham
Coventry, UK   Achim Rosemann

Contents

Contributors

Marco Alessandrini Department of Immunology, Institute for Cellular and Molecular Medicine, SAMRC Extramural Unit for Stem Cell Research and Therapy, Faculty of Health Sciences, University of Pretoria, Pretoria, South Africa

John D. Banja Department of Rehabilitation Medicine, Emory University, Atlanta, GA, USA

Center for Ethics, Emory University, Atlanta, GA, USA

Nicole Baur Department of Sociology, Philosophy and Anthropology, Egenis, Centre for the Study of the Life Sciences, University of Exeter, Exeter, UK

Yasser El-Sherbiny Leeds Institute of Rheumatic and Musculoskeletal Medicine, University of Leeds, Leeds, St James's University Hospital, Leeds, UK

Adrian Ely Science Policy Research Unit (SPRU), University of Sussex, Brighton, UK

Erden Eren Department of Neuroscience, Institute of Health Sciences, Dokuz Eylul University Health Campus, Izmir, Turkey

Izmir Biomedicine and Genome Center, Dokuz Eylul University Health Campus, Izmir, Turkey

Erdogan Pekcan Erkan Department of Neuroscience, Institute of Health Sciences, Dokuz Eylul University Health Campus, Izmir, Turkey

Izmir Biomedicine and Genome Center, Dokuz Eylul University Health Campus, Izmir, Turkey

Sermin Genc Department of Neuroscience, Institute of Health Sciences, Dokuz Eylul University Health Campus, Izmir, Turkey

Izmir Biomedicine and Genome Center, Dokuz Eylul University Health Campus, Izmir, Turkey

Kemal Kursad Genc Department of Neuroscience, Institute of Health Sciences, Dokuz Eylul University Health Campus, Izmir, Turkey

Gero Hütter Cellex GmbH, Dresden, Germany

Christine Hauskeller Department of Sociology, Philosophy and Anthropology, Egenis, Centre for the Study of the Life Sciences, University of Exeter, Exeter, UK

Carlo S. Jackson Department of Immunology, Institute for Cellular and Molecular Medicine, SAMRC Extramural Unit for Stem Cell Research and Therapy, Faculty of Health Sciences, University of Pretoria, Pretoria, South Africa

Li Jiang Kenneth Wang School of Law, Soochow University, Suzhou, Jiangsu, China

Elena Jones Leeds Institute of Rheumatic and Musculoskeletal Medicine, University of Leeds, Leeds, St James's University Hospital, Leeds, UK

Dimitrios Kouroupis Department of Biomedical Research, Foundation for Research and Technology-Hellas, Institute of Molecular Biology and Biotechnology, University of Ioannina, Ioannina, Greece

Tamra Lysaght Centre for Biomedical Ethics, Yong Loo Lin School of Medicine, National University of Singapore, Singapore, Singapore

Paul Martin Department of Sociological Studies, University of Sheffield, Sheffield, UK

Dennis McGonagle Leeds Institute of Rheumatic and Musculoskeletal Medicine, University of Leeds, Leeds, St James's University Hospital, Leeds, UK

Iñigo de Miguel Beriain Department of Public Law, University of the Basque Country, Bilbao, Spain

Fikile M. Mnisi Steve Biko Centre for Bioethics, Witwatersrand University, Johannesburg, South Africa

Michael S. Pepper Department of Immunology, Institute for Cellular and Molecular Medicine, SAMRC Extramural Unit for Stem Cell Research and Therapy, Faculty of Health Sciences, University of Pretoria, Pretoria, South Africa

Phuc Van Pham Stem Cell Institute, University of Science, Vietnam National University, Ho Chi Minh City, Vietnam

Sujatha Raman Institute for Science and Society (ISS), School of Sociology and Social Policy, University of Nottingham, Nottingham, UK

Science and Technology Studies (STS), University of Nottingham, Nottingham, UK

Achim Rosemann Centre for Education Studies, Faculty of Social Sciences, University of Warwick, Coventry, UK

Margaret Sleeboom-Faulkner Centre for Bionetworking, Department of Anthropology, School of Global Studies, Arts C 209, University of Sussex, Brighton, UK

Suli Sui School of Social Sciences and Humanities, Peking Union Medical College, Beijing, China

Shashank S. Tiwari Individualized Treatment Technology Laboratories LLC, Birmingham, AL, USA

Institute for Science and Society (ISS), School of Sociology and Social Policy, University of Nottingham, Nottingham, UK

Barbara von Tigerstrom College of Law, University of Saskatchewan, Saskatoon, SK, Canada

Xiao Nong Wang Institute of Cellular Medicine, University of Newcastle upon Tyne, Newcastle, UK

Xinqing Zhang School of Social Sciences and Humanities, Peking Union Medical College, Beijing, China

About the Editors

Phuc Van Pham received his Ph.D. in Human Physiology from Vietnam National University, Ho Chi Minh City, Vietnam. He is currently a Professor of Biology at University of Science (Vietnam National University Ho Chi Minh City, Vietnam); Director of the Stem Cell Institute; and Vice-Director, Key Laboratory of Cancer Research. He is a longstanding lecturer and translational scientist at the University and is a member of several societies and editorial boards of journals focused on stem cells. Dr. Pham and his colleagues have established one of the first multidisciplinary stem cell centers in Vietnam, and he has successfully launched an array of technological appliances in stem cell isolation. His research interests include stem cell isolation, stem cell therapy, mesenchymal stem cells, cancer stem cells, immunotherapy, and regenerative medicine, and he has published extensively in these areas. After many years of experience as an embryologist, cell biologist, and molecular biologist, collaborating with leading researchers in Singapore, Japan, and the United States, Dr. Pham is a student again, keen to reach beyond the traditional boundaries of biology.

Achim Rosemann is a Research Fellow in the Centre for Education Studies at the Faculty of the Social Sciences, University of Warwick. His research addresses the social, political, regulatory and economic dimensions of life and health science research, with a regional focus on developments in China and East Asia. He has received his PhD in Social Anthropology from the University of Sussex, with a specialization in Science and Technology Studies and Medical Anthropology. He is also a research associate at the Centre for Bionetworking at the University of Sussex.

Part I
Safety

Chapter 1
The Safety of Allogeneic Stem Cell Transplantation

Gero Hütter

1.1 Introduction

Allogeneic stem cell transplantation (SCT) was developed between the 1950s and 1970s and became later a standard treatment for leukemia, hematological cancer, and nonmalignant bone marrow diseases. The procedure is associated with challenging side effects and a high rate of early and late mortality. Aside from relapse of the underlying malignancy, particularly graft-versus-host disease and negative implication of the necessary immune suppression with reactivation of latent infections are the main reasons of failing to cure the patient.

In comparison to these treatment-related threats, possible risks from infectious or noninfectious complications that originated from the stem cell graft itself are almost unnoticed. Here, we report on the safety of stem cell products for the use of hematopoietic reconstitution and discuss possible implications for clinical decisions.

1.2 Stem Cell Sources

1.2.1 Bone Marrow

Stem cells from the bone marrow can be collected after general anesthesia and puncture of the iliac crest. Thereby up to 1.500 ml of bone marrow blood will be aspirated, and by this artificial blood stream, stem cells will be carried away from their niches into the collection bag. The stem cell product content has a similar hematocrit, a lower platelet count, and a higher concentration of leukocytes as

G. Hütter (✉)
Cellex GmbH, Fiedlerstr. 36, 01307 Dresden, Germany
e-mail: g.huetter@cellex.me

© Springer International Publishing AG 2017

P.V. Pham, A. Rosemann (eds.), *Safety, Ethics and Regulations*, Stem Cells in Clinical Applications, DOI 10.1007/978-3-319-59165-0_1

compared to peripheral blood. Because of the high amount of erythrocytes in the product, blood group compatibility has to be considered in every patient.

Because of the lower incident rate of chronical graft-versus-host disease, bone marrow is the preferred stem cell source in infants/children and adults with nonmalignant hematological diseases. In infants and children, special attention to the total volume of the product has to be paid especially in terms of volume overload and AB0 major/minor incompatibility.

1.2.2 *Peripheral Blood Stem Cells*

The technique of stem cell collection from peripheral blood was developed in the 1990s after sufficient electrophoresis machines and the granulocyte colony-forming factor (G-CSF) were available. After a 4-day stimulation period with G-CSF, donor stem cells will be enriched in an external blood circulation procedure via centrifugation and continuous or discontinuous collection. The product contains a high concentration of leukocytes, a considerable plasma portion of, and a low hematocrit usually less than 4%.

1.2.3 *Umbilical Cord Blood*

Cord blood units are collected immediately after delivery and when the placenta is still in utero. Puncture of the umbilical vein is done after desinfection. Some collection sides perform the collection when the placenta is completely developed and suspended on a metal frame. At least 75 ml of cord blood should be obtained for a sufficient preparation. After collection, it is feasible to reduce the amount of erythrocytes after centrifugation. The product will be stored after addition of dimethyl sulfoxide (DMSO) in the gas phase of liquid nitrogen. After this procedure CB units can be safely stored for years and decades.

1.3   Donor Selection and Testing

1.3.1 *HLA Restrictions*

The general availability of donors is restricted by the compatibility of the human lymphocyte antigen (HLA) system which is the major barrier between allogeneic cell sources in terms of immune tolerance and rejection. For stem cell sources from adult donors, the compatibility of 9–10 from 10 HLA alleles (A, B, C, DRB1, and DQB1) is mandatory for a successful transplantation. For cord blood a matching of only 6 HLA alleles may be acceptable.

1.3.2   Related Donor

Looking for siblings in the family of the patient is generally the first step in donor search. The probability in finding an HLA-matched family donor is $1-(3/4)^{n-1}$ where n is the number of children of the parents. If no HLA-matching sibling available, other family members like children or parents can be considered as haploidentical stem cell source. Usually, the criteria of donor clearance are comparable to unrelated donors except the donor's age.

1.3.3   Unrelated Donor

Unrelated donor selection underwent typically four checkpoints in terms of screening for potential risks: 1. at the recruitment stage, 2. during confirmatory typing (CT), 3. during workup (WU)/donor clearance, and 4. at donation.

At the time point of donor recruitment, no test for infectious disease markers (IDM) will be performed. Donor assessment and risk group stratification will only be done by questionnaire and medical history. It may last years or decades after the recruitment that the donor will be activated for a special patient. That is the time point of the stage of CT, where questionnaire and medical history are updated and first laboratory testing on the infectious state of the donor are available. During WU, the donor is examined at the collection center where the medical record is updated, IDMs taken, and physical and additional examinations like ultrasound or chest X-ray will be performed. The validity of IDMs is usually limited to 30 days after blood samples were taken. Any delay concerning time point of donation requires repetition of IDM testing.

Stem cells stay stable and vital for at least 72 h after donation at 4° C as well as at room temperature. Because of this restriction of the storage life, the products usually get released by IDM testing from the day of the WU up to 30 days before. Therefore, there is a residual risk to release potential infectious products, e.g., during the time of seroconversion or in case of occult infection. To avoid this potential risk some collection centers obtain the results of IDM testing at the donation day before releasing the product.

For donor clearance, only IDM test kits which are valid and considered suitable by the responsible health authorities can be used. In Europe all reagents for diagnostic testing must be CE marked (guide to preparation) or for the United States exhibit FDA approval. Tests for donor clearance differ in minor points between the collection sides and are mostly based on the recommendations of national or international guidelines (Table 1.1).

Donor selection follows usually the local and national regulations of blood donation. Additional to a complete medical history and risk evaluation, the crucial crossroad is still the IDM test. It is a well-described observation that blood donors who give their blood repetitively and frequently have lower incidence rates of infectious

Table 1.1 Recommended tests of infectious disease markers to release stem cell products

IDM		European Union	WMDA	FDA[a]
HIV-1/2 ab		X	X	X
HIV NAT		X	X	X
p24			X	
HBV	Anti-HBs		X	
	Anti-HBc	X	X	X[b]
	HBsAg	X	X	X
HBV NAT		X	X	
HCV ab		X	X	X
HCV NAT		X	X	X
Syphilis		X	X	X
HTLV	Type I	X	X	X
	Type II		X	X
CMV				X[b]

(*IDM* infectious disease marker, *FDA* Food and Drug Administration, *WMDA* World Marrow Donor Association) based on Lown et al. (2014), EU Commission Directive 2006/17/EC, and FDA Guidance for Industry: Eligibility Determination for Donors of Human Cells, Tissues, and Cellular and Tissue-Based Products on www.FDA.gov
[a]All tests have to be FDA-licensed
[b]IgG and IgM

Table 1.2 Incidence of positive infectious disease markers (IDM) in a cohort of blood donors

Incidence of positive IDMs screening/100,000 donations[a]				
	HIV	HCV	HBV	Lues
First donation male	11.1	90.4	175.8	45.0
First donation woman	2.3	57.2	78.7	25.0
Multiple donation male	4.1	2.8	1.5	6.2
Multiple donation woman	0.4	2.0	0.5	1.5
Calculated risk of transmission per 1,000,000 transfusions[b]				
	0.1	0.0	0.2	NR

[a]According to Offergeld and coworkers (Offergeld et al. 2012)
[b]Report of the German Inspection Authority 2011/2012 (Paul-Ehrlich Institute, Langen, Germany) (www.pei.de/haemovigilanzbericht)

disease like HBV, HCV, and HIV compared to donors who gave blood once and for the first time (Table 1.2). Because stem cell donors are tested at least twice the time before donation (CT and WU), the estimated relative risk of carrying undetected infections should be attributed to the group of repetitive blood donors. However, during urgent stem cell request, CT and final donor clearance have to be done simultaneously. Then, the stem cell products will be released by only a single IDM test done in maximum 30 days before donation. Taken together, there is a residual risk that during the asymptomatic and occult phase of a viral infection donor suitability is confirmed, but the donor may develop a critical viremia at the day of collection.

1.3.4   Cord Blood

The principal agreement for cord blood donation is usually done before delivery. Therefore, the quality and safety of cord blood products is highly dependent on the effort of the cord blood bank to minimize potential risks. In a survey on maternal-neonatal donor pairs, an evaluation of possible exclusion criteria after donation revealed a high percentage of ineligibilities which were not known at the time point the product was stored (Jefferies et al. 1999).

Generally, not only the mother but also the cord blood unit gets tested for IDM, and the results of both are essential for the release of the product (Kogler et al. 1999). Some cord blood banks try to enhance the safety of their product by recontacting the donors after a couple of months after delivery. Usually, only medical questionnaire will be performed, and no new IDMs collected; potential problematic units will be sorted off the blood bank (Lecchi et al. 2001). The complex relationship of mother-to-child transmission of certain infectious diseases is reviewed in (Berencsi et al. 2013). In general, the burden of common infectious diseases from the herpesvirus group seems to be less frequent in CB units (Behzad-Behbahani et al. 2005).

1.4   Reported Microbiological Transmissions

1.4.1   Transmission of Human Immunodeficiency Virus Type 1 (HIV-1)

After the lessons from the HIV catastrophe in the 1980s, blood products have been become saver and saver every year. Today, transmissions of HIV are only reported sporadically despite the fact that millions of blood donations and transfusions are performed every year. The probability of transmission of HIV has been estimated to lower than 1:4,300,000 (CI 2.39–21.37 $\times$ 10^6) in most West European countries and the United States (Schmidt et al. 2014). Considering this low probability and the fact that up to date approximately 300,000 patients already received an allogeneic transplantation since 1980 and that this number increases around 25,000–45,000 new patients every year, statistically transmission of HIV is thought to be a very uncommon event in stem cell recipients.

Principally, HIV can be transmitted by stem cell transplantation. The first case was reported by Furlini and coworkers in which a young woman with acute lymphoblastic leukemia underwent SCT from her brother, who was found later to be HIV-positive. At the time point of donation no HIV-test was available. Five months after engraftment, HIV antibody testing in the recipient was positive (Furlini et al. 1988). After the introduction of routine HIV antibody and later HIV-RNA testing, there is no further documented transmission of HIV via SCT (Lown et al. 2014). However, there is a residual risk from HIV+ donors before seroconversion harbouring a mutant HIV strains which may escape from routine NAT testing (Delwart et al. 2004; Schmidt et al. 2009).

In a large survey between 2010 and 2012 of the German Red Cross Blood Service screening 2.7 million blood donations, four plasma specimens were found with false-negative HIV-NAT. Two of them had a viral load below usual sensitivity of detection, but two samples had a clear viral load not detected by standard commercial assays. Investigators found deficiencies in the 5'-long terminal repeat (LTR) as target region for NAT testing prevented binding of the used primers and leading into false-negative results (Muller et al. 2013). As a consequence health authorities have now directed to use dual-targeted NAT assays to minimize the risk of transmission of these variants of HIV before seroconversion.

1.4.1.1   Prevention of HIV Transmission by a Viremic Donor

Based on the experiences of serodisconcordant couples, post exposition prophylaxis, and mother-to-child settings, HIV-1 transmission can effectively be prevented by early antiretroviral therapy of the exposed individual. However, compared to those routes of transmission with considerable low amounts of infectious particles, contaminated blood and therefore also contaminated stem cells products have a tremendous higher risk of transmission (transmission risk: contaminated iv needle use 6.7‰ compared to contaminated blood product 90%) (Smith et al. 2005).

Irrespectively, there are reports of preventing successfully HIV transmission after contaminated blood product transfusion by antiretroviral post exposition prophylaxis assuming that this procedure could also be effective in contaminated stem cell sources (Katzenstein et al. 2000).

1.4.2   Transmission of Hepatitis B Virus (HBV)

Transmission of HBV is still a major concern in transfusion medicine and has been reported several times during SCT (Lau et al. 1999). Although current serological testing for hepatitis surface antigen (HBsAg), serological markers like anti-HBs and anti-HBc, and most effective the nucleic acid test for HBV genome have improved the safety of blood products, HBV transmission occurs still with the highest incidence of all serious transfusion-associated viral infections. Several improvements in optimizing the detection test have already been achieved, and the window period length for HBsAg test has decreased from 59 days using enzyme immunoassay (EIA) to 36–38 days using the new EIAs and chemolumynoimmunoassay (ChLIA) tests (Assal et al. 2009; Schreiber et al. 1996). Moreover, HBV NAT testing arises around 21 days after transmission and 15 days before the serologic testing detects the HBV infection (Kleinman et al. 2009).

However, HBV transmission may occur in window period where donors display low viremia. Today, the HBV transmission risk has been significantly reduced mostly by the standard use of DNA-based techniques. It has been estimated that the risk of HBV transmission by occult hepatitis B infection (OBI) is still around 1:116,000–1:150,000 depending on the local prevalence of HBV (Candotti and Allain 2009).

The prevalence of HBV and thereby the risk of transmission by OBI donors are highly dependent on the country from which the graft was collected. The prevalence of anti-HBc variates from nearly 15% in Greece down to 1% and lower in the United Kingdom and the United States (Candotti and Allain 2009; Solves et al. 2014). The absence of surface antigen and a negative NAT for HBV make these anti-HBc donors suitable for stem cell donation. However, low-level viral replication during the acute or chronic phase of HBV infection could be present but undetectable to the routinely used NAT assays. Furthermore, escape mutants can occur undetectable to current HBsAg screening assays.

1.4.2.1   How to Proceed with Anti-HBc-Positive Donors?

Approximately 90% of blood donors carrying anti-HBc also carry anti-HBs indicating a recovery from HBV infection and thereby immunity (Allain et al. 2003). The remaining 10% of donors are attributed as "anti-HBc only" may originate either from recovered infections that have lost detectable anti-HBs or from late stage chronic infections having lost detectable HBsAg. Principally, both groups may harbor latent virus which is competent to replicate and might be transmitted. Usually, "anti-HBc only" will be excluded from blood donation, whereas donors with sufficient (>100 IE) titer of anti-HBs are eligible to donate. Theoretically, these donors have the ability to develop escape mutants between donor clearance and donation irrespectively a sufficient anti-HBs titer. Therefore, products from donors with positive anti-HBc test should be released by the results of IDM (HBV-NAT) testing on the day of donation and not only by the results from donor clearance.

1.4.2.2   Prevention of HBV Infection in Case of HBV Viremic Donors

Recipients of SCT are exposed against several blood products during their course of treatment. Therefore, vaccination against HBV would be an easy way for protection against transmission. However, 57% of the pre-transplant vaccinated recipients loose their immunity after transplantation (Idilman et al. 2003). A considerable high risk in HBV reactivation is reported in recipients with previous infection and seroscare (Knoll et al. 2007). In the case that only a donor with history of HBV infection is available, there are reasonable strategies to prevent transmission. In a case series of 13 patients, Frange and coworkers reported transplantation of units with increased risk of HBV transmission. Four donors were HBs-Ag positive with low but detectable viremia; all other units were HBs-Ag positive but HBV-NAT negative. Recipients were variable pretreated with, e.g., vaccination, immune globulins, specific anti-HBV immune globulins, and lamivudine prophylaxis. All stem cell units were tested negative for HBV-NAT. No one of the 13 recipients developed seroconversion post-transplant (Frange et al. 2014). In a similar setting, another group reported a reduction of the risk of HBV transmission by administrating prophylactic lamivudine and vaccination from 48.0 down to 6.9% (Hui et al. 2005).

1.4.3 Transmission of Hepatitis C Virus (HCV)

Recipients of stem cell transplantation receive frequently blood products, and the cumulative number of transfused units may exceed 100 products over the time of cancer treatment. Therefore, transmission of HCV in these patients was a major problem before routine screening for HCV antibodies and later NAT testing was established (Kolho et al. 1993; Locasciulli et al. 1991). Despite of the risk of transmission from supportive blood products, transmission of HCV by the graft itself was reported frequently in the past. In a case series, Shuhart and coworker reported on 12 patients who tested HCV negative before transplantation and received grafts from HCV-positive donors. In seven donors the NAT testing for HCV was positive, and all recipients of those units were infected with the virus. On the other side, the remaining five recipients who received only the antibody-positive and NAT-negative product stayed free of disease (Shuhart et al. 1994). Today in the era of HCV, NAT testing transmissions of hepatitis C via blood products is an outstanding rare event.

1.4.3.1 Prevention of HCV Transmission

There are some reports where stem cell donors with HCV viral load received antiviral medication to suppress replication and then successfully donate the graft without transmission (Surapaneni et al. 2007). Moreover, a report of a prophylactic administration of peg-interferon in combination with ribavirin to treat the recipient receiving a stem cell product from an viremic donor successfully prevented HCV transmission (Hsiao et al. 2014).

According to the biology of HCV, it has been assumed that the infection is limited to hepatocytes, and transmission would derive from virus particles in the serum. Removing the serum and other components of the blood except of the stem cells by, e.g., CD34+ selection should remove the major source of infectious material. However, as shown by Thomas and coworkers, this method was not sufficient to prevent transmission in a patient receiving CD34 + −selected stem cells even after the stem cell enriched supernatant was washed and tested apparently negative for HCV-NAT (Tomas et al. 1999). One of the reasons for this failure of graft purging to prevent transmission is that HCV can also harbor in cells in CD19+ mononuclear peripheral blood cells which could not completely be removed by CD34 selection (Di Lello et al. 2014; Zehender et al. 1997).

1.4.4   Transmission of HTLV-I + II

HTLV-1 + II infection is very rare in Europe and is mostly limited to risk groups like intravenous drug users. The disease is endemic in Japan, Caribbean countries, some parts of Africa, and South America. The World Marrow Donor Association (WMDA) recommends HTLV antibody testing for all stem cell donation. The recent German guidelines allow to differentiate between donors with high or low risk for HTLV infection and only those how have been exposed or originate from the endemic sides are required to get tested (DOI: 10.3238/arztebl.2014.rl_haematop_sz01). However, transmissions of HTLV have been repetitively reported in recipients of contaminated stem cell products (Kikuchi et al. 2000; Ljungman et al. 1994).

1.4.5   Cytomegalic Virus (CMV) in CMV-Negative Recipients

Despite effective therapy, infection with or reactivation of cytomegalovirus (CMV) remains one of the leading causes of morbidity and mortality after hematopoietic stem cell transplantation (Boeckh 1999). Transmission of CMV by seropositive donors to seronegative recipients of standard blood products has become rare today after these preparations are generally reduced concerning their content of leukocytes (Thiele et al. 2011). In the case of stem cell units from serodisconcordant donors, there is a significant risk of CMV transmission. The cumulative risk for seroconversion is 19% during the first 100 days after transplantation as shown from a survey from the Fred Hutchinson Cancer Research Center (Pergam et al. 2012). Approximately 10% of these seroconverters displayed active CMV infection after transplantation. The risk of transmission was not connected with the stem cell source but with the number of nucleated cells in the graft.

If possible, donors are selected to have the same CMV serology as the patient. However, disconcordant transplantation occurs frequently due to the limited choice of donors. Transplanting CMV-negative stem cells to CMV-positive recipient is associated with a remarkable high risk of CMV reactivation in the recipient, whereas CMV+ donor on CMV+ recipient is without negative impact. Most relevant is the administration of CMV-positive stem cells on CMV-negative recipients which is associated with a significant decreased survival after allogeneic transplantation (Ljungman 2014).

In terms of cord blood units, there is also a low but still existing risk of CMV transmission. In a survey of 1221 CB/donor matches, Albano and coworker were able to identify two transmissions of CMV by CB units (Albano et al. 2006). Both CB preparations were found antibody negative but positive for CMV-DNA. Similar findings have been reported from another group where 0.5% of the cord blood samples derived from only CMV-IgG-positive mothers were also found CMV-NAT positive (Theiler et al. 2006).

Nevertheless, in a multivariate analysis of 753 allogeneic SCT form PBSC, BM, and CB, the CMV viremia was significantly more likely in patients who: 1. were already seropositive for CMV, 2. had acute graft-versus-host disease, and 3. who received T cell-depleted grafts. Graft source did not independently contribute to the risk of CMV infection and did not impact survival after CMV infection indication of a minor influence of the graft source on CMV transmission (Walker et al. 2007).

1.4.6 Possible Risks from Donors with Epstein-Barr Virus (EBV)

Stem cell donors are generally tested for EBV. The major concerns in transplanting EBV-positive stem cell units on EBV-negative recipients are the problem of acute EBV infection during the transplantation and engraftment process and the consecutive development of possible EBV-associated lymphoproliferative disorders. In case of acute infection or spontaneous viremia of the donor, the collection may be shifted to a later time point.

In a small survey of donor/recipient pairs, Sakellariet and coworkers found an occult EBV viremia in several donors. However, this EBV viremia was not associated with EBV reactivation at the donor and not an indicator for a risk of transmission. Furthermore, risk factors like donor EBV seronegativity were not significantly correlated with EBV reactivation in case of EBV positivity at the recipient (Sakellari et al. 2014).

However, other blood products like red blood packages or thrombocyte concentrates are not regularly tested for EBV, and these blood products are depleted but not free from leukocytes. Due to the large number of administered blood packages during transplantation procedure, there is a cumulative risk for transmission of EBV to EBV-native recipients of 13.4% after 60 days in a cohort of SCT recipients (Trottier et al. 2012).

One infrequent but serious complication after stem cell transplantation is the EBV-associated post-transplant lymphoproliferative disorder (PTLD). The origin of the EBV-infected cells in most cases is the donor, and this disease can arise from any stem cell source including UC (Haut et al. 2001; McClain 1997). For treatment of EBV-associated PTLD, there are no established guidelines available. Approaches including high-dose aciclovir, ganciclovir, alpha interferon, cytotoxic drugs, and cytotoxic T-cell therapy have been tested. Today reduction of immuno suppression is the first line treatment, and early use of rituximab as a second option has to be considered. In more aggressive forms of PTLD, upfront chemotherapy may offer a better and more durable response (Singavi et al. 2015).

1.4.7 Transmission of Treponema pallidum

Treponema pallidum is transmitted primarily by sexual contact or during pregnancy. The spirochete is able to pass through intact mucous membranes or compromised skin; parenteral transmission is possible and reported infrequently for blood products and in some rare cases of stem cell products (Naohara et al. 1997). Donors generally underwent sensitive screening for *Treponema pallidum*, and therefore transmissions are uncommon.

1.4.8 Commensal Bacteria

Contamination of commensal bacteria is a known problem in blood products. However, the highest risk for life-threatening events after administration of those units is mostly restricted to blood products stored at room temperature like platelet concentrates. Stem cell products are usually stored and transferred at 4 °C ± 2 °C, and the storage time is limited to 72 h. Contamination of stem cell products from peripheral apheresis is reported usually in around 1% of the collections (Table 1.3) (Kamble et al. 2005; Kelly et al. 2006; Kozlowska-Skrzypczak et al. 2014).

As expected due to the different collection technique, contamination in bone marrow harvest is reported more frequently (Kamble et al. 2005; Vanneaux et al. 2007). Although the surgery is performed aseptically, the technique of repetitive insertion of the bone marrow aspiration needle in the same cut in the donor's skin is associated with a noteworthy risk of carrying germs into the product.

Usually agents from normal skin flora like coagulase-negative *Staphylococcus* and *Propionibacterium acnes* are the most frequent isolated microorganisms. The clinical significance of bacterial contamination seems to be limited: in a survey with 19 contaminated products, no adverse sequelae occurred after infusion, and none of the transplanted patients developed bacteriemia that could have been related to the isolated microorganism (Borecki et al. 1991).

The fact that most of the recipients of allogeneic stem cell transplantation receive routinely antibiotic treatment (e.g., for gut decontamination) may contribute to the low rate of reported and documented transfusion-associated infections.

Table 1.3 Relative frequency of positive sterile testing of stem cell products

Germ	PBSC own data[a]	PBSC[b]	BM[b]	UC[b]
Propionibacterium acnes	53.5%	40.6%	11.2%	10.6%
Micrococcus luteus	21.4%	3.1%	2.8%	0
Staphylococcus spp.	17.8%	40.6%	46.4%	42.7%
Steptococcus spp.	3.5%	6.2%	0	3.9%
Enteroccus spp.	0	0	0	25.7%
Others	3.5%	9.3%	2.8%	17.4%
Overall positive sterile testing	0.9%	0.13–3.1%	6.2–60.0%	4.9–7.5%

PBSC peripheral blood stem cells, *BM* bone marrow, *UC* umbilical cord blood
[a]*N* = 2974
[b]Based on an evaluation of the Paul-Ehrlich Institute, Langen, Germany of 23 German collection centers (http://www.pei.de)

There are few reports of contamination that did not origin from donor as reported from Kassis and coworkers. They found 6 of 45 stem cell units contaminated with a *Mycobacterium mucogenicum*-related pathogen. As the source of this contamination, they found colonized ice cubes which were used during stem cell processing as the responsible origin (Kassis et al. 2007).

Although reports of relevant bacterial transmission are rare, they can also cause severe complication like in the case reported from Kelley and coworkers, where a 36-year-old patient with chronic myelogenous leukemia received a stem cell product contaminated with *Bacillus cereus*. The patient developed life-threatening acute renal failure and disseminated intravasal coagulation (Kelley et al. 2014). Interestingly the stem cell product was transferred from the collection center at room temperature for 31 h.

Storage temperature may have divergent effects on possible bacterial contamination as shown by Hahn and coworkers. They spiked 66 individual bone marrow samples stored at 20–24 °C room temperature or 3–5 °C, respectively. In products spiked with typical bone marrow contaminants (*P. acnes* and *S. epidermidis*) and *E. coli* germ outgrowth was arrested under room temperature. However, under these conditions several pathogenic bacteria (*S. aureus* and *K. pneumoniae*) proliferated dramatically, but these agents have minor relevance in terms of artificial contamination so that authors recommend room temperature for BM storage (Hahn et al. 2014).

Sterility of CB is highly dependent on the training skills of medical personnel during this procedure. For example, the London Blood Bank reported to have lowered the contamination rate from 28% to less than 1% after an intensive training program (Armitage et al. 1999). As expected, the spectrum of germs is different to PBSC and BM, and aside from skin flora, there is also vaginal flora and fecal and environmental germs detectable. Usually, contamination rates between 4 and 5% are reported (Kamble et al. 2005; M-Reboredo et al. 2000). Secondary contamination of the product, for example, after rupture of the collection bag, is uncommon but may occur in cryopreserved packages (Mele et al. 2005).

1.4.9 Other Pathogens

Global traffic, climate change, and immigration modify the occurrence of known infectious disease and may alter the incidence of new health threats. In the past years, new entities of infectious agents have tracked attention like dengue fever, West Nile virus, chikungunya virus, or hepatitis E virus which can all be transmitted by blood products. However, reported cases are mostly limited to countries where these agents are endemic (Stramer 2014). However, transmission of malaria, Chagas disease, and brucellosis have been reported occasionally after SCT (Ertem et al. 2000; Mejia et al. 2012; Villalba et al. 1992). Whereas transmission of WNV is a reasonable risk in endemic areas for blood donation, cases of WNV transmission by SCT have not been published so far. According to this, not every infrequent agent can be tested routinely, but careful donor questionnaire, anamnesis, and travel history may narrow the possible exposures.

In a small case series during the influenza H1N1 pandemic, a Korean group reported on three donors who have been found to be infected during stem cell donation. The virus was detectable in nasal aspirate in all and in two of them in the blood stream. One stem cell product was found positive for influenza H1N1 PCR before administration. The recipients received prophylactic oseltamivir at a dose of 150 mg twice daily for 7 days and did not show symptoms or signs associated with influenza (Lee et al. 2011).

Herpes viruses are general less frequent in CB units as compared to adult donors except for human herpes virus type 6 (HHV-6) which is casually isolated from CB progenitor cells. De Pagter and coworkers reported a case of transmission after CB transplantation by HHV-6, a virus which is able to integrate into chromosomal DNA. Although the recipient was HHV-6 positive before transplantation, transfusion of the cord blood led to a rapid and over months sustained viral load in the patient (de Pagter et al. 2010).

Infection or reactivation with parvovirus B19 (PVB19) after SCT is uncommon but may cause severe complications. In a review of 98 patients after SCT, Eid and coworkers described three cases with fatal PVB19-associated myocarditis and red blood cell anemia in 98.8% of these cases leading to a transplant loss in 10% (Eid et al. 2006). BVB19 can be transmitted by any blood product and has been reported casually for stem cell products (Heegaard and Laub Petersen 2000). Screening for PVB19 DNA is easy and efficient to detect viremia in the donor. However, variants in the PVB19 genome as reported in the genotype 3 are highly associated with chronic anemia after SCT, and viremia is not detectable by using routine PCR testing assays (Knoester et al. 2012).

1.5   Transfusion-Related Complications

1.5.1   Transfusion-Related Acute Lung Injury (TRALI)

TRALI reaction is characterized by acute respiratory distress (occurrence during or within 6 h after the transfusion), dyspnea, hypoxemia, new bilateral lung infiltrations in the chest radiograph, and no evidence of hypervolemia. The reported frequency in transfusion medicine is between 1/1120 and 1/5000 transfusions depending on the plasma portion of the product.

Causatives for this complication are donor-derived antibodies against recepient's leukocyte antigens (HLA class I and II), as well as against recipient's human neutrophil antigen (HNA) system (van Stein et al. 2010). These antibodies are detectable more frequently in female donors but may also be present in male donors (Nguyen et al. 2011). The risk of TRALI reaction after infusion is proportional to the plasma content of the product, and therefore this complication has been reported more frequently in bone marrow products rather than PBSC and has not been reported in CB products so far (Urahama et al. 2003).

As figured out, the typical TRALI originates from donor-derived antibodies. However, TRALI or TRALI-like reaction can also originate from HLA or HNA antibodies from the recipient after exposition against leukocyte-containing products like granulocyte concentrate or stem cell products. Knop and coworker reported a case where the recipient developed a severe and lethal pancytopenia after bone marrow infusion. Causative was that the recipient harbored antibodies against all cell lineages including anti-lymphocyte antibodies with specificity against HLA molecules and anti-neutrophilic antibodies against HNA-2a (NB1, CD177) (Knop et al. 2004).

To reduce the transfusion-related risks of TRALI, a granulocyte agglutination test (GAT) may uncover these cross-reactions. More advanced, very sensitive, and rapid working tests like the automated flow cytometric granulocyte immunofluorescence test (Flow-GIFT) are available and have been tested in clinical setting for both donors and recipients to reduce the risk of TRALI reaction after receiving leukocyte-containing products (Schulze et al. 2011).

1.5.2 Hemolytic Transfusion Reaction (HTR)

AB0 blood group system should be considered during allogeneic SCT for several reasons. Major AB0 incompatibility can cause severe hemolytic transfusion reaction, especially in the case bone marrow is used (Lopez et al. 1998). Furthermore, outcome in AB0 compatible donor-recipient setting is significant better than in minor and major incompatibility (Kimura et al. 2008).

Moreover, most patients receiving SCT are heavily pretreated with transfusions and have an increased risk to develop irregular antibodies. Kim and coworker reported a patient with alloreactive anti-Jk(a) where the donor was Jk(a) antigen positive. To avoid hemolytic transfusion reaction, rituximab was added to the conditioning regimen of fludarabine and melphalan. The patient proceeded to receive a peripheral blood stem cell transplant from a matched unrelated donor with no adverse events (Kim et al. 2013).

Not only major but also minor hemolytic transfusion reaction caused by the plasma portion of the transplant has been frequently reported. Several transplant centers request for plasma diluted PBSC to decrease the cell concentration and thereby enhance the vitality of the stem cells during transport. For collection centers the benefits and risks of a freehanded dilution with plasma have to be balanced and discussed with the transplant center (Akkok et al. 2013).

Interestingly, the type of conditioning regimen of the patient may reduce the risk of minor hemolytic transfusion reactions. GVHD prophylaxis with MTX reduces the risk of hemolytic reactions after receiving minor ABO-mismatched PBPC grafts (Worel et al. 2002). Transfusion reaction may occur delayed but massively like in the case reported from Salmon and coworkers: a donor (Rh D-positive) received an 0 Rh D-positive graft. Hemolysis developed on day 7 after transfusion with rapid and complete hemolysis of the recipient's erythrocytes. Causative for this dramatic development was an unusual high anti-A antibody titer in the donor's graft (Salmon et al. 1999). In any case, careful monitoring of hemolysis parameters during the first 15 days after SCT is mandatory.

1.6 Risk of Transmission of Donor-Derived Noninfectious Diseases

1.6.1 *Possible Long-Time Effects*

Donors for PBSC are generally pretreated with recombinant G-CSF. Based on large observational studies in healthy stem cell donors after G-CSF administration, there is no increased risk for malignancies or autoimmune diseases in this group (Holig et al. 2009). Additionally, in the past there were some concerns that G-CSF administration could alter the genome of the consecutive collected stem cells. However large follow-up trials in healthy donors after peripheral stem cell collection have revealed no increased risk for secondary malignancies. Moreover, in-depth analysis of the genome of stem cells after PBSC in terms of genomic stability and epigenetic alteration were all negative indicating no sustained effect of the stimulation (Leitner et al. 2014; Shapira et al. 2003).

1.6.2 *Donor-Derived Malignancies*

Donor-derived myelodysplastic syndrome/acute leukemia following allogeneic SCT is rare and has been reported since 1971 in about 50 cases (Reichard et al. 2006). The mechanism of the phenomenon is poorly understood, and in some cases a preexisting unrecognized chromosomal aberration of the donor cells can be causative (Dickson et al. 2014). In a survey of 2390 engrafted patients, the incidence of donor-derived MDS/leukemia is not attributed to a stem cell source (PBSC, BM, and UC) (Dietz et al. 2014).

Whereas donor-derived leukemia occurs sporadically, donor-derived solid tumors are extremely rare but also reported, for example in a case of a male patient with acute myeloid leukemia who received one HLA mismatch unrelated graft from a female donor and was 9 years later diagnosed with a well-differentiated adenocarcinoma of the bile duct and underwent pancreaticoduodenectomy. Fluorescence in situ hybridization analysis revealed female patterns of the tumor cells, which suggested that the tumor cells are originated from the donor (Haruki et al. 2014). However, evidence of donor-derived cells within tumorous tissues is not always evidentiary that these donor cells are causative for the tumor. In a large survey of suspected post-transplant donor-derived malignancies, Worthley and coworkers took a closer look on the donor-derived cells in the neoplastic tissue and identified these cells as surrounding and reactive myofibroblasts and not as tumor cells. However, the role of these identified donor cells concerning tumorigenesis remains unclear (Avital et al. 2007; Worthley et al. 2009).

1.7 Conclusion

Enormous efforts have been reached today concerning the safety of blood products after introduction of sensitive test assays. Furthermore, donor screening and donor clearance has been improved and is getting more and more standardized worldwide. Taken together, transmission of infectious diseases by stem cell preparations is quite uncommon. However, there is still some residual risk, e.g., for hepatitis B infection or bursts of endemic or pandemic agents; the risk is not substantial different to other blood products. However, these other blood products are regularly free or at least reduced from leukocytes which is, as a matter of course, not possible in stem cell preparations and therefore may harbor additional risk from cell-based or cell-integrated diseases.

Prevention of stem cell-based impairment of the recipient should be attributed in first line to the donor registry and the collection center. Secondary the transplant units are in duty to monitor the recipient carefully during engraftment to evaluate potential health threats at the earliest point of time. Most of the possible transmission originated from stem cell products can successfully be prevented by post-exposure prophylaxis.

References

Akkok CA, Haugaa H, Galgerud A et al (2013) Severe hemolytic transfusion reaction due to anti-A1 following allogeneic stem cell transplantation with minor ABO incompatibility. Transfus Apher Sci 48:63–66

Albano MS, Taylor P, Pass RF et al (2006) Umbilical cord blood transplantation and cytomegalovirus: posttransplantation infection and donor screening. Blood 108:4275–4282

Allain JP, Candotti D, Soldan K et al (2003) The risk of hepatitis B virus infection by transfusion in Kumasi, Ghana. Blood 101:2419–2425

Armitage S, Warwick R, Fehily D et al (1999) Cord blood banking in London: the first 1000 collections. Bone Marrow Transplant 24:139–145

Assal A, Barlet V, Deschaseaux M et al (2009) Sensitivity of two hepatitis B virus, hepatitis C virus (HCV), and human immunodeficiency virus (HIV) nucleic acid test systems relative to hepatitis B surface antigen, anti-HCV, anti-HIV, and p24/anti-HIV combination assays in seroconversion panels. Transfusion 49:301–310

Avital I, Moreira AL, Klimstra DA et al (2007) Donor-derived human bone marrow cells contribute to solid organ cancers developing after bone marrow transplantation. Stem Cells 25:2903–2909

Behzad-Behbahani A, Pouransari R, Tabei SZ et al (2005) Risk of viral transmission via bone marrow progenitor cells versus umbilical cord blood hematopoietic stem cells in bone marrow transplantation. Transplant Proc 37:3211–3212

Berencsi G, Csire M, Szomor K et al (2013) Virological risks and benefits of cord blood banking. In: Vanostrand L (ed) Cord blood banks and banking, ethical issues and risks/benefits. Nova Biomedical, New York

Boeckh M (1999) Current antiviral strategies for controlling cytomegalovirus in hematopoietic stem cell transplant recipients: prevention and therapy. Transpl Infect Dis 1:165–178

Borecki IB, Rice T, Bouchard C et al (1991) Commingling analysis of generalized body mass and composition measures: the Quebec family study. Int J Obes 15:763–773

Candotti D, Allain JP (2009) Transfusion-transmitted hepatitis B virus infection. J Hepatol 51:798–809

De Pagter PJ, Virgili A, Nacheva E et al (2010) Chromosomally integrated human herpesvirus 6: transmission via cord blood-derived unrelated hematopoietic stem cell transplantation. Biol Blood Marrow Transplant 16:130–132

Delwart EL, Kalmin ND, Jones TS et al (2004) First report of human immunodeficiency virus transmission via an RNA-screened blood donation. Vox Sang 86:171–177

Di Lello FA, Culasso AC, Parodi C et al (2014) New evidence of replication of hepatitis C virus in short-term peripheral blood mononuclear cell cultures. Virus Res 191:1–9

Dickson MA, Papadopoulos EB, Hedvat CV et al (2014) Acute myeloid leukemia arising from a donor derived premalignant hematopoietic clone: a possible mechanism for the origin of leukemia in donor cells. Leuk Res Rep 3:38–41

Dietz AC, Defor TE, Brunstein CG et al (2014) Donor-derived myelodysplastic syndrome and acute leukaemia after allogeneic haematopoietic stem cell transplantation: incidence, natural history and treatment response. Br J Haematol 166:209–212

Eid AJ, Brown RA, Patel R et al (2006) Parvovirus B19 infection after transplantation: a review of 98 cases. Clin Infect Dis 43:40–48

Ertem M, Kurekci AE, Aysev D et al (2000) Brucellosis transmitted by bone marrow transplantation. Bone Marrow Transplant 26:225–226

Frange P, Leruez-Ville M, Neven B et al (2014) Safety of hematopoietic stem cell transplantation from hepatitis B core antibodies-positive donors with low/undetectable viremia in HBV-naive children. Eur J Clin Microbiol Infect Dis 33:545–550

Furlini G, Re MC, Bandini G et al (1988) Antibody response to human immunodeficiency virus after infected bone marrow transplant. Eur J Clin Microbiol Infect Dis 7:664–666

Hahn S, Sireis W, Hourfar K et al (2014) Effects of storage temperature on hematopoietic stability and microbial safety of BM aspirates. Bone Marrow Transplant 49:338–348

Haruki K, Shiba H, Futagawa Y et al (2014) Successfully-treated advanced bile duct cancer of donor origin after hematopoietic stem cell transplantation by pancreaticoduodenectomy: a case report. Anticancer Res 34:3789–3792

Haut PR, Kovarik P, Shaw PH et al (2001) Detection of EBV DNA in the cord blood donor for a patient developing Epstein-Barr virus-associated lymphoproliferative disorder following mismatched unrelated umbilical cord blood transplantation. Bone Marrow Transplant 27:761–765

Heegaard ED, Laub Petersen B (2000) Parvovirus B19 transmitted by bone marrow. Br J Haematol 111:659–661

Holig K, Kramer M, Kroschinsky F et al (2009) Safety and efficacy of hematopoietic stem cell collection from mobilized peripheral blood in unrelated volunteers: 12 years of single-center experience in 3928 donors. Blood 114:3757–3763

Hsiao HH, Liu YC, Wang HC et al (2014) Hepatitis C transmission from viremic donors in hematopoietic stem cell transplant. Transpl Infect Dis 16:1003–1006

Hui CK, Lie A, Au WY et al (2005) Effectiveness of prophylactic anti-HBV therapy in allogeneic hematopoietic stem cell transplantation with HBsAg positive donors. Am J Transplant 5:1437–1445

Idilman R, Ustun C, Karayalcin S et al (2003) Hepatitis B virus vaccination of recipients and donors of allogeneic peripheral blood stem cell transplantation. Clin Transpl 17:438–443

Jefferies LC, Albertus M, Morgan MA et al (1999) High deferral rate for maternal-neonatal donor pairs for an allogeneic umbilical cord blood bank. Transfusion 39:415–419

Kamble R, Pant S, Selby GB et al (2005) Microbial contamination of hematopoietic progenitor cell grafts-incidence, clinical outcome, and cost-effectiveness: an analysis of 735 grafts. Transfusion 45:874–878

Kassis I, Oren I, Davidson S et al (2007) Contamination of peripheral hematopoeitic stem cell products with a mycobacterium mucogenicum-related pathogen. Infect Control Hosp Epidemiol 28:755–757

Katzenstein TL, Dickmeiss E, Aladdin H et al (2000) Failure to develop HIV infection after receipt of HIV-contaminated blood and postexposure prophylaxis. Ann Intern Med 133:31–34

Kelley JM, Onderdonk AB, Kao G (2014) Bacillus Cereus septicemia attributed to a matched unrelated bone marrow transplant. Transfusion 53:394–397

Kelly M, Roy DC, Labbe AC et al (2006) What is the clinical significance of infusing hematopoietic cell grafts contaminated with bacteria? Bone Marrow Transplant 38:183–188

Kikuchi H, Ohtsuka E, Ono K et al (2000) Allogeneic bone marrow transplantation-related transmission of human T lymphotropic virus type I (HTLV-I). Bone Marrow Transplant 26:1235–1237

Kim MY, Chaudhary P, Shulman IA et al (2013) Major non-ABO incompatibility caused by anti-Jk(a) in a patient before allogeneic hematopoietic stem cell transplantation. Immunohematology 29:11–14

Kimura F, Sato K, Kobayashi S et al (2008) Impact of AB0-blood group incompatibility on the outcome of recipients of bone marrow transplants from unrelated donors in the Japan marrow donor program. Haematologica 93:1686–1693

Kleinman SH, Lelie N, Busch MP (2009) Infectivity of human immunodeficiency virus-1, hepatitis C virus, and hepatitis B virus and risk of transmission by transfusion. Transfusion 49:2454–2489

Knoester M, Von Dem Borne PA, Vossen AC et al (2012) Human parvovirus B19 genotype 3 associated with chronic anemia after stem cell transplantation, missed by routine PCR testing. J Clin Virol 54:368–370

Knoll A, Boehm S, Hahn J et al (2007) Long-term surveillance of haematopoietic stem cell recipients with resolved hepatitis B: high risk of viral reactivation even in a recipient with a vaccinated donor. J Viral Hepat 14:478–483

Knop S, Bux J, Kroeber SM et al (2004) Fatal immune-mediated pancytopenia and a TRALI-like syndrome associated with high titers of recipient-type antibodies against donor-derived peripheral blood cells after allogeneic bone marrow transplantation following dose reduced conditioning. Haematologica 89:ECR12

Kogler G, Somville T, Gobel U et al (1999) Haematopoietic transplant potential of unrelated and related cord blood: the first six years of the EUROCORD/NETCORD Bank Germany. Klin Padiatr 211:224–232

Kolho E, Ruutu P, Ruutu T (1993) Hepatitis C infection in BMT patients. Bone Marrow Transplant 11:119–123

Kozlowska-Skrzypczak M, Bembnista E, Kubiak A et al (2014) Microbial contamination of peripheral blood and bone marrow hematopoietic cell products and environmental contamination in a stem cell bank: a single-center report. Transplant Proc 46:2873–2876

Lau GK, Lee CK, Liang R (1999) Hepatitis B virus infection and bone marrow transplantation. Crit Rev Oncol Hematol 31:71–76

Lecchi L, Rebulla P, Ratti I et al (2001) Outcomes of a program to evaluate mother and baby 6 months after umbilical cord blood donation. Transfusion 41:606–610

Lee SH, Cheuh H, Yoo KH et al (2011) Hematopoietic stem cell transplantation from a related donor infected with influenza H1N1 2009. Transpl Infect Dis 13:548–550

Leitner GC, Faschingbauer M, Wenda S et al (2014) Administration of recombinant human granulocyte-colony-stimulating factor does not induce long-lasting detectable epigenetic alterations in healthy donors. Transfusion 54:3121–3126

Ljungman P (2014) The role of cytomegalovirus serostatus on outcome of hematopoietic stem cell transplantation. Curr Opin Hematol 21:466–469

Ljungman P, Lawler M, Asjo B et al (1994) Infection of donor lymphocytes with human T lymphotrophic virus type 1 (HTLV-I) following allogeneic bone marrow transplantation for HTLV-I positive adult T-cell leukaemia. Br J Haematol 88:403–405

Locasciulli A, Bacigalupo A, Vanlint MT et al (1991) Hepatitis C virus infection in patients undergoing allogeneic bone marrow transplantation. Transplantation 52:315–318

Lopez A, De La Rubia J, Arriaga F et al (1998) Severe hemolytic anemia due to multiple red cell alloantibodies after an ABO-incompatible allogeneic bone marrow transplant. Transfusion 38:247–251

Lown RN, Philippe J, Navarro W et al (2014) Unrelated adult stem cell donor medical suitability: recommendations from the World marrow donor Association clinical working group committee. Bone Marrow Transplant 49:880–886

Mcclain KL (1997) Immunodeficiency states and related malignancies. Cancer Treat Res 92:39–61

Mejia R, Booth GS, Fedorko DP et al (2012) Peripheral blood stem cell transplant-related plasmodium falciparum infection in a patient with sickle cell disease. Transfusion 52:2677–2682

Mele L, Dallavalle FM, Verri MG et al (2005) Safety control of peripheral blood progenitor cell processing-eight year-survey of microbiological contamination and bag ruptures in a single institution. Transfus Apher Sci 33:269–274

M-Reboredo N, Diaz A, Castro A et al (2000) Collection, processing and cryopreservation of umbilical cord blood for unrelated transplantation. Bone Marrow Transplant 26:1263–1270

Muller B, Nubling CM, Kress J et al (2013) How safe is safe: new human immunodeficiency virus type 1 variants missed by nucleic acid testing. Transfusion 53:2422–2430

Naohara T, Suzuki G, Masauzi N et al (1997) Positive seroconversion syphilis in a patient with acute lymphocytic leukemia after allogeneic bone marrow transplantation. Rinsho Ketsueki 38:228–230

Nguyen XD, Dengler T, Schulz-Linkholt M et al (2011) A novel tool for high-throughput screening of granulocyte-specific antibodies using the automated flow cytometric granulocyte immunofluorescence test (Flow-GIFT). ScientificWorldJournal 11:302–309

Offergeld R, Ritter S, Hamouda O (2012) HIV-, HCV-, HBV- und Syphilis surveillance unter Blutspendern in Deutschland 2008–2010. Bundesgesundheitsbl 55:907–913

Pergam SA, Xie H, Sandhu R et al (2012) Efficiency and risk factors for CMV transmission in seronegative hematopoietic stem cell recipients. Biol Blood Marrow Transplant 18:1391–1400

Reichard KK, Zhang QY, Sanchez L et al (2006) Acute myeloid leukemia of donor origin after allogeneic bone marrow transplantation for precursor T-cell acute lymphoblastic leukemia: case report and review of the literature. Am J Hematol 81:178–185

Sakellari I, Papalexandri A, Batsis D et al (2014) EBV detected viral load in peripheral blood of donors at the time of transplant has no impact on EBV-lymphoprliferative diesease development in recipients. Bone Marrow Transplant 49:342–343

Salmon JP, Michaux S, Hermanne JP et al (1999) Delayed massive immune hemolysis mediated by minor ABO incompatibility after allogeneic peripheral blood progenitor cell transplantation. Transfusion 39:824–827

Schmidt M, Korn K, Nubling CM et al (2009) First transmission of human immunodeficiency virus type 1 by a cellular blood product after mandatory nucleic acid screening in Germany. Transfusion 49:1836–1844

Schmidt M, Geilenkeuser WJ, Sireis W et al (2014) Emerging pathogens—how safe is blood? Transfus Med Hemother 41:10–17

Schreiber GB, Busch MP, Kleinman SH et al (1996) The risk of transfusion-transmitted viral infections. The retrovirus epidemiology donor study. N Engl J Med 334:1685–1690

Schulze TJ, Nguyen XD, Stötzer F et al (2011) The flow cytometric granulocyte immunofluorescence test (Flow-GIFT) in donor and patient screening is able to improve the safety of granolucyte transfusions. Onkologie 34:868

Shapira MY, Kaspler P, Samuel S et al (2003) Granulocyte colony stimulating factor does not induce long-term DNA instability in healthy peripheral blood stem cell donors. Am J Hematol 73:33–36

Shuhart MC, Myerson D, Childs BH et al (1994) Marrow transplantation from hepatitis C virus seropositive donors: transmission rate and clinical course. Blood 84:3229–3235

Singavi AK, Harrington AM, Fenske TS (2015) Post-transplant lymphoproliferative disorders. Cancer Treat Res 165:305–327

Smith DK, Grohskopf LA, Black RJ et al (2005) Antiretroviral postexposure prophylaxis after sexual, injection-drug use, or other nonoccupational exposure to HIV in the United States: recommendations from the U.S. Department of Health and Human Services. MMWR Recomm Rep 54:1–20

Solves P, Mirabet V, Alvarez M (2014) Hepatitis B transmission by cell and tissue allografts: how safe is safe enough? World J Gastroenterol 20:7434–7441

Stramer SL (2014) Current perspectives in transfusion-transmitted infectious diseases: emerging and re-emerging infections. ISBT Sci Ser 9:30–36

Surapaneni SN, Hari P, Knox J et al (2007) Suppressive anti-HCV therapy for prevention of donor to recipient transmission in stem cell transplantation. Am J Gastroenterol 102:449–451

Theiler RN, Caliendo AM, Pargman S et al (2006) Umbilical cord blood screening for cytomegalovirus DNA by quantitative PCR. J Clin Virol 37:313–316

Thiele T, Kruger W, Zimmermann K et al (2011) Transmission of cytomegalovirus (CMV) infection by leukoreduced blood products not tested for CMV antibodies: a single-center prospective study in high-risk patients undergoing allogeneic hematopoietic stem cell transplantation (CME). Transfusion 51:2620–2626

Tomas JF, Rodriguez-Inigo E, Bartolome J et al (1999) Transplantation of allogeneic CD34-selected peripheral stem cells does not prevent transmission of hepatitis C virus from an infected donor. Bone Marrow Transplant 24:109–112

Trottier H, Buteau C, Robitaille N et al (2012) Transfusion-related Epstein-Barr virus infection among stem cell transplant recipients: a retrospective cohort study in children. Transfusion 52:2653–2663

Urahama N, Tanosaki R, Masahiro K et al (2003) TRALI after the infusion of marrow cells in a patient with acute lymphoblastic leukemia. Transfusion 43:1553–1557

Van Stein D, Beckers EA, Sintnicolaas K et al (2010) Transfusion-related acute lung injury reports in the Netherlands: an observational study. Transfusion 50:213–220

Vanneaux V, Fois E, Robin M et al (2007) Microbial contamination of BM products before and after processing: a report of incidence and immediate adverse events in 257 grafts. Cytotherapy 9:508–513

Villalba R, Fornes G, Alvarez MA et al (1992) Acute Chagas' disease in a recipient of a bone marrow transplant in Spain: case report. Clin Infect Dis 14:594–595

Walker CM, Van Burik JA, De for Te E et al (2007) Cytomegalovirus infection after allogeneic transplantation: comparison of cord blood with peripheral blood and marrow graft sources. Biol Blood Marrow Transplant 13:1106–1115

Worel N, Greinix HT, Keil F et al (2002) Severe immune hemolysis after minor ABO-mismatched allogeneic peripheral blood progenitor cell transplantation occurs more frequently after nonmyeloablative than myeloablative conditioning. Transfusion 42:1293–1301

Worthley DL, Ruszkiewicz A, Davies R et al (2009) Human gastrointestinal neoplasia-associated myofibroblasts can develop from bone marrow-derived cells following allogeneic stem cell transplantation. Stem Cells 27:1463–1468

Zehender G, Meroni L, De Maddalena C et al (1997) Detection of hepatitis C virus RNA in CD19 peripheral blood mononuclear cells of chronically infected patients. J Infect Dis 176:1209–1214

Chapter 2
Induced Pluripotent Stem Cell Therapy and Safety Concerns in Age-Related Chronic Neurodegenerative Diseases

Erden Eren, Erdogan Pekcan Erkan, Sermin Genc, and Kemal Kursad Genc

Abbreviations

6-OHDA	6-Hydroxydopamine
AD	Alzheimer's disease
AKT	RAC-alpha serine/threonine-protein kinase
APOE	Apolipoprotein E
APP	Amyloid precursor protein
BBB	Blood-brain barrier
BiP	Binding immunoglobulin protein
BSI	β-secretase inhibitor
ChAT	Choline acetyltransferase
c-Myc	Cytoplasmic Myc protein
CNS	Central nervous system
CNV	Copy number variation
Crispr/Cas9	Clustered regularly interspaced short palindromic repeat/CAS9 RNA-guided nucleases
DHA	Docosahexaenoic acid
DJ-1	PARK7
ECM	Extracellular matrix

E. Eren • E.P. Erkan • S. Genc
Department of Neuroscience, Institute of Health Sciences, Dokuz Eylul University Health Campus, Inciralti, 35340 Izmir, Turkey

Izmir Biomedicine and Genome Center, Dokuz Eylul University Health Campus, Inciralti, 35340 Izmir, Turkey

K.K. Genc (✉)
Department of Neuroscience, Institute of Health Sciences, Dokuz Eylul University Health Campus, Inciralti, 35340 Izmir, Turkey
e-mail: kkursadgenc@hotmail.com

© Springer International Publishing AG 2017
P.V. Pham, A. Rosemann (eds.), *Safety, Ethics and Regulations*, Stem Cells in Clinical Applications, DOI 10.1007/978-3-319-59165-0_2

ER	Endoplasmic reticulum
ES	Embryonic stem cells
FAD	Familial form of Alzheimer's disease
FDA	Food and Drug Administration
GABA	Gamma-aminobutyric acid
G-CSF	Granulocyte colony-stimulating factor
GSK-3	Activated glycogen synthase kinase 3
GSK-3	Glycogen synthase kinase 3
GWAS	Genome-wide associated studies
hES	Human embryonic stem cells
hiPSCs	Human-induced pluripotent stem cells
HLA	Human leukocyte antigen
HUVEC	Human umbilical vein endothelial cells
IFN-γ	Interferon gamma
IKBKAP	Inhibitor of kappa light polypeptide gene enhancer in B-cells, kinase complex-associated protein
iNSC	Induced neural stem cells
iPSC	Induced pluripotent stem cells
Klf4	Kruppel-like factor 4
LRRK2	Leucine-rich repeat kinase 2
LV	Lentivirus
MAO	Monoamine oxidase
MEFs	Mouse embryonic fibroblasts
MMP	Mitochondrial membrane permeabilization
MPTP	1-Methyl-4-phenyl-1,2,3,6-tetrahydropyridine
mtDNA	Mitochondrial DNA
NaB	Sodium butyrate
NAB2	*N*-aryl benzimidazole
NHP	Nonhuman primate
NMDA	*N*-methyl-D-aspartate
NOS	Nitric acid synthase
NPC	Neural progenitor cells
Nrf2	Nuclear factor (erythroid-derived 2)-like 2
NTFs	Neurofibrillary tangles
Oct3/4	Octamer 3/4
PCR	Polymerase chain reaction
PD	Parkinson's disease
PDAPP	Promoter-driven amyloid precursor protein
PDGF	Platelet-derived growth factor
PINK1	PTEN-induced putative kinase 1
PLG	Polylactide-co-glycolide
PSEN1	Presenilin 1
PSEN2	Presenilin 2
ROS	Reactive oxygen species
SAD	Sporadic form of Alzheimer's disease

SNCA   Synuclein, Alpha
Sox2   SRY-box containing gene 2
TALEN   Transcription activator-like effector nucleases
VEGF   Vascular endothelial growth factor
VPA   Valproic acid
VPS35   Vacuolar protein sorting-associated protein 35
VSV-G   Vesicular stomatitis virus G
ZFN   Zinc finger nucleases

2.1 Overview of Age-Related Chronic Neurodegenerative Diseases

2.1.1 Alzheimer's Disease

Alzheimer's disease (AD) was identified more than 100 years ago, and it is considered as the most common type of dementia (Alzheimer's Association 2015). AD is known as a progressive disease that affects primarily memory and cognitive and functional abilities (Nussbaum and Ellis 2003). Approximately 8 million new cases are recorded each year, and it is estimated that this number will reach 115 million by the end of 2050 (Jindal et al. 2014). Additionally, family members and other unpaid caregivers provided 17.9 billion hours of care in the USA (Alzheimer's Association 2015). However, the prevalence of AD around the world is changing depending on diagnostic criteria and other factors such as ethnicity, age, etc. (Hendrie et al. 2001; Hy and Keller 2000). Yet, definite diagnosis can only be done post-mortem; there are several studies, which are being studied to identify novel biomarkers for earlier diagnosis.

The pathological hallmarks of AD are loss of neurons in the hippocampus and extracellular senile plaques consisting of β-amyloid peptides and neurofibrillary tangles (NFTs), which are composed of hyperphosphorylated form of microtubule protein tau (De-Paula et al. 2012).

β-Amyloid is produced by cleavage of amyloid precursor protein (APP) with α-, β-, and γ-secretases. Normally, the cleavage of APP with first α-secretase and then γ-secretase occurs in the non-amyloidogenic pathway. However, involvement of β-secretase results in the formation of longer C-terminal fragment (C99), which contains amyloidogenic amino acid sequence. Further cleavage with γ-secretase yields β-amyloid peptides. β-Amyloid $_{(1-42)}$ is the most toxic form of amyloid oligomers, and it can aggressively accumulate in the extracellular niche, leading to neuronal cell death.

Another hallmark of AD is hyperphosphorylation of tau protein. Tau is a microtubule-associated protein, which binds to α- and β-tubulins for their stabilization. Additionally, its phosphorylation state is important in stabilization. Yet, its abnormal phosphorylation leads inability to bind tubulins that result in destabilization of microtubules and finally cell death.

AD have two types: sporadic form (SAD) and familial form (FAD). Mutations in three different genes result in familial AD. These genes are APP, which encodes amyloid precursor protein, and presenilin 1 and 2 (*PSEN1* and *PSEN2*), which encode parts of gamma-secretase family proteins. Individuals with mutations in one of those genes are likely to develop AD. However, familial AD constitutes only 1–2% of all AD cases.

SAD constitutes the vast majority of the disease. Since it is considered late onset, it develops in individuals >65 years old. Younger individuals can also develop AD before the age of 65, but this is rare. There are certain risk genes that may cause to develop AD. Apolipoprotein E (APOE) gene is the most validated risk gene, which has three different alleles. If an individual has at least one copy of apoε4 allele, the risk for developing AD is 3- to12-folds higher (Alzheimer's Association 2015).

2.1.2 Parkinson's Disease

Parkinson's disease (PD) is an idiopathic and chronic neurodegenerative disease that primarily affects motor functions. It is the second most common neurodegenerative disease, which was named in 1800s in the honor of James Parkinson. PD is not common in younger adults aged below 40 (Beitz 2014). The prevalence of the disease varies with increasing age. Additionally, several factors affect the prevalence of PD; these includes geographical location, sex, and age (Pringsheim et al. 2014). Moreover, several studies have shown that exposure to exogenous toxins, genetic background, inflammation, and their combinations can increase the chance of developing PD (Bartels and Leenders 2009).

PD is characterized by the loss of dopaminergic neurons in the substantia nigra and the formation of Lewy bodies. Together with the formation of cellular inclusions, these findings represent the hallmarks of pathophysiology of PD (Davie 2008). Lewy bodies contain neurofilamentous proteins along with the proteins that are responsible for proteolysis including ubiquitin, a heat shock protein. Mutations in α-synuclein are responsible for familial PD. However, mutations in *parkin* gene can cause parkinsonism, without the formation of Lewy bodies. Furthermore, *LRRK 2* gene is known to cause sporadic, idiopathic, or familial PD (Davie 2008). Genome-wide association studies (GWAS) have revealed that mutations in different genes may cause PD development. Those include *SNCA*, *VPS35*, *PINK1*, and *DJ-1* in addition to genes that are mentioned above. Mutations in *SNCA*, *LRRK2*, and *VPS35* genes are known as an autosomal dominant cause of PD. Furthermore, mutations in *parkin*, *DJ-1*, and *PINK1* genes are an autosomal recessive form of PD and accounted for early-onset parkinsonism (Bonifati 2014).

Clinical diagnosis is based on some physical changes and requires accurate anamnesis. Those characteristics include rest tremor, rigidity, bradykinesia, and postural instability. Additionally, some other clinical signs are worth to give attention such as problems in handwriting and reduced facial expression (Hughes et al. 1992). Since, it's shown that Lewy bodies first accumulate in the olfactory bulb, reduced sense of smell cannot be ruled out for more accurate diagnosis (Hawkes 1995).

2.2 Induced Pluripotent Stem Cells Technology

Embryonic stem (ES) cells are capable of differentiating into cells of all three germ layers. They are able to proliferate indefinitely, while preserving their pluripotency. Furthermore, they hold great promise to treat neurodegenerative diseases, such as AD and PD. However, ethical concerns have emerged about the use of ES cells, since they are found in the inner mast of the blastocysts in addition to tissue rejection problems (Vazin and Freed 2010).

To overcome those issues regarding ES cells, new ways have to be found to produce stem cells while maintaining their pluripotency and self-renewal capabilities. In 2006, Yamanaka and his coworkers found a new way to obtain from somatic cells, and his work was granted the Nobel Prize in Physiology or Medicine 2012. They found that using four transcription factors known as "Yamanaka factors" can reprogram mouse embryonic and adult fibroblasts into pluripotent stem cells. These factors are octamer 3/4 (Oct3/4), SRY-box containing gene 2 (Sox2), cytoplasmic Myc protein (c-Myc), and Kruppel-like factor 4 (Klf4) (Takahashi and Yamanaka 2006).

In 2007, Yamanaka and his coworkers move a step to further their work and use adult human fibroblasts to produce induced pluripotent stem cells (iPSCs) by using the same defined transcription factors (Takahashi et al. 2007). Since then, a great number of studies have been done to develop iPSCs technology. Furthermore, human somatic cells were reprogrammed with Oct4, Sox2, Nanog, and LIN28 (Yu et al. 2007). For this purpose, different reprogramming factors, small compounds, mRNAs, and proteins are being used to enhance efficiency for the generation of iPSCs. Moreover, different delivery methods and sources are being examined.

Apart from fibroblasts, various cell types are being used to generate iPSCs, since reprogramming the efficiency and quality of iPSCs differs among different cells. To date, different cells have been used as source for iPSCs, such as primary hepatocytes, exfoliated renal epithelial cells, umbilical cord and peripheral blood cells, keratinocytes (Raab et al. 2014), pancreatic β cells (Stadtfeld et al. 2008), melanocytes (Utikal et al. 2009), neural cells (Kim et al. 2008), and adipose tissue cells (Sugii et al. 2010). Furthermore, human umbilical vein endothelial cells (HUVEC) are reprogrammed into iPSCs and differentiated to astrocytes and neurons (Haile et al. 2015). The choice of cell origin for reprogramming depends on several factors including reprogramming efficiency, availability, invasiveness, and methods to be used (Durnaoglu et al. 2011). Fibroblasts are still the first choice for iPSCs reprogramming studies. There are some disadvantages to start with fibroblasts. First, fibroblasts are obtained from the skin by punch biopsy. This procedure is very painful and has some risks such as bleeding and infection. Other disadvantages are longtime period and efficiency. The whole reprogramming takes a long time (5 weeks), and the efficiency is quite low compared with keratinocytes. Peripheral blood is another source for iPSCs generation. The donor should be prepared with G-CSF injection and then CD34$^+$ cells isolated via 4 h of apheresis. The convenient alternative source is keratinocyte. It is possible to obtain keratinocytes easily from scalp hair. In addition, these cells can be reprogrammed faster, and the method has higher efficiency.

2.2.1 *Reprogramming Methods*

There are several methods to deliver reprogramming factors into the cells. These methods can be classified according to the vector type: viral vector based, naked DNA based, and non-DNA based (de Lazaro et al. 2014).

2.2.1.1 Viral Vector-Based Methods

Retroviruses are the most used and known method for reprogramming. This method contains higher risk of immunogenicity, and the integration of reprogramming factors into genome can be a problem for further applications of iPSCs. Moreover, lentiviruses (LV) are used for low-efficiency problems of retroviruses, since they can only transduce dividing cells, and they also have enhanced tropism owing to vesicular stomatitis virus G (VSV-G) pseudotyping. Integration can still be an issue for LV transduction (Hu 2014). However, the use of excisable transgenes with LV vectors may overcome this problem (Sommer et al. 2010). Another method involves adenoviruses, which make transgene-free iPSCs possible; yet it has a low expression of reprogramming factors and higher integration frequency than naked plasmid DNA. *Sendai virus* (SeV)-based vectors are DNA-free vectors, and a single vector can contain all four reprogramming factors. So, efficiency is much higher compared to using four different vectors. In addition, genomic integration does not occur, and removal of virus particles is much easier. Thus, safety and immunogenicity can be established. Furthermore, alphaviruses are used to deliver RNA replicons. However, integration can occur due to cDNA conversion in the target cells.

2.2.1.2 Naked DNA-Based Methods

Other method involves the use of naked DNA. For this purpose PiggyBac transposons, plasmids, and episomal plasmids are used. Using plasmids for the delivery of reprogramming factors requires repeated transfection steps, and also current transfection methods are inadequate for reprogramming. Nucleofection can be used, but this method requires a relatively high number of cells, as it results in significant cell death. Nevertheless, some random integration can occur. Using episomal plasmids can be delivered by using *Epstein-Barr virus*, which can replicate in human cells. With this method, the quality of iPSCs is high and has lower immunogenicity. However, efficiency is very low, and additional reprogramming factors are required. Furthermore, polycistronic sequences can also be used. By this way, integration into genome can be reduced, and every infected cell receives all four factors, but packing into viral particles is hard due to a larger size of plasmid. In addition to polycistronic sequences, Lox sequences can be added using 2A sequences. Using Cre recombinase, excision of integrated sequences can be easy. The PiggyBac transposon system is a more advanced version of plasmid system, since it provides excision without any genomic alteration and lower immunogenicity. Yet, an extra excision step is required, and imperfect excision may occur (Hu 2014).

2.2.1.3 Non-DNA-Based Methods

Non-DNA-based methods include the use of synthetic mRNAs, miRNA mimics, and small compounds. mRNA transfection can induce innate immune response through toll-like receptors, which leads to severe cytotoxicity. However, synthetic mRNAs can bypass innate responses and allow the generation of transgene-free iPSCs. This method has higher efficiency and low toxicity. Additionally, functionality of mRNAs is higher due to translation in the cytoplasm and proper posttranslational modifications. The main disadvantage of this system is that expression time is low (about 2–3 days), and repeated transfection is needed.

miRNAs are a class of short, noncoding RNAs, which regulate their target mRNAs by binding to the 3′ untranslated regions (UTRs), 5′ UTRs, or open reading frames (ORFs). miRNAs have key regulatory functions, starting from the embryonic development and extending to cellular differentiation and growth. Thus, it is not surprising that miRNAs are associated with pluripotency of stem cells. Earlier studies have demonstrated the requirement for ES cell-specific miRNA signatures for self-renewal and differentiation of ES cells (Kanellopoulou et al. 2005; Jia et al. 2013). Afterwards, several studies have shown that miRNAs can be used as reprogramming factors. Advantages of this system include the ease of their synthesis, non-integrating nature of miRNAs, and controllable administration. Furthermore, miRNA expression in the cytoplasm is relatively longer, and less transfection is needed (Hu 2014).

iPSCs generation can also be done using proteins of four reprogramming factors. Thus, there is no need for any exogenous genetic material that can cause integration into genome. However, this technique requires permeabilization of cell membranes prior to the delivery of proteins. So, there are a few techniques to achieve this problem. One of them is the usage of cell-penetrating peptides, which contain high amount of basic amino acids. These peptides can be linked with C-terminus of four reprogramming factors. These fusion proteins can be produced in *E. coli* or HEK293 cell line. Furthermore, nuclear localization signal peptide can be fused to reprogramming factors, and this provides minimalization of lysosomal degradation of proteins (Li et al. 2014).

Small compounds are also used to generate iPSCs. These molecules enhance the efficiency of iPSCs generation. For this purpose, histone deacetylation, demethylation, and methyltransferase inhibitors are used. Furthermore, signaling pathway inhibitors (e.g., glycogen synthase kinase 3, GSK-3) and epigenetic modulators can also be used. The purpose is to increase efficiency and generate iPSCs without using genetic materials. One of these compounds is the valproic acid (VPA), which inhibits histone deacetylation. It enhances iPSC generation and can be used as replacement for c-Myc. Furthermore, this chemical can improve efficiency as far as 1000-fold. Another chemical used for this purpose is sodium butyrate (NaB). It can be used throughout the whole process, and reprogramming efficiency is increased (Revilla et al. 2015). Moreover, lithium also has increased the efficiency of reprogramming of both mouse endothelial fibroblasts (MEFs) and HUVECs (Wang et al. 2011; Wu et al. 2013; Masuda et al. 2013). In addition, vitamin C can reduce senescence state partially. Sodium chloride can reduce nearly all demethylation levels via hyperosmosis. Ascorbic acid and GSK3-β inhibitor can facilitate reprogramming as well (Revilla et al. 2015). Besides, Hou et al. showed that VPA, CHIR99021, 616,452, tranylcypromine, forskolin, 3-deazaneplanocin A, 2-methyl-5-hydroxytryptamine hydrochloride, and D4476 are used to generate iPSCs from mouse somatic cells (Hou et al. 2013).

2.3 Applications of iPSCs

Animal models are used to understand the mechanisms of neurodegenerative diseases and to screen potential drugs and seeking therapeutic strategies. However, generating models that accurately mimic the disease as in human physiology is a problem, since there are differences in species, cell-line specificity, and lack of brain complexity (Wan et al. 2014). Furthermore, there are no models for rare diseases, and using animal models to observe disease progression remains difficult and raises some ethical issues regarding using too much animals. Moreover, screening for new drugs and performing toxicity tests for available drugs are time-consuming. An additional challenge is to obtain cells (e.g., neurons) from living individuals. In this context, Daley's group developed disease-specific iPSC models for the first time. They use fibroblasts and bone marrow mesenchymal cells to generate disease-specific iPSCs including PD, Huntington's disease, and Down syndrome (Park et al. 2008).

Using iPSCs technology to establish disease models has its own advantages. This technology allows us to model diseases more accurately. Therefore, it can provide insight into the mechanistic basis of the diseases and leads to discovery of new effective treatment strategies. Furthermore, high-throughput chemical screening with iPSCs allows predicting more accurate drug-induced toxicity. Additionally, cell replacement therapies with patient-specific iPSCs are the ultimate goal, and it can develop our current personalized medicine strategies for various diseases.

2.3.1 Disease Modeling

There are inherent differences between the nervous systems of rodents and humans, and difference in life spans of those species may also cause inability to serve as appropriate AD and PD models. Modeling sporadic and familial AD with iPSCs provides understanding the mechanisms of AD pathology and establishing new drug testing platforms. Recent studies revealed that human-generated iPSCs could be used for disease modeling. Initial studies were focused on familial AD mutations, since these are more homogenized and well characterized (Doege and Abeliovich 2014).

First AD-specific iPSCs were produced from the skin fibroblast of familial AD patients with *PSEN1* and *PSEN2* mutations (Yagi et al. 2011). Then iPSC-derived neurons from different familial AD mutations (*APP*, *PSEN1*, and *PSEN2*) have been generated to study the pathogenesis of the disease (Table 2.1). Woodruff and colleagues used transcription activator-like effector nucleases (TALENs) to introduce ΔE9 PSEN1 mutations, whether the mutation reduced γ-secretase activity in iPSC-derived neural cells (Woodruff et al. 2013).

Additionally, tau phosphorylation was also increased in iPSC-derived neurons from patients with familial AD (Israel et al. 2012). In another study, iPSCs, which are reprogrammed from fibroblasts of one daughter and father, carrying APP London mutation (V717I), differentiated into forebrain neurons, and they showed AD-like phenotypes and increased levels of $A\beta_{(42)}$ and $A\beta_{(38)}$ and t-tau and p-tau. Interestingly, they found an alteration in γ-secretase cleavage site (Muratore et al. 2014). Heterogenic phenotypes were seen in iPSC-derived neurons. For example, increased phosphorylated tau levels were not seen in the neurons carrying *PSEN1* or *PSEN2* mutation (Yagi et al. 2011).

In addition to AD phenotypes, differential gene expression changes were seen in iPSC-derived neurons from patients with familial AD. iPSC-derived neuron with different PSEN1 mutations have shown that ten different genes have been upregulated, and four genes have been downregulated along with increased generation of $A\beta_{(42)}/A\beta_{(40)}$ (Yagi et al. 2011; Sproul et al. 2014; Liu et al. 2014). Furthermore, in one study, iPSC-derived cortical neurons have increased the endoplasmic reticulum and oxidative stress, and also accumulated Aβ oligomers are prone to proteolysis (Kondo et al. 2013).

Apart from these models, iPSC-derived neurons can be obtained from sporadic AD patients who carry the risk gene allele Apoϵ4 and others (Table 2.1). iPSC-derived cholinergic neurons carrying Apoϵ4 allele showed elevated $A\beta_{(42)}/A\beta_{(40)}$ ratio, increased calcium levels within the cytoplasm upon glutamate exposure, and sensitivity for neurotoxic stimuli (Duan et al. 2014). iPSC-derived neurons generated from sporadic form of AD showed increased levels of $A\beta_{(1-40)}$ and phosphor-tau$_{(Thr231)}$ levels along with activated GSK-3β. Yet, one of them has increased Aβ levels in neurons and increased ER and oxidative stress also (Israel et al. 2012; Kondo et al. 2013).

PD is the other common neurodegenerative disease. There are many studies which are used in iPSC-derived dopaminergic neurons to model PD. Despite of familial PD cases compromising 5–10% of total PD cases (Nishimura and Takahashi 2013), most of the iPSCs are derived from fibroblasts of patients who have familial, while a few studies use iPSCs from sporadic PD patients. Patient-derived iPSC, which have different mutations in *LRRK2*, *PINK1*, *SCNA*, and *PARK2*, can differentiate into dopaminergic neurons as control iPSCs. These iPSC-derived dopaminergic neurons show various phenotypes including increased oxidative stress, increased α-synuclein expression and elevated mitochondrial gene expressions, etc. (Table 2.2). These findings are consistent with non-iPSCs models and brain autopsies (Lee et al. 2012b). Moreover, other studies have revealed novel phenotypes, which are worth to investigate. For instance, increased monoamine oxidase (MAO) activity was observed in *PARK2* mutant iPSC-derived dopaminergic neurons, and cells showed increased dopamine release and decreased uptake (Jiang et al. 2012). Furthermore, Ryan et al. reported that A53T α-synuclein mutant

Table 2.1 Disease modeling studies with patient-specific iPSCs in Alzheimer's disease

Gene	Genomic alteration	Reprogramming delivery methods	iPSC-derived cell types	Phenotype	Reference
APP	Duplication	Retroviral (*OCT4, SOX2, KLF4, c-MYC*, and ± *EGFP*)	Neuron (glutamergic and GABAergic)	Increased Aβ40, p-tau(Thr 231) and aGSK-3β	Israel et al. (2012)
	E693Δ	Episomal vector (*SOX2, KLF4, OCT4, L-MYC, LIN28*, and small hairpin RNA for p53)	Cortical neuron, astrocyte	Aβ oligomers accumulated in iPSC-derived neurons, increased ER, and oxidative stress	Kondo et al. (2013)
	V717 L	Episomal vector (*SOX2, KLF4, OCT4, L-MYC, LIN28*, and small hairpin RNA for p53)	Cortical neuron, astrocyte	Intracellular Aβ oligomers were not detected, increased extracellular Aβ42/Aβ40 ratio	Kondo et al. (2013)
	V717I	Lentiviruses (Oct4, SOX2, cMYC, and KLF4)	Forebrain neurons	Alteration in initial cleavage site of γ-secretase, increased Aβ42 and Aβ38, increased total and p-tau	Muratore et al. (2014)
	Trisomy 21	Retroviral (*OCT4, SOX2, cMYC, KLF4*, and *NANOG*)	Neuron	**Increased Aβ peptide production and aggregation,** hyperphosphorylation, and redistribution of tau (localized to cell bodies and dendrites)	Shi et al. (2012)

PS 1	A246E	Retroviral (OCT4, SOX2, KLF4, LIN28, and NANOG)	Neuron	Increased Aβ secretion, no tau accumulation	Yagi et al. (2011)
		Retroviral (Oct4, KLF4, SOX2, and c-Myc)	Neuron	Elevated levels of Aβ42, increased Aβ42/Aβ40 ratio	Liu et al. (2014)
		Retroviral (Oct4, KLF4, SOX2, and c-Myc)	Basal forebrain cholinergic	Not increased Aβ42/Aβ40 ratio	Duan et al. (2014)
		Retroviral (Oct4, KLF4, SOX2, and c-Myc)	Neuron	Increased generation of Aβ42/40, ten upregulated genes (*ASB9*, *BIK*, *C7orf16*, *NDP*, *NLRP2*, *PLP1*, *SLC45A2*, *TBX2*, *TUBB4*, *ZNF300*), and four downregulated genes (*ADM2*, *FLJ35024*, *MT2A*, *PTGS2*)	Sproul et al. (2014)
		Episomal vector (OCT4, SOX2, NANOG, LIN28), (OCT4, SOX2, SV40LT, KLF4), or (OCT4, SOX2, MYC, KLF4)	Neuron	Elevated levels of Aβ42, increased Aβ42/Aβ40 ratio	Mahairaki et al. (2014)
	M146 L	Retroviral (Oct4, KLF4, SOX2, and c-Myc)	Neuron	Increased generation of Aβ42/40, ten upregulated genes (*ASB9*, *BIK*, *C7orf16*, *NDP*, *NLRP2*, *PLP1*, *SLC45A2*, *TBX2*, *TUBB4*, *ZNF300*), and four downregulated genes (*ADM2*, *FLJ35024*, *MT2A*, *PTGS2*)	Sproul et al. (2014)
		Retroviral (Oct4, KLF4, SOX2, and c-Myc)	Neuron	Elevated levels of Aβ42, increased Aβ42/Aβ40 ratio	Liu et al. (2014)
	H163R	Retroviral (Oct4, KLF4, SOX2, and c-Myc)	Neuron	Elevated levels of Aβ42, increased Aβ42/Aβ40 ratio	Liu et al. (2014)
	ΔE9	Retroviral from normal human fibroblast (*OCT4*, *SOX2*, *KLF4*, *c-MYC*, and ± *EGFP*), mutation generated with TALENs	Neuron	Decreased Aβ40, increased Aβ42/Aβ40 ratio	Woodruff et al. (2013)
	L166P	Retroviral	Neuron	Partial loss of γ-secretase function, decreased production of endogenous Aβ40, and an increased Aβ42/β40 ratio	Koch et al. (2012)

(continued)

Table 2.1 (continued)

Gene	Genomic alteration	Reprogramming delivery methods	iPSC-derived cell types	Phenotype	Reference
PS2	N1411	Retroviral (OCT4, SOX2, KLF4, LIN28, and NANOG)	Neuron	Increased Aβ secretion, no tau accumulation	Yagi et al. (2011)
Sporadic	*ApoE3/E4* genotypes	Retroviral (Oct4, KLF4, SOX2, and c-Myc)	Basal forebrain cholinergic neurons	Increased Aβ42/40 ratios, increased sensitivity to neurotoxic stimuli, increased cytoplasmic calcium levels upon glutamate exposure	Duan et al. (2014)
	None	Retroviral (*OCT4, SOX2, KLF4, c-MYC,* and ± *EGFP)*	Neuron	One patient iPSC shows increased Aβ40, p-tau(Thr 231), and active glycogen synthase kinase-3β, one patient iPSC is normal	Israel et al. (2012)
	None	Episomal vector (*SOX2, KLF4, OCT4, L-MYC, LIN28,* and small hairpin RNA for p53)	Cortical neuron, astrocyte	Aβ oligomers accumulated in iPSC-derived neurons, increased ER and oxidative stress	Kondo et al. (2013)
	None	Retroviral (Oct4, KLF4, SOX2, and c-Myc)	Neuron	N/A	Yagi et al. (2011)
	None	Retroviral (Oct4, KLF4, SOX2, and c-Myc)	Neuron	Positive for neuronal cell markers (PAX6, NESTIN, and b-TUBULIN III), expressed p-tau and GSK3B, down-regulation of AD-related genes	Hossini et al. (2015)

Table 2.2 Disease modeling studies with patient-specific iPSCs in Parkinson's disease

Gene	Genomic alteration	Initial cell	Reprogramming methods	iPSC-derived cell types	Phenotype	Genetic correction	Reference
SNCA	Triplication	Fibroblast	Retroviral (Oct4, KLF4, SOX2, and c-Myc)	Dopaminergic neuron	Increased α-synuclein expression, gene expression profile is similar with hES cell	N/A	Devine et al. (2011)
					Increased α-synuclein expression, overexpression of markers of oxidative stress (DNAJA1, HMOX2, UCHL1, HSPB1, and MAO-A), and sensitivity to peroxide-induced oxidative stress	N/A	Byers et al. (2011)
					Reduced proliferative capacity in NPCs, altered cellular energy balance and decreased mitochondrial function, delayed protein import and increased protein aggregation, increased cellular stress and reactive oxygen species	Lentiviral SNCA gene knockdown	Flierl et al. (2014)
					Impaired mitochondrial integrity and permeability transition, Caspase activation and apoptosis	N/A	Chung et al. (2013)
	A53T (G209A)	Fibroblast	Lentiviral vectors (Oct4, KLF4, SOX2 c-Myc, and Nanog)	Dopaminergic neuron	Higher number of CNVs, ZFN correction leads loss of expression of the mutated A53T (G209A) transcript, the genetic repair did not affect differentiation capacity of iPSc	ZFN	Soldner et al. (2011)
			Lentiviral vectors (Oct4, KLF4, SOX2, and c-Myc)	Dopaminergic neuron	α-synuclein aggregation, increased oxidative and nitrosative stress, increased toxin mediated apoptosis	ZFN	Ryan et al. (2013)
			Lentiviral vectors (Oct4, KLF4, SOX2, and c-Myc)	Dopaminergic neuron	Endoplasmic reticulum form of glucoserebrosidase and nicastin accumulated	ZFN	Chung et al. (2013)

(continued)

Table 2.2 (continued)

Gene	Genomic alteration	Initial cell	Reprogramming methods	iPSC-derived cell types	Phenotype	Genetic correction	Reference
PARK2	Heterozygous exon 3 and 5 deletion	Fibroblast	Lentiviral vectors (Oct4, KLF4, c-Myc, and Nanog)	Dopaminergic neuron	Increased dopamine release, decreased dopamine uptake, elevated oxidative stress, increased MAO transcript	Lentiviral overexpression of parkin	Jiang et al. (2012)
					Reduce the length and complexity of neuronal processes, reduce microtubule stability	Lentiviral overexpression of parkin	Ren et al. (2015)
	Exon 3 deletion	Fibroblast	Lentiviral vectors (Oct4, KLF4, c-Myc, and Nanog)	Dopaminergic neuron	Increased dopamine release, decreased dopamine uptake, elevated oxidative stress, increased MAO transcript	Lentiviral overexpression of parkin	Jiang et al. (2012)
					Reduce the length and complexity of neuronal processes, reduce microtubule stability	Lentiviral overexpression of parkin	Ren et al. (2015)
	Exon 2–4 deletion	Fibroblast	Retroviral (Oct4, KLF4, SOX2, and c-Myc)	Dopaminergic neuron	Increased oxidative stress and enhanced activity of Nrf2, abnormal mitochondrial morphology	N/A	Imaizumi et al. (2012)
	Exon 6–7 deletion						
	V324A	Fibroblast	*Sendai virus* (KLF4, SOX2, and c-Myc)	Dopaminergic neuron	Progerin overexpression causes increased apoptosis, enhanced dendrite shortening, and reduced AKT activation	N/A	Miller et al. (2013)
	Heterozygous R275W				Increased in condensed nuclei and caspase-3, progerin overexpression leads loss of dendrite length	N/A	Miller et al. (2013)
	40 bp deletion in exon 3 of one allele and exons 5 and 6 deletion in the other allele	Fibroblast	Retroviral (Oct4, KLF4, SOX2, and c-Myc) and episomal (OCT3/4, SOX2, KLF4, L-MYC, LIN28, and shRNA for TP53)	NPC	Mn exposure was associated with significantly higher reactive oxygen species	N/A	Aboud et al. (2012)

	R42P and exon 3 deletion	Fibroblast	*Sendai virus* (Oct4, KLF4, SOX2, and c-Myc)	Dopaminergic neuron	Impaired dopaminergic differentiation on the percentage of TH+ neurons was decreased, reduced mitochondrial volume fraction, increased expression of mitochondrial genes and cell death genes	N/A	Shaltouki et al. (2015)
	Exons 3–4 deletion 1-bp deletion 255A						
	R275W						
	R42P						
PINK1	Q456X	Fibroblast	Retroviral (Oct4, KLF4, SOX2, and c-Myc)	Dopaminergic neuron	Impaired stress induced mitochondrial translocation of parkin, increased mitochondrial copy number, and upregulation of PGC-1α	Lentiviral overexpression of PINK1	Seibler et al. (2011)
			Retroviral (Oct4, KLF4, SOX2, and c-Myc)	Dopaminergic neuron	Vulnerability to cell stressors, oxidative stress	N/A	Cooper et al. (2012)
			Sendai virus (KLF4, SOX2, and c-Myc)	Dopaminergic neuron	With progerin overexpression causes increased apoptosis, enlarged mitochondria	N/A	Miller et al. (2013)
	V170G	Fibroblast	Retroviral (Oct4, KLF4, SOX2, and c-Myc)	Dopaminergic neuron	Impaired recruitment of lentivirally expressed parkin to mitochondria, increased mitochondrial copy number, and upregulation of PGC-1α	Lentiviral overexpression of PINK1	Seibler et al. (2011)
					Detectable mitophagy when parkin gene is overexpressed in fibroblast but not iPSC derived neurons	Lentiviral overexpression of PINK1	Rakovic et al. (2013)

(continued)

Table 2.2 (continued)

Gene	Genomic alteration	Initial cell	Reprogramming methods	iPSC-derived cell types	Phenotype	Genetic correction	Reference
LRRK2	G2019S	Fibroblast, keratinocytes	Retroviral (Oct4, KLF4, SOX2, and c-Myc) or (Oct4, KLF4, and SOX2)	Dopaminergic neuron	Increased expression of key oxidative stress-response genes and a-synuclein protein; increased sensitivity to caspase-3 activation and cell death caused by exposure to stress agents (hydrogen peroxide, MG-132, and 6-hydroxydopamine), decreased tau expression, and increased α-synuclein, 6-OHDA sensitivity	N/A	Nguyen et al. (2011)
		Fibroblast	Retroviral (Oct4, KLF4, SOX2, and c-Myc)	Neural cells	Decreased oxygen consumption rate, dysfunctional mitochondrial mobility, sensitive to valinomycin and concanamycin A	N/A	Cooper et al. (2012)
		Fibroblast	Retroviral (Oct4, KLF4, SOX2, and c-Myc)	Dopaminergic neuron	LMX1A overexpression efficient way for generate dopaminergic neurons	N/A	Sanchez-Danes et al. (2012a)
		Fibroblast, keratinocytes	Retroviral (Oct4, KLF4, SOX2, and c-Myc)	Dopaminergic neuron	Reduced numbers of neurites, impaired autophagy	N/A	Sanchez-Danes et al. (2012b)
		Fibroblast	Retroviral (Oct4, KLF4, SOX2, and c-Myc)	NPCs	Increased susceptibility to proteasomal stress and passage-dependent deficiencies in nuclear-envelope organization, clonal expansion, and neuronal differentiation	Gene targeting with HDAdV	Liu et al. (2012)
		Fibroblast	Retroviral (Oct4, KLF4, SOX2, and c-Myc)	Dopaminergic neuron	Impaired autophagy, increased α-synuclein accumulation	N/A	Orenstein et al. (2013)

	Fibroblast	Retroviral (Oct4, KLF4, SOX2, and c-Myc)	Dopaminergic neuron	Neurite shortening and sensitivity to neurotoxins (rotenone, 6-OHDA), tau upregulation, dysregulation of CPNE8, MAP7, UHRF2, ANXA1, and CADPS2, correction decreased sensitivity of neurons to rotenone	ZFN	Reinhardt et al. (2013)
	Fibroblast	Retroviral (Oct4, KLF4, SOX2, and c-Myc)	Dopaminergic neuron	Altered mitochondrial morphology, less mitochondria loss of MMP, increased mitoROS, and decreased ATP levels	N/A	Su and Qi (2013)
	Fibroblast	Retroviral (Oct4, KLF4, SOX2, and c-Myc)	NPC, mature neural cell	Increased mtDNA damage	ZFN	Sanders et al. (2014)
R1441C	Fibroblast	Retroviral (Oct4, KLF4, SOX2, and c-Myc)	Neural cells	Decreased oxygen consumption rate, dysfunctional mitochondrial mobility, sensitive to valinomycin and concanamycin A	N/A	Cooper et al. (2012)
	Fibroblast	Retroviral (Oct4, KLF4, SOX2, and c-Myc)	NPC, mature neural cell	Increased mtDNA damage	ZFN	Sanders et al. (2014)

(continued)

Table 2.2 (continued)

Gene	Genomic alteration	Initial cell	Reprogramming methods	iPSC-derived cell types	Phenotype	Genetic correction	Reference
GBA	N370S/84GG insertion	Fibroblast	Retroviral (Oct4, KLF4, SOX2, and c-Myc)	Dopaminergic neuron	Lysosomal protein degradation, causes accumulation of α-synuclein and results in neurotoxicity	N/A	Mazzulli et al. (2011)
	L444P N370S	Fibroblast	lentiviral vectors (Oct4, KLF4, SOX2, and c-Myc)	Dopaminergic neuron, astrocyte, oligodendrocyte	Very low GC enzymatic activity	N/A	Panicker et al. (2012)
Sporadic	None	Fibroblast	Retroviral (Oct4, KLF4, SOX2, and c-Myc)	N/A	N/A	N/A	Park et al. (2008)
	None	Fibroblast, keratinocytes	Retroviral (Oct4, KLF4, SOX2, and c-Myc)	Dopaminergic neuron	Reduced numbers of neurites, impaired autophagy	N/A	Sanchez-Danes et al. (2012b)
	None	Fibroblast	lentiviral vectors (Oct4, KLF4, SOX2, and c-Myc) or (Oct4, KLF4, and SOX2)	Dopaminergic neuron	Differentiation efficiency similar with control. TH+ and TUJ + neurons generated	N/A	Soldner et al. (2009)

iPSC-derived cells nitrosative/oxidative stress resulted in S-nitrosylation of myocyte enhancer factor 2C (MEF2C) (Ryan et al. 2013). Neural progenitor cells (NPC) differentiated from both healthy and PARK2 mutation-carrying individuals without any PD manifestations showed that manganese (Mn) treatment did not result in any difference between groups. Yet, Mn treatments caused increased reactive oxygen species (ROS) levels in mutated iPSC-derived NPCs (Aboud et al. 2012). Ren et al. have shown that iPSC-derived neurons carrying parkin mutation have decreased microtubule stability and shorter neurite length. Further, overexpression of parkin gene restores microtubule stability and complexity of neural processes (Ren et al. 2015).

Not all studies are related with familial PD mutations. iPSC-derived dopaminergic neurons generated from sporadic PD patient cells have revealed that their phenotype is similar to those found in familial PD. However, dopaminergic neurons generated from sporadic PD-derived iPSCs need to be cultured for a long time in cell culture to be able to show PD-related phenotype (Sanchez-Danes et al. 2012b).

Taken together, these findings are accelerating the research on neurodegenerative disease and lead to new translational approaches such as high-throughput drug screening. In spite of new developments, there are still major concerns to overcome before using iPSC technology.

2.3.2 Drug Screening and Testing

The current drug discovery methods are time-consuming and expensive, as well as failure rate is higher due to serious side effects such as cardiotoxicity and hepatotoxicity. Approximately 90% of the drugs are not able to reach the market. Additionally, 30% of the drugs are given up due to side effects and lack of efficiency in clinical trials (Singh et al. 2015). Furthermore, safety data come from animal models, and interpreting the results is not efficient due to species-specific differences. Using human cell-based toxicity test can overcome these problems, since organ-specific cells can be used for high-throughput toxicity screening, while ethical concerns and time-consuming procedures of animal usage are emerging (Heilker et al. 2014).

Current treatments for AD include cholinesterase inhibitors, which are used to treat mild to moderate AD, and *N*-methyl-D-aspartate (NMDA) receptor antagonists which are used to treat moderate to severe AD. However, these treatment strategies rely on only improving symptoms. To date, there are no drugs that can reverse neuronal loss and stop the cognitive decline in AD. Several experimental therapy options exist; these include immunotherapies, which target to enhance Aβ clearance such as bapineuzumab, solanezumab, and intravenous immunoglobulins. Moreover, gamma-secretase inhibitors and modulators have been also tested in clinical trials. However, these studies have failed to pass phase II and III trials. Herbal supplements such as docosahexaenoic acid (DHA) have the potential as drugs in the treatment of AD, and they need to be investigated further to provide as symptomatic treatment option. Conclusively, there are also significant amount of potential drug failures in late-stage clinical trials, yet these failures may alter the future of novel therapy options (Berk and Sabbagh 2013).

Despite the advancements in PD treatment, there is no drug that can cure PD completely. The major challenge for this problem is that molecular mechanism of PD pathology remains unknown and primary cause of dopaminergic neuron loss is also not known. Several drugs have been found to be effective in animal models; however, they failed in clinical trials due to the aforementioned problems related to animal models. Also, doses of drugs used in clinical trials may not be effective. Furthermore, there is no drug used for neuroprotection. Coenzyme Q, green tea, creatine, and minocycline have no effect on disease progression. Currently used drugs focus on the improvement of symptoms as in AD treatment. Levodopa (best known antiparkinsonian drug), dopamine agonists, glutamate antagonists, MAO B inhibitors, and catechol-O-methyltransferase (COMT) inhibitors are used to improve symptoms such as dyskinesia (Stocchi 2014).

Currently, experimental drugs are being evaluated using iPSC-derived neurons and dopaminergic neurons in both AD and PD models (Table 2.3). Kondo and colleagues used β-secretase inhibitor (BSI), DHA, NSC23766 (Rac1 inhibitor), and dibenzoylmethane (DBM14–26) to examine the effects on familial and sporadic AD iPSC-derived neurons. The authors found no change in the levels of Aβ oligomers; on the other hand, they found that DHA reduced ROS generation as well as cleaved caspase-4 and peroxiredoxin-4 in neurons. Furthermore, high dose of DBM14–26, NSC23766, or DHA treatment elevated binding immunoglobulin protein (BiP) levels. In familial AD mutant iPSC-derived neurons, long-term DHA treatment increased cell viability. However, the same treatment did not alter cell survival of sporadic AD iPSC-derived neurons (Kondo et al. 2013). Furthermore, Israel et al. used β and γ-secretase inhibitors to examine the relationship among amyloid-β, p-tau, and GSK-3β. Both inhibitors reduced A$\beta_{(1-40)}$. However, only β-secretase inhibitors (βSi-II and OM99–2) significantly reduced aGSK-3β and p-tau/total tau (Israel et al. 2012). Moreover, compound E, a γ-secretase inhibitor, was used to examine its effects on a different mutation carrying iPSC-derived neurons in AD model. Compound E reduced both A$\beta_{(42)}$ and A$\beta_{(40)}$ levels in both PSEN1 and PSEN2 mutations carrying iPSC-derived neurons, and they further used compound W (selective Aβ lowering agent) and found reduced A$\beta_{(42)}$/A$\beta_{(40)}$ ratio in iPSC-derived neurons (Yagi et al. 2011). Another study has shown that compound E treatment reduces p-tau levels in AD-iPSC-derived neurons (Hossini et al. 2015).

Experimental drugs are also being evaluated in iPSC-based PD models. Cooper et al. have used coenzyme Q_{10}, rapamycin, and LRRK2 inhibitor (GW 5074) in iPSC-derived neurons and found that coenzyme Q_{10} reduced cell vulnerability to valinomycin and concanamycin. However, rapamycin did not change cell death induced by concanamycin. Furthermore, GW 5074 reduced cell death by valinomycin but not concanamycin (Cooper et al. 2012). Moreover, N-arylbenzimidazole (NAB2) reversed the PD phenotype such as mitochondrial dysfunction in A53T mutation carrying iPSC-derived neurons (Chung et al. 2013). In another study, they used L-NAME, a nitric oxide synthase (NOS) inhibitor, to prevent S-nitrosylation for the enhancement of MEF2C-PGC1α pathway. They found that L-NAME pretreatment partially recovered pesticide-induced cell death (Ryan et al. 2013).

Table 2.3 Drug testing using iPSCs derived from in Alzheimer's disease and Parkinson's disease patients

Disease	Molecules	Genetic of iPS	Results	Reference
Alzheimer's disease	β-secretase inhibitor (BSI)	APP (E693Δ)	Intracellular Aβ oligomers decreased, inhibited secretion of Aβ40 from control neural cells, decreased the amount of binding immunoglobulin protein, and cleaved caspase-4	Kondo et al. (2013)
	β-secretase inhibitor II	APP duplication Sporadic AD	Decreased Aβ40, aGSK-3β, p-tau/total tau	Israel et al. (2012)
	OM99–22 (β-secretase inhibitor)	APP duplication Sporadic AD	Decreased Aβ40, aGSK-3β, p-tau/total tau	Israel et al. (2012)
	Docosahexaenoic acid (DHA)	APP (E693Δ)	Decreased binding immunoglobulin protein, cleaved caspase-4, peroxiredoxin-4, ROS, not altered Aβ oligomer	Kondo et al. (2013)
	Docosahexaenoic acid (DHA)	Sporadic AD	decreased binding immunoglobulin protein, peroxiredoxin-4	Kondo et al. (2013)
	DBM14–26	APP (E693Δ)	Increased binding immunoglobulin protein	Kondo et al. (2013)
	NSC23766 (Rac 1 inhibitor)	APP (E693Δ)	Increased binding immunoglobulin protein	Kondo et al. (2013)
	DAPT (gamma-secretase inhibitor)	APP (V717I)	Decreased Aβ38, Aβ40, and Aβ42	Muratore et al. (2014)
		Trisomy 21	Decreased Aβ42 and Aβ40	Shi et al. (2012)
		PS1 (A246E) *ApoE3/E4* genotypes	Variable responses (increased secretion Aβ40 at 100 nM and 200 nM one *ApoE3/E4* genotypes iPSC and decreased the other *ApoE3/E4* genotypes iPSC; inhibited Aβ40 production in all iPSC at 400 nM and 800 nM)	Duan et al. (2014)
		PS1 (L166P)	Secretion of both Aβ40 and Aβ 42 was strongly reduced	Koch et al. (2012)
		APP duplication Sporadic AD	Decreased Aβ40	Israel et al. (2012)

(continued)

Table 2.3 (continued)

Disease	Molecules	Genetic of iPS	Results	Reference
	CPD-E (gamma-secretase inhibitor)	APP duplication Sporadic AD	Decreased Aβ40	Israel et al. (2012)
	Compound E (gamma-secretase inhibitor)	PS1 (A246E), PS2 (N1411)	Decreased Aβ42 and Aβ40	Yagi et al. (2011)
		ApoE3/E4 genotypes	Variable responses (200 pm concentration of compound E increased A β40, 400 pm concentration of compound E decreased Aβ40 in other iPSC line)	Duan et al. (2014)
		Sporadic	Reduced p-Tau in one iPSC but did not result in any reduction of p-tau in other iPSC; p-GSK-3β and GSK-3β expression did not alter	Hossini et al. (2015)
	Compound W (selective Abeta1–42 lowering agent)	PS1 (A246E), PS2 (N1411)	Decreased in the ratio Aβ42 to Aβ40 in neurons	Yagi et al. (2011)
	Semagacestat (gamma-secretase inhibitor)	PS1 (A246E)	Decreased Aβ42 and Aβ40, the ratio Aβ42 to Aβ40	Liu et al. (2014)
	Gamma-secretase modulator-4	PS1 (A246E, M146, H163R) PS2 (N1411)	Decreased Aβ42 and Aβ40 not Aβ38 and total Aβ	Liu et al. (2014)
	Ibuprofen	PS1 (L166P)	Reduced the secretion of Aβ42 and Aβ42/Aβ40	Koch et al. (2012)
	Indomethacin	PS1 (L166P)	Reduced the secretion of Aβ42 and Aβ42/Aβ40	Koch et al. (2012)

Parkinson's disease	Coenzyme Q10	PINK1 (Q456X), LRRK2 (G2019S, R1441C)	Reduced cell vulnerability to low-dose valinomycin and concanamycin A	Cooper et al. (2012)
	Rapamycin		Reduced cell death induced by valinomycin but not concanamycin A in LRRK2 mutant cells. PINK1 mutant cells unresponsive to valinomycin	
	GW 5074 (LRRK2 inhibitor)		Reduced cell death induced by valinomycin but not concanamycin A in PINK1 mutant cells; reduced cell death induced both valinomycin and concanamycin A	
	Y-27632 (ROCK inhibitor)	LRRK2 (G2019S)	No effect on hydrogen peroxide, exacerbate cell death induced by MG-132	Nguyen et al. (2011)
	N-arylbenzimidazole (NAB2)	SCNA A53T (G209A)	Reversed pathologic phenotypes	Chung et al. (2013)
	L-NAME (NOS inhibitor)	SCNA A53T	Rescue from paraquat and MANEB-induced death	Ryan et al. (2013)
	P110 (inhibitor of fission dynamin-related protein 1	LRRK2 (G2019S)	Reduced mitochondrial impairment, lysosomal hyperactivity, and neurite shortening	Su and Qi (2013)

Currently, animal models are used for drug screening and toxicity tests. Yet these systems are merely imperfect replicas of human system. Molecules which are found to be toxic in one animal species may not be toxic for another species. Furthermore, newly discovered drugs should be tested on human cells or the human itself; since it is not exactly possible, we need a system to mimic the conditions in human physiology. iPSC-derived cells are ideal sources for drug screening and toxicity testing, and they represent diseases more accurately in vitro and facilitate drug discovery efforts. Developing more robust and reliable differentiation techniques will improve the application of iPSCs to drug development for different diseases. Also, it is possible to assess how an individual may respond to certain drugs. Furthermore, significant risks and costs of early-stage clinical trials can be avoided. Thus, development of new drugs and patient-specific treatments can increase easily (Qi et al. 2014; Ko and Gelb 2014).

Lee et al. used patient-specific iPSC-derived neural crest precursor cells for drug screening. They performed high-throughput assay which consists of 6912 compounds. They found eight candidate compounds to rescue *IKBKAP* expression, and one of them was SKF-86466 which induces *IKBKAP* expression (Lee et al. 2012a). Furthermore, a low-throughput assay composed of 44 compounds was performed by using iPSC-derived dopaminergic neurons in MPP^+ and rotenone toxicity. They found that 16 of 44 compounds showed a neuroprotective effect (Peng et al. 2013). Moreover, 3.313 drugs were screened using iPSC-derived hepatocytes, yielding 263 hit compounds, 42 of which are approved by the Food and Drug Administration (FDA). Further screening of these 42 compounds showed that 5 compounds were found to be consistent by showing a similar effect on four different patient-specific iPSC-derived hepatocytes (Choi et al. 2013).

Therefore, it is possible to say that iPSC-based drug screening provides a safer, cost-effective, and faster way to develop new drugs and testing existing drugs for toxicity and side effects. Additionally, comparing healthy and diseased iPSC-derived mature cells provides valuable information about disease mechanisms, and novel molecular therapeutic targets can be found in a dish (Giri and Bader 2015).

2.3.3 Cell Replacement Therapy

The first transplantation of iPSC-derived cells was used for the treatment of humanized sickle cell anemia mouse model (Hanna et al. 2007). iPSC-derived cell replacement is a new alternative for the treatment of neurodegenerative diseases. Several groups transplanted iPSC-derived cells into the brain in preclinical animal models of neurodegenerative diseases including AD and PD. Transplanted cells can survive in the different brain regions and provide functional recovery (Table 2.4). Human clinical studies using iPSCs in neurodegenerative disease have not started yet.

The important advantages of autologous iPSCs for transplantation therapy are absence of immune rejection risk and ethical problems. On the other hand, a previous study has shown that autologous undifferentiated iPSCs elicit a very strong

Table 2.4 Cell therapy using iPSCs in preclinical models of Alzheimer's disease and Parkinson's disease

	Animal	Model	iPS source	Cell numbers	Route of administration	Evaluation time	Results	Reference
AD	Mice	APP	hiPSC cell lines, 201B7 and 253G1	2×10^5	Hippocampus		Spatial memory improved, choline acetyltransferase (ChAT) positive cholinergic human neurons and vesicle GABA transporter (VGAT) positive GABAergic human neurons detected in the hippocampus	Fujiwara et al. (2013)
PD	Rat	6-OHDA	hiPSC from PD patient	1×10^5	Intrastriatal	16 week	Behavioral recovery, transplanted cell survive and functional	Hargus et al. (2010)
	Rat	6-OHDA	iPSC cell line (IMR90, clone 4)	1×10^6	Intrastriatal	6 week	No behavioral recovery, but hiPSC mDA progenitor cells survive long term and many develop into bona fide mDA neurons	Cai et al. (2010)

(continued)

Table 2.4 (continued)

Animal	Model	iPS source	Cell numbers	Route of administration	Evaluation time	Results	Reference
Rat	6-OHDA	Autologous fibroblast	7.5×10^5	Intrastriatal	8 weeks	Dramatic functional recovery in amphetamine-induced rotation scores, abundant TH+ cells	Rhee et al. (2011)
			2.5×10^5	Intrastriatal	8 weeks	Moderate but significant level of functional recovery The graft size was more appropriate and uniform, without irregular proliferating mass or rosette formation	
			2.5×10^5 (terminal differentiated)	Intrastriatal	8 weeks	No behavioral recovery, no TH+ cells were readily detectable in the graft. Severe cell death was evident from the central area of the grafts	
Rat	6-OHDA	Embryonic mouse fibroblast	1×10^5 with DHA	Intrastriatal	4 months	DHA induces the differentiation of functional dopaminergic precursors and improves the abnormal behavior	Chang et al. (2012)
Rat	6-OHDA	Mice iNSC derived with transdifferentiation from Sertoli cells	8×10^5—Lmx1a	Intrastriatal	10 weeks	Decreased ipsilateral rotations, a few transplanted cells survive	Wu et al. (2015)
Macaca mulatta monkeys	MPTP	Autologous fibroblast	1×10^5	Caudate nucleus, putamen and the substantia nigra	6 months	iPSC-derived neural progenitors survive for up to 6 months and differentiate into neurons, astrocytes, and myelinating oligodendrocytes	Emborg et al. (2013)

immune response with high lymphocytic infiltration and elevated interferon-gamma (IFN-γ), granzyme B, and perforin intragraft (de Almeida et al. 2014). Generation of autologous iPSCs is a time-consuming process and delays iPSC-based cell therapies. Therefore, allogenic iPSC-derived cells provide a useful strategy for transplantation therapy in acute brain disorders such as stroke. It requires well-characterized human leukocyte antigen (HLA)-typed iPSC lines and their biobanking. Even so, small difference in culture conditions alters gene expressions (Newman and Cooper 2010). The differentiation stage of iPSCs is an important point for the successful outcome in transplantation therapy. Differentiated neurons are less immunogenic and tumorigenic and do not require differentiation factors. Due to the ability of NSCs to differentiate into a variety of cell types, including neurons, astrocytes, and oligodendrocytes, they become a favorable cell type in cell replacement therapy. However, undifferentiated iPSCs could cause teratoma formation in the transplanted brain region (Kawai et al. 2010).

The most critical factor affecting the success of stem cell therapy is the route of administration. Intravenous and intraperitoneal routes are the easiest ways; however, the number of cells that reach the brain is limited (Li et al. 2015). Additionally, blood-brain barrier (BBB) also limits iPSCs to cross to the brain. More specific methods should be found for delivering iPSCs into the brain. Direct intrastriatal and intranigral routes have been successfully used for the transplantation of iPSC-derived dopaminergic neurons in animal model of PD (Nishimura and Takahashi 2013). Intracerebral injection could be a more specific way, but it is invasive and carries tissue injury risk throughout the route of administration (Martinez-Morales et al. 2013). Intracerebroventricular route may help widespread distribution of iPSCs into the CNS, but it is also an invasive route for stem cell delivery (Li et al. 2015). Intranasal route is an easy and noninvasive delivery method for stem cell therapy. In addition, it is suitable for repeated administration (Li et al. 2015). This route was used for delivering other types of stem cells, but not for iPSCs delivery, even in animal experiments.

Genome editing methods allow the correction of mutations in iPSCs from individuals carrying mutations. Zinc finger nucleases (ZFNs) are the first used genome editing method for mutation correction. It has some disadvantages such as off-target effects and cell toxicity (Velasco et al. 2014). Similar to ZFNs, TALENs also generate double-strand breaks at target site in the genome (Gupta and Musunuru 2014). The advantages of TALENs include easier design, low levels of off-target effects, and toxicity; however, the size of TALENs limits their use in stem cell therapy. The third genome editing tool is clustered regularly interspaced short palindromic repeat/CAS9 RNA-guided nucleases (CRISPR/CAS9), which has gained attention due to easier design, high success rate, low cost, and side effects (Velasco et al. 2014).

Before proceeding to clinical trials, the benefits and safety of iPSCs-derived cell transplantation should be evaluated in in vivo animal studies. Toxin-based animal models of PD, especially the 6-OHDA model, were used for iPSCs transplantation studies (Table 2.4). Intrastriatal delivery method was preferred due to regional localization of degenerative neurons. Most studies have confirmed long-term survival of transplanted dopaminergic neurons, which provide behavioral recovery (Hargus et al. 2010; Swistowski et al. 2010; Rhee et al. 2011; Chang et al. 2012; Wu et al. 2015). Although there was a high concentration of iPSC-derived dopaminergic neuron transplantation, functional recovery was not observed in short-term periods (Cai et al. 2010). Additionally, transplantation of terminally differentiated cells results in ineffective engraftment (Rhee et al. 2011). The available iPSCs transplantation methods still need to be improved before clinical trial. For instance, selection of subtypes of neurons using cell sorting may increase transplantation success (Doi et al. 2014). The first in vivo iPSC-derived neuronal precursor cell transplantation study for AD was carried out in platelet-derived growth factor (PDGF) promoter-driven amyloid precursor protein (PDAPP) transgenic mice (Fujiwara et al. 2013). iPSC-derived cholinergic neurons were transplanted into bilateral hippocampus of the 10-week-old PDAPP mice. Transplanted neurons survive and show cholinergic and GABAergic phenotypes in the recipient mouse brain 45 days after transplantation. Additionally, transplantation of the neurons restored spatial memory dysfunction of PDAPP mice.

Nonhuman primates (NHPs) have anatomical and functional similarities compared to humans. In addition, gene expression profile in the brain is also similar between NHPs and humans (Verdier et al. 2015). These similarities make NHPs useful animal models to study neurodegenerative diseases. NHP models enable the monitoring of long-term outcome of the transplanted cells (Qiu et al. 2013). Moreover, these models provide considerable information about aging process in the brain due to display characteristics similar to human aging. AD-related pathological findings, including Aβ accumulation, tau phosphorylation, and atrophy, also naturally occur in NHPs (Verdier et al. 2015). They are also valuable animals for Parkinson's disease studies. While 1-methyl-4-phenyl-1,2,3,6-tetrahydropyridine (MPTP) does not show toxicity in rats, it leads to loss of dopaminergic neurons of the substantia nigra with PD symptoms in NHPs (Capitanio and Emborg 2008). In addition to toxin models, mutant A53T α-synuclein over-expressing monkey and transgenic A53T monkey have been generated as NHPs animal models of PD (Eslamboli et al. 2007; Niu et al. 2015). NHPs have been used for iPSCs transplantation studies for PD. Dopaminergic neurons derived from human iPSCs were transplanted to the putamen of MPTP-lesioned cynomolgus monkeys, and transplanted cells survived in the monkey brain for 6 months (Kikuchi et al. 2011). Autologous fibroblast of *Macaca mulatta* monkeys was used for the generation of iPSCs in NHPs model of PD, and iPSC-derived neural progenitors survived 6 months and differentiated into neurons and glial cells (Emborg et al. 2013).

The first human clinical study using iPSC derivatives was started for the treatment of patients suffering from age-related macular degeneration (Okano and Yamanaka 2014). The results of this study have not been reported yet. There are also several planned clinical trials of iPSCs-based therapies (Okano and Yamanaka 2014). But, there is no record on web pages of clinical trials for iPSCs-derived cell replacement therapy (clinicaltrials.gov, Accessed 04 May 2015). Clinical studies should be started with safety and side effects analysis.

2.4   Future Directions: Challenges and Advancements

2.4.1   Limitations in iPSC Generation and Potential Solutions

2.4.1.1   Integration into Genome

One of the major limitations of iPSC technology is low efficiency. The efficiency is as low as 0.001% with current methods. Besides, the current methods are slow and require the overexpression of multiple transcription factors at the same time. Retroviral systems are still the most used methods for generating iPSCs. This system uses transduction of reprogramming factors into host genome, which may result in random integration into genome, karyotype abnormalities, and copy number variations. Further, this problem can also affect differentiation efficiency. Furthermore, major goals in iPSC generation are avoiding genomic integration with developing more efficient non-integrative method. Moreover, the elimination of residual transgene expression and reactivation of reprogramming factors should be achieved (Hu 2014). Another problem is that individual iPSC clones have differences among them even if they are generated from the same individual. This might affect differentiation efficiency. Gender and usage of integration factors may cause this problem. If iPSCs are reprogrammed from cells of a female person, cells have X chromosome inactivation which can lead to altered expression of cognition and brain development-related genes. If integrating method was used, there might be incomplete transgene silencing (Zhao et al. 2014). For avoiding this problem, proper PCR screening, excisable vector, or non-integrative methods can be used that are mentioned in the reprogramming of iPSCs. However, these non-integrating methods are not perfect and have disadvantages such as lower efficiency. There are various ways to excise those integration transgenes, but these methods have their own disadvantages such as micro-deletions in genomic DNA. Lastly, high-quality iPSCs must be met along with high efficient reprogramming.

2.4.1.2   Epigenetic Memory

Epigenetic memory could be a problem in differentiation and reprogramming. A study showed that iPSCs showed significant reprogramming variability independent of reprogramming technique including epigenetic memory of somatic cells and abnormal DNA methylation after reprogramming (Lister et al. 2011). However, reprogramming cells can erase epigenetic signature but imperfectly. This can affect reprogramming and differentiation capacity. Furthermore, disease-related epigenetic signature can be required for showing disease phenotype, so removal of epigenetic signature may result in the loss of disease phenotype (Doege and Abeliovich 2014).

2.4.1.3 Differentiation and Purity

The use of iPSCs in disease modeling has various limitations and problems. Currently, there are no standardized and optimized differentiation protocols for a given cell type. In additionally, the available protocols are time-consuming and inefficient. For instance, iPSC differentiation into dopaminergic neurons takes about 21–70 days, and the efficiency differs with respect to techniques used (Badger et al. 2014).

After differentiation, cell population must be validated. For this purpose, immunochemistry and PCR methods are used for cell-specific markers. In addition to those, in-depth analysis of neuron functionality can be performed via calcium imaging, dopamine release, and electrophysiological properties such as patch clamp method (Badger et al. 2014). Further, differentiation yields heterogeneous population because of maturation at different time points. Hu et al. has shown that regardless of cell origin which iPSCs are derived from, differentiation is highly variable and not efficient. They investigated PAX6+ levels and found that PAX6+ expression is variable among different clones and clones that were generated from iPSCs reprogrammed in the same fibroblasts (Hu et al. 2010). Selecting clones with the same differentiation capacity may eliminate this variation. Furthermore, disease phenotype can be mixed with abnormal phenotypes (Zhao et al. 2014). Moreover, current methods are unable to generate specific cell type in reliable and high amounts. Differentiation results in mixed cell types. For example, differentiation neurons from iPSCs results in cell population composed of neurons and glial cells (Kondo et al. 2013). However, if the cell type of interest is unknown for a given disease, analyzing multiple cell types at once will be an opportunity, since state-of-the-art methods can be used for single cell analysis.

In addition to those problems, most of the studies focused on cell-autonomous models. While this approach is acceptable in first steps, the developing brain is not working in that way. Notably, cell-cell interactions are required for proper modeling, but it is hard to study in differentiation, synapse formation, etc. Studying diseases at the network level may be required for adequately modeling diseases. For this purpose, complex cellular interactions are needed rather than single-cell analysis (REF).

Furthermore, aging is another problem which researchers are faced, since developing AD and PD takes decades in humans. However, current iPSC-dependent models in culture take a couple of months. Fibroblasts taken from elder individuals have shown aging markers, whereas iPSCs reprogrammed from these fibroblasts did not show any of those markers (Miller et al. 2013). Furthermore, progerin, which is associated with premature aging syndrome "progeria," overexpression in iPSCs provided aging-associated marker expression (Miller et al. 2013), and progerin expression may promote degenerative phenotypes in iPSC-derived models. Moreover, it is proposed that environmental factors (toxins, nutrion stress, etc.) may promote aging in culture (Doege and Abeliovich 2014).

2.4.1.4 Control Groups for Research

Selection of control groups in iPSC-derived disease models is another problem. Proper control groups are required for revealing disease mechanism and/or drug screening. They can be generated from the same age, gender, and ethnic group of healthy controls. However, they have different genetic background along with different risk factor exposure (Santostefano et al. 2015). Ideally, isogenic control groups should be used. To be able to obtain such cells, mutations in iPSCs should be corrected via proper gene editing tools which are ZNFs, TALENs, and Crispr/Cas9 system (Xu and Zhong 2013).

2.4.2 Safety Concerns for Clinical Grade iPSCs

2.4.2.1 Tumorigenicity, Immunogenicity, and Genomic Instability

Tumorigenicity is one of the major concerns in the usage of iPSCs, since oncogenes are used for reprogramming somatic cells into iPSC. Furthermore, the potential presence of undifferentiated iPSCs can also cause tumor formation after transplantation. It is shown that reprogramming without oncogenes (c-Myc and Klf4) can reduce tumor formation in mice. Alternative methods can be used to achieve this problem such as non-viral methods mentioned earlier (Sect. 2.2.1).

Genomic aberrations can still occur in iPSCs regardless how they were generated. In most cases, reprogramming does not result in alterations in the karyotype, but in some instances, it is possible to see abnormalities in karyotype. Furthermore, when genomic stability is looked closer, it can be seen that subkaryotypic changes can happen during reprogramming or in prolonged subculture periods. Copy number variation (CNV) can be detected in iPSC lines. After CNV analysis, early passages of iPSCs can have deletions in tumor suppressor genes, and early passages tend to have more deletions than late passages. On the other hand, amplifications in oncogenes tend to occur in late passages. Furthermore, mutations in exons (i.e., protein-coding regions) can occur, and most of these mutations are acquired during reprogramming or in culture of iPSCs. Mutations are maintained during culture of iPSCs. All in all, genomic instability of iPSCs, whether gained during reprogramming or in culture, can affect the quality of iPSCs in clinical use (Martins-Taylor and Xu 2012).

Another safety concern in iPSC-based therapy is the issue of immunogenicity. Generally, autologous cell therapy is generally considered as immune safe. However, Zhao and colleagues showed that iPSC cell therapy can induce immune response in syngeneic recipients. Furthermore, they showed that after iPSC implantation, immune rejection occurred via T-cell infiltration. Moreover, they identified two genes (*Hormad1* and *Zg16*), which were expressed in iPSC-teratoma, and these genes directly contribute immunogenicity of iPSC derivatives (Zhao et al. 2011). Immunogenicity is essential in iPSC-based cell therapy. Using autologous iPSC-derived cell can bypass immune rejection. However, autologous iPSC confer some practical problems. The most important one is that it is time-consuming. Further, clone selection, differentiation, and characterization require more time. If donors have mutations in their genome, correction of mutated gene or genes is a must, and this requires additional time.

2.4.2.2 Biobanking

iPSC banking is an essential matter in terms of both research and clinical applications. It should assure scientific reproducibility in iPSC researches. Furthermore, iPSC lines can have genomic and epigenetic variations, their quality must be checked carefully, and genotyping is necessary for providing required genotypic iPSC lines for scientific researches (Stacey et al. 2013). Additionally, iPSC biobanks can meet cell demands in cell replacement therapies. Patient-specific iPSCs can be used in cell replacement therapies. However, their generation is time-consuming and expensive. Additionally, such cells can have genetic defects which have to be corrected. Moreover, the quality of such cells can be an issue. iPSC biobanks can provide high quality and low chance of immune rejection iPSCs. Furthermore, iPSC supply of these banks cannot be depleted due to the nature of iPSCs. Biobanks, which are based on HLA matching, can be used to avoid immune rejection problems. HLA is a polymorphic gene inherited with monogenic dominant Mendelian manner. Overall, more than 2558 possible HLA classes (HLA I and II) exist. However, according to one estimate, 150 lines are sufficient to match with 90% of England population, and more diverse population may need more than 150 lines. Therefore, it can be said that the creation of HLA-matched-based banks with sufficient HLA matching which represent different geographical population can ease iPSC-based therapies (Solomon et al. 2015).

2.4.2.3 Clearance of Animal Products

Contamination with animal products is an important issue in terms of producing clinical-grade iPSCs and biobanking. The use of animal products in reprogramming and differentiating iPSCs contains the risk of unknown pathogens, exogenous antigens, etc., because there would be unpredictable risks to humans. Animal product-free culture medium is also important. Furthermore, animal-derived MEFs are used for feeder layer. Alternatives to mouse-derived feeder layer and human-derived feeder layer were developed. However, producing them is time- and effort consuming. Matrigel can be used in human iPSCs generation, yet Matrigel is derived from Engelbreth-Holm-Swarm mouse. Other alternatives including recombinant proteins, CellStart®, and synthetic polymers can be used instead of Matrigel (Seki and Fukuda 2015).

2.4.3   Recent Biological and Biotechnological Advancements

2.4.3.1   Alternative Strategies for Reprogramming

iPSC generation includes nuclear reprogramming. These iPSCs are in a transient pluripotent state which are susceptible to chromosomal aberrations. The generation of iPSC results in almost complete epigenetic memory erasure. Furthermore, methods for generation and differentiation of iPSCs are time-consuming and expensive. There are alternative ways to achieve those problems. Direct reprogramming is one of them. Epigenetic memory is not completely erased after reprogramming. Moreover, iPSC generation is performed in vitro, but direct reprogramming can be performed in both in vitro and in vivo (Amamoto and Arlotta 2013), for instance, overexpression of Pax6 in astrocytes isolated from postnatal cerebral cortex of mice differentiated into neurons (Heins et al. 2002). However, some cell types are not always feasible when human physiology is considered. This problem leads the scientist into more lineage distant cell types such as skin fibroblasts. Vierbuchen et al. showed that mouse tail fibroblast could be directly differentiated into neurons with three distinct transcription factors, which are Brn2, Ascl1, and Myt1l (Vierbuchen et al. 2010). Thus, these methods can provide time-efficient patient-specific cells for both research and clinical applications. Apart from these, in vivo reprogramming is a developing area. The advantages of it include cells residing in their native tissue, having low tumorigenesis risk, and having new cells autologous in origin. Furthermore, delivery of transcription factor into specific cell type requires virus mediated transfer. However, it can have unknown consequences (Heinrich et al. 2015).

2.4.3.2   Three-Dimensional (3D) and Organoid Cultures

iPSCs are capable to receive early developmental signals. Thus, when specific signal is given to iPSCs, different cell type can be differentiated autonomously via interacting each other and the environment. Two-dimensional cultures are limited to deliver full potential of iPSCs. Using 3D culture with functional biomaterials create more relevant physiological environment (Shao et al. 2015). Using human midbrain-derived neural progenitor cells, 3D cell culture has been established and called as 3D neurospheres containing functional dopaminergic neurons, oligodendrocytes, and astrocytes (Simao et al. 2015). Sophisticated 3D organoid cultures can also be used to mimic this differentiation process. Lancaster et al. were able to generate 3D organoid culture with human iPSCs. This system includes cerebral cortex (Lancaster et al. 2013). Additionally, similar systems have been developed by various groups (Dye et al. 2015; Beauchamp et al. 2015).

2.4.3.3 Biotechnological Strategies

Biomaterials

The efficiency of iPSC expansion and differentiation process can be improved by controlling the microenvironment. For this purpose, biomaterials, which are designed to interact with cells, can be used. Nano- or microparticles can control reprogramming factors as well as modulating epigenetic state of iPSCs. Biomaterials can be designed to deliver reprogramming factors more safely and efficiently. Furthermore, they can increase the efficiency of iPSC derivation by controlling the duration of exposure to extracellular matrix (Tong et al. 2015). Moreover, artificial transcription factors can be generated such as NanoScript which was designed as a platform for mimicking transcription factor domains. It contained nuclear localization signal, DNA binding domain, and activation domain (Patel et al. 2014). Additionally, factors can be integrated into scaffolds, and their differential release can be adjusted. For instance, PLGA-based scaffold was used for differential release of both vascular endothelial growth factor (VEGF) and PDGF. VEGF was mixed with polymer for rapid release, and PDGF was pre-encapsulated with polylactide-co-glycolide (PLG) for extended release (Richardson et al. 2001).

Bioprinting

Three-dimensional bioprinting is also biomaterial-based iPSC-derived organ systems. Functional tissues and organs can be produced by using biological elements such as cells via 3D bioprinting. These tissues/organs can further be transplanted. Spatial control of layers can be controlled, and desired biological properties can be obtained. Furthermore, with nanoscale resolution of bioprinting, it is possible to construct closely mimicking physiological properties of a desired tissue/organ. Moreover, this technique diminishes vascularization and innervation problems in 3D culture systems (Tong et al. 2015). Kolesky et al. showed that they printed intricate and heterogeneous tissue construct supplied with vasculature, extracellular matrix (ECM), and multiple cell types by their novel 3D printing method (Kolesky et al. 2014).

2.5 Conclusion

iPSCs technology by Shinya Yamanaka and colleagues in 2006 was a groundbreaking invention. Since then, numerous advancements have been made in the field of iPSC research. This development can provide the treatment of incurable diseases. Consequently, iPSCs can overcome immune rejection problems and problems faced with ES cells such as ethical issues. Although, there are some disadvantages of this technology, numerous studies are being conducted for achieving those obstacles. Recent technical advances have provided to overcome safety issues in clinical use of iPSCs. However, these methods are still in their infancy. Further investigations will provide much safer and efficient ways to the use of iPSCs in clinical applications.

References

Aboud AA, Tidball AM, Kumar KK, Neely MD, Ess KC, Erikson KM, Bowman AB (2012) Genetic risk for Parkinson's disease correlates with alterations in neuronal manganese sensitivity between two human subjects. Neurotoxicology 33(6):1443–1449. doi:10.1016/j.neuro.2012.10.009

Alzheimer's Association (2015) 2015 Alzheimer's disease facts and figures. Alzheimers Dement 11(3):332–384. doi:10.1016/j.jalz.2015.02.003

Amamoto R, Arlotta P (2013) Reshaping the brain: direct lineage conversion in the nervous system. F1000Prime Rep 5:33. doi:10.12703/P5-33

Badger JL, Cordero-Llana O, Hartfield EM, Wade-Martins R (2014) Parkinson's disease in a dish—using stem cells as a molecular tool. Neuropharmacology 76(Pt A):88–96. doi:10.1016/j.neuropharm.2013.08.035

Bartels AL, Leenders KL (2009) Parkinson's disease: the syndrome, the pathogenesis and pathophysiology. Cortex 45(8):915–921. doi:10.1016/j.cortex.2008.11.010

Beauchamp P, Moritz W, Kelm JM, Ullrich ND, Agarkova I, Anson BD, Suter TM, Zuppinger C (2015) Development and characterization of a scaffold-free 3D spheroid model of induced pluripotent stem cell-derived human cardiomyocytes. Tissue Eng Part C Methods. doi:10.1089/ten.TEC.2014.0376

Beitz JM (2014) Parkinson's disease: a review. Front Biosci 6:65–74

Berk C, Sabbagh MN (2013) Successes and failures for drugs in late-stage development for Alzheimer's disease. Drugs Aging 30(10):783–792. doi:10.1007/s40266-013-0108-6

Bonifati V (2014) Genetics of Parkinson's disease—state of the art, 2013. Parkinsonism Relat Disord 20(Suppl 1):S23–S28. doi:10.1016/S1353-8020(13)70009-9

Byers B, Cord B, Nguyen HN, Schule B, Fenno L, Lee PC, Deisseroth K, Langston JW, Pera RR, Palmer TD (2011) SNCA triplication Parkinson's patient's iPSC-derived DA neurons accumulate alpha-synuclein and are susceptible to oxidative stress. PLoS One 6(11):e26159. doi:10.1371/journal.pone.0026159

Cai J, Yang M, Poremsky E, Kidd S, Schneider JS, Iacovitti L (2010) Dopaminergic neurons derived from human induced pluripotent stem cells survive and integrate into 6-OHDA-lesioned rats. Stem Cells Dev 19(7):1017–1023. doi:10.1089/scd.2009.0319

Capitanio JP, Emborg ME (2008) Contributions of non-human primates to neuroscience research. Lancet 371(9618):1126–1135. doi:10.1016/S0140-6736(08)60489-4

Chang YL, Chen SJ, Kao CL, Hung SC, Ding DC, Yu CC, Chen YJ, Ku HH, Lin CP, Lee KH, Chen YC, Wang JJ, Hsu CC, Chen LK, Li HY, Chiou SH (2012) Docosahexaenoic acid promotes dopaminergic differentiation in induced pluripotent stem cells and inhibits teratoma formation in rats with Parkinson-like pathology. Cell Transplant 21(1):313–332. doi:10.3727/0963689 11X580572

Choi SM, Kim Y, Shim JS, Park JT, Wang RH, Leach SD, Liu JO, Deng C, Ye Z, Jang YY (2013) Efficient drug screening and gene correction for treating liver disease using patient-specific stem cells. Hepatology 57(6):2458–2468. doi:10.1002/hep.26237

Chung CY, Khurana V, Auluck PK, Tardiff DF, Mazzulli JR, Soldner F, Baru V, Lou Y, Freyzon Y, Cho S, Mungenast AE, Muffat J, Mitalipova M, Pluth MD, Jui NT, Schule B, Lippard SJ, Tsai LH, Krainc D, Buchwald SL, Jaenisch R, Lindquist S (2013) Identification and rescue of

E. Eren et al.

alpha-synuclein toxicity in Parkinson patient-derived neurons. Science 342(6161):983–987. doi:10.1126/science.1245296

Cooper O, Seo H, Andrabi S, Guardia-Laguarta C, Graziotto J, Sundberg M, McLean JR, Carrillo-Reid L, Xie Z, Osborn T, Hargus G, Deleidi M, Lawson T, Bogetofte H, Perez-Torres E, Clark L, Moskowitz C, Mazzulli J, Chen L, Volpicelli-Daley L, Romero N, Jiang H, Uitti RJ, Huang Z, Opala G, Scarffe LA, Dawson VL, Klein C, Feng J, Ross OA, Trojanowski JQ, Lee VM, Marder K, Surmeier DJ, Wszolek ZK, Przedborski S, Krainc D, Dawson TM, Isacson O (2012) Pharmacological rescue of mitochondrial deficits in iPSC-derived neural cells from patients with familial Parkinson's disease. Sci Transl Med 4(141):141ra190. doi:10.1126/scitranslmed.3003985

Davie CA (2008) A review of Parkinson's disease. Br Med Bull 86:109–127. doi:10.1093/bmb/ldn013

de Almeida PE, Meyer EH, Kooreman NG, Diecke S, Dey D, Sanchez-Freire V, Hu S, Ebert A, Odegaard J, Mordwinkin NM, Brouwer TP, Lo D, Montoro DT, Longaker MT, Negrin RS, Wu JC (2014) Transplanted terminally differentiated induced pluripotent stem cells are accepted by immune mechanisms similar to self-tolerance. Nat Commun 5:3903. doi:10.1038/ncomms4903

de Lazaro I, Yilmazer A, Kostarelos K (2014) Induced pluripotent stem (iPS) cells: a new source for cell-based therapeutics? J Control Release 185:37–44. doi:10.1016/j.jconrel.2014.04.011

De-Paula VJ, Radanovic M, Diniz BS, Forlenza OV (2012) Alzheimer's disease. Subcell Biochem 65:329–352. doi:10.1007/978-94-007-5416-4_14

Devine MJ, Ryten M, Vodicka P, Thomson AJ, Burdon T, Houlden H, Cavaleri F, Nagano M, Drummond NJ, Taanman JW, Schapira AH, Gwinn K, Hardy J, Lewis PA, Kunath T (2011) Parkinson's disease induced pluripotent stem cells with triplication of the alpha-synuclein locus. Nat Commun 2:440. doi:10.1038/ncomms1453

Doege CA, Abeliovich A (2014) Dementia in a dish. Biol Psychiatry 75(7):558–564. doi:10.1016/j.biopsych.2014.01.007

Doi D, Samata B, Katsukawa M, Kikuchi T, Morizane A, Ono Y, Sekiguchi K, Nakagawa M, Parmar M, Takahashi J (2014) Isolation of human induced pluripotent stem cell-derived dopaminergic progenitors by cell sorting for successful transplantation. Stem Cell Reports 2(3):337–350. doi:10.1016/j.stemcr.2014.01.013

Duan L, Bhattacharyya BJ, Belmadani A, Pan L, Miller RJ, Kessler JA (2014) Stem cell derived basal forebrain cholinergic neurons from Alzheimer's disease patients are more susceptible to cell death. Mol Neurodegener 9:3. doi:10.1186/1750-1326-9-3

Durnaoglu S, Genc S, Genc K (2011) Patient-specific pluripotent stem cells in neurological diseases. Stem Cells Int 2011:212487. doi:10.4061/2011/212487

Dye BR, Hill DR, Ferguson MA, Tsai YH, Nagy MS, Dyal R, Wells JM, Mayhew CN, Nattiv R, Klein OD, White ES, Deutsch GH, Spence JR (2015) In vitro generation of human pluripotent stem cell derived lung organoids. Elife 4. doi:10.7554/eLife.05098

Emborg ME, Liu Y, Xi J, Zhang X, Yin Y, Lu J, Joers V, Swanson C, Holden JE, Zhang SC (2013) Induced pluripotent stem cell-derived neural cells survive and mature in the nonhuman primate brain. Cell Rep 3(3):646–650. doi:10.1016/j.celrep.2013.02.016

Eslamboli A, Romero-Ramos M, Burger C, Bjorklund T, Muzyczka N, Mandel RJ, Baker H, Ridley RM, Kirik D (2007) Long-term consequences of human alpha-synuclein overexpression in the primate ventral midbrain. Brain 130(Pt 3):799–815. doi:10.1093/brain/awl382

Flierl A, Oliveira LM, Falomir-Lockhart LJ, Mak SK, Hesley J, Soldner F, Arndt-Jovin DJ, Jaenisch R, Langston JW, Jovin TM, Schule B (2014) Higher vulnerability and stress sensitivity of neuronal precursor cells carrying an alpha-synuclein gene triplication. PLoS One 9(11):e112413. doi:10.1371/journal.pone.0112413

Fujiwara N, Shimizu J, Takai K, Arimitsu N, Saito A, Kono T, Umehara T, Ueda Y, Wakisaka S, Suzuki T, Suzuki N (2013) Restoration of spatial memory dysfunction of human APP transgenic mice by transplantation of neuronal precursors derived from human iPS cells. Neurosci Lett 557(Pt B):129–134

Giri S, Bader A (2015) A low-cost, high-quality new drug discovery process using patient-derived induced pluripotent stem cells. Drug Discov Today 20(1):37–49. doi:10.1016/j. drudis.2014.10.011

Gupta RM, Musunuru K (2014) Expanding the genetic editing tool kit: ZFNs, TALENs, and CRISPR-Cas9. J Clin Invest 124(10):4154–4161. doi:10.1172/JCI72992

Haile Y, Nakhaei-Nejad M, Boakye PA, Baker G, Smith PA, Murray AG, Giuliani F, Jahroudi N (2015) Reprogramming of HUVECs into induced pluripotent stem cells (HiPSCs), generation and characterization of HiPSC-derived neurons and astrocytes. PLoS One 10(3):e0119617. doi:10.1371/journal.pone.0119617

Hanna J, Wernig M, Markoulaki S, Sun CW, Meissner A, Cassady JP, Beard C, Brambrink T, Wu LC, Townes TM, Jaenisch R (2007) Treatment of sickle cell anemia mouse model with iPS cells generated from autologous skin. Science 318(5858):1920–1923. doi:10.1126/science.1152092

Hargus G, Cooper O, Deleidi M, Levy A, Lee K, Marlow E, Yow A, Soldner F, Hockemeyer D, Hallett PJ, Osborn T, Jaenisch R, Isacson O (2010) Differentiated Parkinson patient-derived induced pluripotent stem cells grow in the adult rodent brain and reduce motor asymmetry in Parkinsonian rats. Proc Natl Acad Sci U S A 107(36):15921–15926. doi:10.1073/pnas.1010209107

Hawkes CM (1995) Diagnosis and treatment of Parkinson's disease. Anosmia is a common finding. BMJ 310(6995):1668

Heilker R, Traub S, Reinhardt P, Scholer HR, Sterneckert J (2014) iPS cell derived neuronal cells for drug discovery. Trends Pharmacol Sci 35(10):510–519. doi:10.1016/j.tips.2014.07.003

Heinrich C, Spagnoli FM, Berninger B (2015) In vivo reprogramming for tissue repair. Nat Cell Biol 17(3):204–211. doi:10.1038/ncb3108

Heins N, Malatesta P, Cecconi F, Nakafuku M, Tucker KL, Hack MA, Chapouton P, Barde YA, Gotz M (2002) Glial cells generate neurons: the role of the transcription factor Pax6. Nat Neurosci 5(4):308–315. doi:10.1038/nn828

Hendrie HC, Ogunniyi A, Hall KS, Baiyewu O, Unverzagt FW, Gureje O, Gao S, Evans RM, Ogunseyinde AO, Adeyinka AO, Musick B, Hui SL (2001) Incidence of dementia and Alzheimer disease in 2 communities: Yoruba residing in Ibadan, Nigeria, and African Americans residing in Indianapolis, Indiana. JAMA 285(6):739–747

Hossini AM, Megges M, Prigione A, Lichtner B, Toliat MR, Wruck W, Schroter F, Nuernberg P, Kroll H, Makrantonaki E, Zoubouliss CC, Adjaye J (2015) Induced pluripotent stem cell-derived neuronal cells from a sporadic Alzheimer's disease donor as a model for investigating AD-associated gene regulatory networks. BMC Genomics 16:84. doi:10.1186/s12864-015-1262-5

Hou P, Li Y, Zhang X, Liu C, Guan J, Li H, Zhao T, Ye J, Yang W, Liu K, Ge J, Xu J, Zhang Q, Zhao Y, Deng H (2013) Pluripotent stem cells induced from mouse somatic cells by small-molecule compounds. Science 341(6146):651–654. doi:10.1126/science.1239278

Hu BY, Weick JP, Yu J, Ma LX, Zhang XQ, Thomson JA, Zhang SC (2010) Neural differentiation of human induced pluripotent stem cells follows developmental principles but with variable potency. Proc Natl Acad Sci U S A 107(9):4335–4340. doi:10.1073/pnas.0910012107

Hu K (2014) All roads lead to induced pluripotent stem cells: the technologies of iPSC generation. Stem Cells Dev 23(12):1285–1300. doi:10.1089/scd.2013.0620

Hughes AJ, Daniel SE, Kilford L, Lees AJ (1992) Accuracy of clinical diagnosis of idiopathic Parkinson's disease: a clinico-pathological study of 100 cases. J Neurol Neurosurg Psychiatry 55(3):181–184

Hy LX, Keller DM (2000) Prevalence of AD among whites: a summary by levels of severity. Neurology 55(2):198–204. doi:10.1212/wnl.55.2.198

Imaizumi Y, Okada Y, Akamatsu W, Koike M, Kuzumaki N, Hayakawa H, Nihira T, Kobayashi T, Ohyama M, Sato S, Takanashi M, Funayama M, Hirayama A, Soga T, Hishiki T, Suematsu M, Yagi T, Ito D, Kosakai A, Hayashi K, Shouji M, Nakanishi A, Suzuki N, Mizuno Y, Mizushima N, Amagai M, Uchiyama Y, Mochizuki H, Hattori N, Okano H (2012) Mitochondrial dysfunction associated with increased oxidative stress and alpha-synuclein accumulation in PARK2 iPSC-derived neurons and postmortem brain tissue. Mol Brain 5:35. doi:10.1186/1756-6606-5-35

Israel MA, Yuan SH, Bardy C, Reyna SM, Mu Y, Herrera C, Hefferan MP, Van Gorp S, Nazor KL, Boscolo FS, Carson CT, Laurent LC, Marsala M, Gage FH, Remes AM, Koo EH, Goldstein LS (2012) Probing sporadic and familial Alzheimer's disease using induced pluripotent stem cells. Nature 482(7384):216–220. doi:10.1038/nature10821

Jia W, Chen W, Kang J (2013) The functions of microRNAs and long non-coding RNAs in embryonic and induced pluripotent stem cells. Genomics Proteomics Bioinformatics 11(5):275–283. doi:10.1016/j.gpb.2013.09.004

Jiang H, Ren Y, Yuen EY, Zhong P, Ghaedi M, Hu Z, Azabdaftari G, Nakaso K, Yan Z, Feng J (2012) Parkin controls dopamine utilization in human midbrain dopaminergic neurons derived from induced pluripotent stem cells. Nat Commun 3:668. doi:10.1038/ncomms1669

Jindal H, Bhatt B, Sk S, Singh Malik J (2014) Alzheimer disease immunotherapeutics: then and now. Hum Vaccin Immunother 10(9):2741–2743. doi:10.4161/21645515.2014.970959

Kanellopoulou C, Muljo SA, Kung AL, Ganesan S, Drapkin R, Jenuwein T, Livingston DM, Rajewsky K (2005) Dicer-deficient mouse embryonic stem cells are defective in differentiation and centromeric silencing. Genes Dev 19(4):489–501. doi:10.1101/gad.1248505

Kawai H, Yamashita T, Ohta Y, Deguchi K, Nagotani S, Zhang X, Ikeda Y, Matsuura T, Abe K (2010) Tridermal tumorigenesis of induced pluripotent stem cells transplanted in ischemic brain. J Cereb Blood Flow Metab 30(8):1487–1493. doi:10.1038/jcbfm.2010.32

Kikuchi T, Morizane A, Doi D, Onoe H, Hayashi T, Kawasaki T, Saiki H, Miyamoto S, Takahashi J (2011) Survival of human induced pluripotent stem cell-derived midbrain dopaminergic neurons in the brain of a primate model of Parkinson's disease. J Parkinsons Dis 1(4):395–412. doi:10.3233/JPD-2011-11070

Kim JB, Zaehres H, Wu G, Gentile L, Ko K, Sebastiano V, Arauzo-Bravo MJ, Ruau D, Han DW, Zenke M, Scholer HR (2008) Pluripotent stem cells induced from adult neural stem cells by reprogramming with two factors. Nature 454(7204):646–650. doi:10.1038/nature07061

Ko HC, Gelb BD (2014) Concise review: drug discovery in the age of the induced pluripotent stem cell. Stem Cells Transl Med 3(4):500–509. doi:10.5966/sctm.2013-0162

Koch P, Tamboli IY, Mertens J, Wunderlich P, Ladewig J, Stuber K, Esselmann H, Wiltfang J, Brustle O, Walter J (2012) Presenilin-1 L166P mutant human pluripotent stem cell-derived neurons exhibit partial loss of gamma-secretase activity in endogenous amyloid-beta generation. Am J Pathol 180(6):2404–2416. doi:10.1016/j.ajpath.2012.02.012

Kolesky DB, Truby RL, Gladman AS, Busbee TA, Homan KA, Lewis JA (2014) 3D bioprinting of vascularized, heterogeneous cell-laden tissue constructs. Adv Mater 26(19):3124–3130. doi:10.1002/adma.201305506

Kondo T, Asai M, Tsukita K, Kutoku Y, Ohsawa Y, Sunada Y, Imamura K, Egawa N, Yahata N, Okita K, Takahashi K, Asaka I, Aoi T, Watanabe A, Watanabe K, Kadoya C, Nakano R, Watanabe D, Maruyama K, Hori O, Hibino S, Choshi T, Nakahata T, Hioki H, Kaneko T, Naitoh M, Yoshikawa K, Yamawaki S, Suzuki S, Hata R, Ueno S, Seki T, Kobayashi K, Toda T, Murakami K, Irie K, Klein WL, Mori H, Asada T, Takahashi R, Iwata N, Yamanaka S, Inoue H (2013) Modeling Alzheimer's disease with iPSCs reveals stress phenotypes associated with intracellular Abeta and differential drug responsiveness. Cell Stem Cell 12(4):487–496. doi:10.1016/j.stem.2013.01.009

Lancaster MA, Renner M, Martin CA, Wenzel D, Bicknell LS, Hurles ME, Homfray T, Penninger JM, Jackson AP, Knoblich JA (2013) Cerebral organoids model human brain development and microcephaly. Nature 501(7467):373–379. doi:10.1038/nature12517

Lee G, Ramirez CN, Kim H, Zeltner N, Liu B, Radu C, Bhinder B, Kim YJ, Choi IY, Mukherjee-Clavin B, Djaballah H, Studer L (2012a) Large-scale screening using familial dysautonomia induced pluripotent stem cells identifies compounds that rescue IKBKAP expression. Nat Biotechnol 30(12):1244–1248. doi:10.1038/nbt.2435

Lee Y, Dawson VL, Dawson TM (2012b) Animal models of Parkinson's disease: vertebrate genetics. Cold Spring Harb Perspect Med 2(10) doi:10.1101/cshperspect.a009324

Li X, Zhang P, Wei C, Zhang Y (2014) Generation of pluripotent stem cells via protein transduction. Int J Dev Biol 58(1):21–27. doi:10.1387/ijdb.140007XL

Li YH, Feng L, Zhang GX, Ma CG (2015) Intranasal delivery of stem cells as therapy for central nervous system disease. Exp Mol Pathol 98(2):145–151. doi:10.1016/j.yexmp.2015.01.016

Lister R, Pelizzola M, Kida YS, Hawkins RD, Nery JR, Hon G, Antosiewicz-Bourget J, O'Malley R, Castanon R, Klugman S, Downes M, Yu R, Stewart R, Ren B, Thomson JA, Evans RM, Ecker JR (2011) Hotspots of aberrant epigenomic reprogramming in human induced pluripotent stem cells. Nature 471(7336):68–73. doi:10.1038/nature09798

Liu GH, Qu J, Suzuki K, Nivet E, Li M, Montserrat N, Yi F, Xu X, Ruiz S, Zhang W, Wagner U, Kim A, Ren B, Li Y, Goebl A, Kim J, Soligalla RD, Dubova I, Thompson J, Yates J 3rd, Esteban CR, Sancho-Martinez I, Izpisua Belmonte JC (2012) Progressive degeneration of human neural stem cells caused by pathogenic LRRK2. Nature 491(7425):603–607. doi:10.1038/nature11557

Liu Q, Waltz S, Woodruff G, Ouyang J, Israel MA, Herrera C, Sarsoza F, Tanzi RE, Koo EH, Ringman JM, Goldstein LS, Wagner SL, Yuan SH (2014) Effect of potent gamma-secretase modulator in human neurons derived from multiple presenilin 1-induced pluripotent stem cell mutant carriers. JAMA Neurol 71(12):1481–1489. doi:10.1001/jamaneurol.2014.2482

Mahairaki V, Ryu J, Peters A, Chang Q, Li T, Park TS, Burridge PW, Talbot CC Jr, Asnaghi L, Martin LJ, Zambidis ET, Koliatsos VE (2014) Induced pluripotent stem cells from familial Alzheimer's disease patients differentiate into mature neurons with amyloidogenic properties. Stem Cells Dev 23(24):2996–3010. doi:10.1089/scd.2013.0511

Martinez-Morales PL, Revilla A, Ocana I, Gonzalez C, Sainz P, McGuire D, Liste I (2013) Progress in stem cell therapy for major human neurological disorders. Stem Cell Rev 9(5):685–699. doi:10.1007/s12015-013-9443-6

Martins-Taylor K, Xu RH (2012) Concise review: genomic stability of human induced pluripotent stem cells. Stem Cells 30(1):22–27. doi:10.1002/stem.705

Masuda S, Wu J, Hishida T, Pandian GN, Sugiyama H, Izpisua Belmonte JC (2013) Chemically induced pluripotent stem cells (CiPSCs): a transgene-free approach. J Mol Cell Biol 5(5):354–355. doi:10.1093/jmcb/mjt034

Mazzulli JR, Xu YH, Sun Y, Knight AL, McLean PJ, Caldwell GA, Sidransky E, Grabowski GA, Krainc D (2011) Gaucher disease glucocerebrosidase and alpha-synuclein form a bidirectional pathogenic loop in synucleinopathies. Cell 146(1):37–52. doi:10.1016/j.cell.2011.06.001

Miller JD, Ganat YM, Kishinevsky S, Bowman RL, Liu B, Tu EY, Mandal PK, Vera E, Shim JW, Kriks S, Taldone T, Fusaki N, Tomishima MJ, Krainc D, Milner TA, Rossi DJ, Studer L (2013) Human iPSC-based modeling of late-onset disease via progerin-induced aging. Cell Stem Cell 13(6):691–705. doi:10.1016/j.stem.2013.11.006

Muratore CR, Rice HC, Srikanth P, Callahan DG, Shin T, Benjamin LN, Walsh DM, Selkoe DJ, Young-Pearse TL (2014) The familial Alzheimer's disease APPV717I mutation alters APP processing and tau expression in iPSC-derived neurons. Hum Mol Genet 23(13):3523–3536. doi:10.1093/hmg/ddu064

Newman AM, Cooper JB (2010) Lab-specific gene expression signatures in pluripotent stem cells. Cell Stem Cell 7(2):258–262. doi:10.1016/j.stem.2010.06.016

Nguyen HN, Byers B, Cord B, Shcheglovitov A, Byrne J, Gujar P, Kee K, Schule B, Dolmetsch RE, Langston W, Palmer TD, Pera RR (2011) LRRK2 mutant iPSC-derived DA neurons demonstrate increased susceptibility to oxidative stress. Cell Stem Cell 8(3):267–280. doi:10.1016/j.stem.2011.01.013

Nishimura K, Takahashi J (2013) Therapeutic application of stem cell technology toward the treatment of Parkinson's disease. Biol Pharm Bull 36(2):171–175

Niu Y, Guo X, Chen Y, Wang CE, Gao J, Yang W, Kang Y, Si W, Wang H, Yang SH, Li S, Ji W, Li XJ (2015) Early Parkinson's disease symptoms in alpha-synuclein transgenic monkeys. Hum Mol Genet 24(8):2308–2317. doi:10.1093/hmg/ddu748

Nussbaum RL, Ellis CE (2003) Alzheimer's disease and Parkinson's disease. N Engl J Med 348(14):1356–1364. doi:10.1056/NEJM2003ra020003

Okano H, Yamanaka S (2014) iPS cell technologies: significance and applications to CNS regeneration and disease. Mol Brain 7:22. doi:10.1186/1756-6606-7-22

Orenstein SJ, Kuo SH, Tasset I, Arias E, Koga H, Fernandez-Carasa I, Cortes E, Honig LS, Dauer W, Consiglio A, Raya A, Sulzer D, Cuervo AM (2013) Interplay of LRRK2 with chaperone-mediated autophagy. Nat Neurosci 16(4):394–406. doi:10.1038/nn.3350

Panicker LM, Miller D, Park TS, Patel B, Azevedo JL, Awad O, Masood MA, Veenstra TD, Goldin E, Stubblefield BK, Tayebi N, Polumuri SK, Vogel SN, Sidransky E, Zambidis ET, Feldman RA (2012) Induced pluripotent stem cell model recapitulates pathologic hallmarks of Gaucher disease. Proc Natl Acad Sci U S A 109(44):18054–18059. doi:10.1073/pnas.1207889109

Park IH, Arora N, Huo H, Maherali N, Ahfeldt T, Shimamura A, Lensch MW, Cowan C, Hochedlinger K, Daley GQ (2008) Disease-specific induced pluripotent stem cells. Cell 134(5):877–886. doi:10.1016/j.cell.2008.07.041

Patel S, Jung D, Yin PT, Carlton P, Yamamoto M, Bando T, Sugiyama H, Lee KB (2014) NanoScript: a nanoparticle-based artificial transcription factor for effective gene regulation. ACS Nano 8(9):8959–8967. doi:10.1021/nn501589f

Peng J, Liu Q, Rao MS, Zeng X (2013) Using human pluripotent stem cell-derived dopaminergic neurons to evaluate candidate Parkinson's disease therapeutic agents in MPP+ and rotenone models. J Biomol Screen 18(5):522–533. doi:10.1177/1087057112474468

Pringsheim T, Jette N, Frolkis A, Steeves TD (2014) The prevalence of Parkinson's disease: a systematic review and meta-analysis. Mov Disord 29(13):1583–1590. doi:10.1002/mds.25945

Qi SD, Smith PD, Choong PF (2014) Nuclear reprogramming and induced pluripotent stem cells: a review for surgeons. ANZ J Surg 84(6):417–423. doi:10.1111/ans.12419

Qiu Z, Farnsworth SL, Mishra A, Hornsby PJ (2013) Patient-specific induced pluripotent stem cells in neurological disease modeling: the importance of nonhuman primate models. Stem Cells Cloning 6:19–29. doi:10.2147/SCCAA.S34798

Raab S, Klingenstein M, Liebau S, Linta L (2014) A comparative view on human somatic cell sources for iPSC generation. Stem Cells Int 2014:768391. doi:10.1155/2014/768391

Rakovic A, Shurkewitsch K, Seibler P, Grunewald A, Zanon A, Hagenah J, Krainc D, Klein C (2013) Phosphatase and tensin homolog (PTEN)-induced putative kinase 1 (PINK1)-dependent ubiquitination of endogenous Parkin attenuates mitophagy: study in human primary fibroblasts and induced pluripotent stem cell-derived neurons. J Biol Chem 288(4):2223–2237. doi:10.1074/jbc.M112.391680

Reinhardt P, Schmid B, Burbulla LF, Schondorf DC, Wagner L, Glatza M, Hoing S, Hargus G, Heck SA, Dhingra A, Wu G, Muller S, Brockmann K, Kluba T, Maisel M, Kruger R, Berg D, Tsytsyura Y, Thiel CS, Psathaki OE, Klingauf J, Kuhlmann T, Klewin M, Muller H, Gasser T, Scholer HR, Sterneckert J (2013) Genetic correction of a LRRK2 mutation in human iPSCs links parkinsonian neurodegeneration to ERK-dependent changes in gene expression. Cell Stem Cell 12(3):354–367. doi:10.1016/j.stem.2013.01.008

Ren Y, Jiang H, Hu Z, Fan K, Wang J, Janoschka S, Wang X, Ge S, Feng J (2015) Parkin mutations reduce the complexity of neuronal processes in iPSC-derived human neurons. Stem Cells 33(1):68–78. doi:10.1002/stem.1854

Revilla A, Gonzalez C, Iriondo A, Fernandez B, Prieto C, Marin C, Liste I (2015) Current advances in the generation of human iPS cells: implications in cell-based regenerative medicine. J Tissue Eng Regen Med. doi:10.1002/term.2021

Rhee YH, Ko JY, Chang MY, Yi SH, Kim D, Kim CH, Shim JW, Jo AY, Kim BW, Lee H, Lee SH, Suh W, Park CH, Koh HC, Lee YS, Lanza R, Kim KS, Lee SH (2011) Protein-based human iPS cells efficiently generate functional dopamine neurons and can treat a rat model of Parkinson disease. J Clin Invest 121(6):2326–2335. doi:10.1172/JCI45794

Richardson TP, Peters MC, Ennett AB, Mooney DJ (2001) Polymeric system for dual growth factor delivery. Nat Biotechnol 19(11):1029–1034. doi:10.1038/nbt1101-1029

Ryan SD, Dolatabadi N, Chan SF, Zhang X, Akhtar MW, Parker J, Soldner F, Sunico CR, Nagar S, Talantova M, Lee B, Lopez K, Nutter A, Shan B, Molokanova E, Zhang Y, Han X, Nakamura T, Masliah E, Yates JR 3rd, Nakanishi N, Andreyev AY, Okamoto S, Jaenisch R, Ambasudhan R, Lipton SA (2013) Isogenic human iPSC Parkinson's model shows nitrosative stress-induced dysfunction in MEF2-PGC1alpha transcription. Cell 155(6):1351–1364. doi:10.1016/j. cell.2013.11.009

Sanchez-Danes A, Consiglio A, Richaud Y, Rodriguez-Piza I, Dehay B, Edel M, Bove J, Memo M, Vila M, Raya A, Izpisua Belmonte JC (2012a) Efficient generation of A9 midbrain dopaminergic neurons by lentiviral delivery of LMX1A in human embryonic stem cells and induced pluripotent stem cells. Hum Gene Ther 23(1):56–69. doi:10.1089/hum.2011.054

Sanchez-Danes A, Richaud-Patin Y, Carballo-Carbajal I, Jimenez-Delgado S, Caig C, Mora S, Di Guglielmo C, Ezquerra M, Patel B, Giralt A, Canals JM, Memo M, Alberch J, Lopez-Barneo J, Vila M, Cuervo AM, Tolosa E, Consiglio A, Raya A (2012b) Disease-specific phenotypes in dopamine neurons from human iPS-based models of genetic and sporadic Parkinson's disease. EMBO Mol Med 4(5):380–395. doi:10.1002/emmm.201200215

Sanders LH, Laganiere J, Cooper O, Mak SK, Vu BJ, Huang YA, Paschon DE, Vangipuram M, Sundararajan R, Urnov FD, Langston JW, Gregory PD, Zhang HS, Greenamyre JT, Isacson O, Schule B (2014) LRRK2 mutations cause mitochondrial DNA damage in iPSC-derived neural cells from Parkinson's disease patients: reversal by gene correction. Neurobiol Dis 62:381–386. doi:10.1016/j.nbd.2013.10.013

Santostefano KE, Hamazaki T, Biel NM, Jin S, Umezawa A, Terada N (2015) A practical guide to induced pluripotent stem cell research using patient samples. Lab Invest 95(1):4–13. doi:10.1038/labinvest.2014.104

Seibler P, Graziotto J, Jeong H, Simunovic F, Klein C, Krainc D (2011) Mitochondrial Parkin recruitment is impaired in neurons derived from mutant PINK1 induced pluripotent stem cells. J Neurosci 31(16):5970–5976. doi:10.1523/JNEUROSCI.4441-10.2011

Seki T, Fukuda K (2015) Methods of induced pluripotent stem cells for clinical application. World J Stem Cells 7(1):116–125. doi:10.4252/wjsc.v7.i1.116

Shaltouki A, Sivapatham R, Pei Y, Gerencser AA, Momcilovic O, Rao MS, Zeng X (2015) Mitochondrial alterations by PARKIN in dopaminergic neurons using PARK2 patient-specific and PARK2 knockout isogenic iPSC lines. Stem Cell Rep. doi:10.1016/j.stemcr.2015.02.019

Shao Y, Sang J, Fu J (2015) On human pluripotent stem cell control: the rise of 3D bioengineering and mechanobiology. Biomaterials 52:26–43. doi:10.1016/j.biomaterials.2015.01.078

Shi Y, Kirwan P, Smith J, MacLean G, Orkin SH, Livesey FJ (2012) A human stem cell model of early Alzheimer's disease pathology in down syndrome. Sci Transl Med 4(124):124ra129. doi:10.1126/scitranslmed.3003771

Simao D, Pinto C, Piersanti S, Weston A, Peddie CJ, Bastos AE, Licursi V, Schwarz SC, Collinson LM, Salinas S, Serra M, Teixeira AP, Saggio I, Lima PA, Kremer EJ, Schiavo G, Brito C, Alves PM (2015) Modeling human neural functionality in vitro: three-dimensional culture for dopaminergic differentiation. Tissue Eng A 21(3–4):654–668. doi:10.1089/ten.TEA.2014.0079

Singh VK, Kalsan M, Kumar N, Saini A, Chandra R (2015) Induced pluripotent stem cells: applications in regenerative medicine, disease modeling, and drug discovery. Front Cell Develop Biol 3:2. doi:10.3389/fcell.2015.00002

Soldner F, Hockemeyer D, Beard C, Gao Q, Bell GW, Cook EG, Hargus G, Blak A, Cooper O, Mitalipova M, Isacson O, Jaenisch R (2009) Parkinson's disease patient-derived induced pluripotent stem cells free of viral reprogramming factors. Cell 136(5):964–977. doi:10.1016/j. cell.2009.02.013

Soldner F, Laganiere J, Cheng AW, Hockemeyer D, Gao Q, Alagappan R, Khurana V, Golbe LI, Myers RH, Lindquist S, Zhang L, Guschin D, Fong LK, Vu BJ, Meng X, Urnov FD, Rebar EJ, Gregory PD, Zhang HS, Jaenisch R (2011) Generation of isogenic pluripotent stem cells differing exclusively at two early onset Parkinson point mutations. Cell 146(2):318–331. doi:10.1016/j.cell.2011.06.019

Solomon S, Pitossi F, Rao MS (2015) Banking on iPSC—is it doable and is it worthwhile. Stem Cell Rev 11(1):1–10. doi:10.1007/s12015-014-9574-4

Sommer CA, Sommer AG, Longmire TA, Christodoulou C, Thomas DD, Gostissa M, Alt FW, Murphy GJ, Kotton DN, Mostoslavsky G (2010) Excision of reprogramming transgenes improves the differentiation potential of iPS cells generated with a single excisable vector. Stem Cells 28(1):64–74. doi:10.1002/stem.255

Sproul AA, Jacob S, Pre D, Kim SH, Nestor MW, Navarro-Sobrino M, Santa-Maria I, Zimmer M, Aubry S, Steele JW, Kahler DJ, Dranovsky A, Arancio O, Crary JF, Gandy S, Noggle SA (2014) Characterization and molecular profiling of PSEN1 familial Alzheimer's disease iPSC-derived neural progenitors. PLoS One 9(1):e84547. doi:10.1371/journal.pone.0084547

Stacey GN, Crook JM, Hei D, Ludwig T (2013) Banking human induced pluripotent stem cells: lessons learned from embryonic stem cells? Cell Stem Cell 13(4):385–388. doi:10.1016/j.stem.2013.09.007

Stadtfeld M, Brennand K, Hochedlinger K (2008) Reprogramming of pancreatic beta cells into induced pluripotent stem cells. Curr Biol 18(12):890–894. doi:10.1016/j.cub.2008.05.010

Stocchi F (2014) Therapy for Parkinson's disease: what is in the pipeline? Neurotherapeutics 11(1):24–33. doi:10.1007/s13311-013-0242-1

Su YC, Qi X (2013) Inhibition of excessive mitochondrial fission reduced aberrant autophagy and neuronal damage caused by LRRK2 G2019S mutation. Hum Mol Genet 22(22):4545–4561. doi:10.1093/hmg/ddt301

Sugii S, Kida Y, Kawamura T, Suzuki J, Vassena R, Yin YQ, Lutz MK, Berggren WT, Izpisua Belmonte JC, Evans RM (2010) Human and mouse adipose-derived cells support feeder-independent induction of pluripotent stem cells. Proc Natl Acad Sci U S A 107(8):3558–3563. doi:10.1073/pnas.0910172106

Swistowski A, Peng J, Liu Q, Mali P, Rao MS, Cheng L, Zeng X (2010) Efficient generation of functional dopaminergic neurons from human induced pluripotent stem cells under defined conditions. Stem Cells 28(10):1893–1904. doi:10.1002/stem.499

Takahashi K, Yamanaka S (2006) Induction of pluripotent stem cells from mouse embryonic and adult fibroblast cultures by defined factors. Cell 126(4):663–676. doi:10.1016/j.cell.2006.07.024

Takahashi K, Tanabe K, Ohnuki M, Narita M, Ichisaka T, Tomoda K, Yamanaka S (2007) Induction of pluripotent stem cells from adult human fibroblasts by defined factors. Cell 131(5):861–872. doi:10.1016/j.cell.2007.11.019

Tong Z, Solanki A, Hamilos A, Levy O, Wen K, Yin X, Karp JM (2015) Application of biomaterials to advance induced pluripotent stem cell research and therapy. EMBO J 34(8):987–1008. doi:10.15252/embj.201490756

Utikal J, Polo JM, Stadtfeld M, Maherali N, Kulalert W, Walsh RM, Khalil A, Rheinwald JG, Hochedlinger K (2009) Immortalization eliminates a roadblock during cellular reprogramming into iPS cells. Nature 460(7259):1145–1148. doi:10.1038/nature08285

Vazin T, Freed WJ (2010) Human embryonic stem cells: derivation, culture, and differentiation: a review. Restor Neurol Neurosci 28(4):589–603. doi:10.3233/RNN-2010-0543

Velasco I, Salazar P, Giorgetti A, Ramos-Mejia V, Castano J, Romero-Moya D, Menendez P (2014) Concise review: generation of neurons from somatic cells of healthy individuals and neurological patients through induced pluripotency or direct conversion. Stem Cells 32(11):2811–2817. doi:10.1002/stem.1782

Verdier JM, Acquatella I, Lautier C, Devau G, Trouche S, Lasbleiz C, Mestre-Frances N (2015) Lessons from the analysis of nonhuman primates for understanding human aging and neurodegenerative diseases. Front Neurosci 9:64. doi:10.3389/fnins.2015.00064

Vierbuchen T, Ostermeier A, Pang ZP, Kokubu Y, Sudhof TC, Wernig M (2010) Direct conversion of fibroblasts to functional neurons by defined factors. Nature 463(7284):1035–1041. doi:10.1038/nature08797

Wan W, Cao L, Kalionis B, Xia S, Tai X (2015) Applications of induced pluripotent stem cells in studying the neurodegenerative diseases. Stem Cells Int 2015:382530. doi:10.1155/2015/382530

Wang Q, Xu X, Li J, Liu J, Gu H, Zhang R, Chen J, Kuang Y, Fei J, Jiang C, Wang P, Pei D, Ding S, Xie X (2011) Lithium, an anti-psychotic drug, greatly enhances the generation of induced pluripotent stem cells. Cell Res 21(10):1424–1435. doi:10.1038/cr.2011.108

Woodruff G, Young JE, Martinez FJ, Buen F, Gore A, Kinaga J, Li Z, Yuan SH, Zhang K, Goldstein LS (2013) The presenilin-1 DeltaE9 mutation results in reduced gamma-secretase activity, but not total loss of PS1 function, in isogenic human stem cells. Cell Rep 5(4):974–985. doi:10.1016/j.celrep.2013.10.018

Wu J, Sheng C, Liu Z, Jia W, Wang B, Li M, Fu L, Ren Z, An J, Sang L, Song G, Wu Y, Xu Y, Wang S, Chen Z, Zhou Q, Zhang YA (2015) Lmx1a enhances the effect of iNSCs in a PD model. Stem Cell Res 14(1):1–9. doi:10.1016/j.scr.2014.10.004

Wu YL, Pandian GN, Ding YP, Zhang W, Tanaka Y, Sugiyama H (2013) Clinical grade iPS cells: need for versatile small molecules and optimal cell sources. Chem Biol 20(11):1311–1322. doi:10.1016/j.chembiol.2013.09.016

Xu XH, Zhong Z (2013) Disease modeling and drug screening for neurological diseases using human induced pluripotent stem cells. Acta Pharmacol Sin 34(6):755–764. doi:10.1038/aps.2013.63

Yagi T, Ito D, Okada Y, Akamatsu W, Nihei Y, Yoshizaki T, Yamanaka S, Okano H, Suzuki N (2011) Modeling familial Alzheimer's disease with induced pluripotent stem cells. Hum Mol Genet 20(23):4530–4539. doi:10.1093/hmg/ddr394

Yu J, Vodyanik MA, Smuga-Otto K, Antosiewicz-Bourget J, Frane JL, Tian S, Nie J, Jonsdottir GA, Ruotti V, Stewart R, Slukvin II, Thomson JA (2007) Induced pluripotent stem cell lines derived from human somatic cells. Science 318(5858):1917–1920. doi:10.1126/science.1151526

Zhao P, Luo Z, Tian W, Yang J, Ibanez DP, Huang Z, Tortorella MD, Esteban MA, Fan W (2014) Solving the puzzle of Parkinson's disease using induced pluripotent stem cells. Exp Biol Med 239(11):1421–1432. doi:10.1177/1535370214538588

Zhao T, Zhang ZN, Rong Z, Xu Y (2011) Immunogenicity of induced pluripotent stem cells. Nature 474(7350):212–215. doi:10.1038/nature10135

Chapter 3
Clinical Safety and Applications of Stem Cell Gene Therapy

Carlo S. Jackson, Marco Alessandrini, and Michael S. Pepper

3.1 Introduction

Gene therapy is becoming increasingly recognized as a potentially important new treatment regimen. Gene therapy is performed by the introduction of genetic material into patients' cells to obtain a specific therapeutic effect (Gould and Favorov 2003). This is accomplished by either eliminating diseased genes or introducing functional genes. Initially, monogenic disorders such as adenosine deaminase severe combined immunodeficiency (ADA-SCID) were considered for gene therapy (Mullen et al. 1996). Subsequently, more complex diseases have been addressed, including the disruption of the human CCR5 gene and HIV replication genes using RNA interference to prevent viral replication in HIV-positive patients (DiGiusto et al. 2010).

The first gene therapy clinical trial was performed in the 1990s (Mullen et al. 1996), and the first success was reported in 2002 for the treatment of X-linked SCID (Hacein-Bey-Abina et al. 2002). However, the excitement was soon dampened when two patients developed leukemia. This was a major setback and resulted in a loss of enthusiasm for the field. However, in-depth investigation into the cause of death and attempts to improve vector safety have led to safer techniques (Check 2002).

The use of genetically manipulated cells that are terminally differentiated only provides a transient therapeutic effect due to the inability of these cells to replace cells that die. Genetically manipulated stem cells on the other hand have the ability to populate a niche in the body and self-renew. This provides a potential long-term supply of cells that can express the therapeutic gene(s) (Burnett et al. 2012). Pluripotent and adult stem cells are the two major categories of stem cell types.

C.S. Jackson • M. Alessandrini • M.S. Pepper (✉)
Department of Immunology, Institute for Cellular and Molecular Medicine, SAMRC
Extramural Unit for Stem Cell Research and Therapy, Faculty of Health Sciences, University of Pretoria, Pretoria, South Africa
e-mail: Michael.Pepper@up.ac.za

© Springer International Publishing AG 2017
P.V. Pham, A. Rosemann (eds.), *Safety, Ethics and Regulations*, Stem Cells in Clinical Applications, DOI 10.1007/978-3-319-59165-0_3

Pluripotent cells include embryonic stem (ES) cells or induced pluripotent stem (iPS) cells. Adult stem cells include hematopoietic, mesenchymal, and epithelial stem cells and are found in various tissues where they ensure a steady number of mature cells.

Hematopoietic stem cells (HSCs) were the first stem cell type to be used for gene therapy purposes given that they had been extensively used in the clinical setting for transplantation purposes (Hacein-Bey-Abina et al. 2002). Bone marrow-derived HSCs, harvested either directly or through mobilization into the peripheral blood, are the most utilized source. The population of HSCs within a harvested sample is relatively rare but can be identified and enriched from this mixed population with cell surface markers such as CD34 (Weissman and Shizuru 2008). Therapeutic genes can be introduced into stem cells with the use of transfer vectors. Virus vectors are the most commonly used, and gene transduction is most efficient using these techniques (Neff et al. 1997). The original virus backbone used was the γ-retrovirus vector, and later adenovirus vectors were adopted (Gould and Favorov 2003). A major drawback of using these vectors is the potential for the therapeutic gene to be inserted into or close to endogenous genes and thereby to modify the activity of these genes (Hacein-Bey-Abina et al. 2003). Typically, the γ-retrovirus vector was found to frequently insert therapeutic genes into or close to oncogenes causing insertional mutagenesis (Blumenthal et al. 2007). A further problem with these vectors is that primitive HSCs, which are mostly quiescent or slow-dividing cells, are not easily transduced (Horn et al. 2015). Lentivirus vectors are currently the vector of choice, and a series of improvements have been made to increase safety and efficacy (Gould and Favorov 2003). Gene therapy has rapidly expanded to include the use of various vector types and the treatment of numerous hematological disorders. Recently, genetically manipulated stem cells types, other than HSCs, have also been explored for the treatment of non-hematologic disorders.

3.2 Applications

The introduction of therapeutic transgenes into target cells or tissues can be accomplished using either a direct delivery strategy or cell-based delivery (Fig. 3.1) (Mohit and Rafati 2013). With the direct delivery strategy, the therapeutic gene is cloned into a plasmid construct. This plasmid construct can be directly administered to the target site for a transient therapeutic effect or packaged into delivery vehicles such as adenoviruses or lentiviruses to ensure a long-term effect. The virus vectors used in this strategy are usually replication competent and are targeted to specifically enter and replicate in the desired cell type (Mohit and Rafati 2013). The use of replication-competent viral vectors in patients faces various challenges, and the preferred delivery strategy is cell-based. In such cases, adult stem cells, ES cells, or iPS cells can be used. Adult stem cells, isolated from the patient or donor and propagated in the laboratory, are currently used in clinical trials and have proven to be safe (Table 3.1). Genetically modified stem cells are reintroduced into patients to

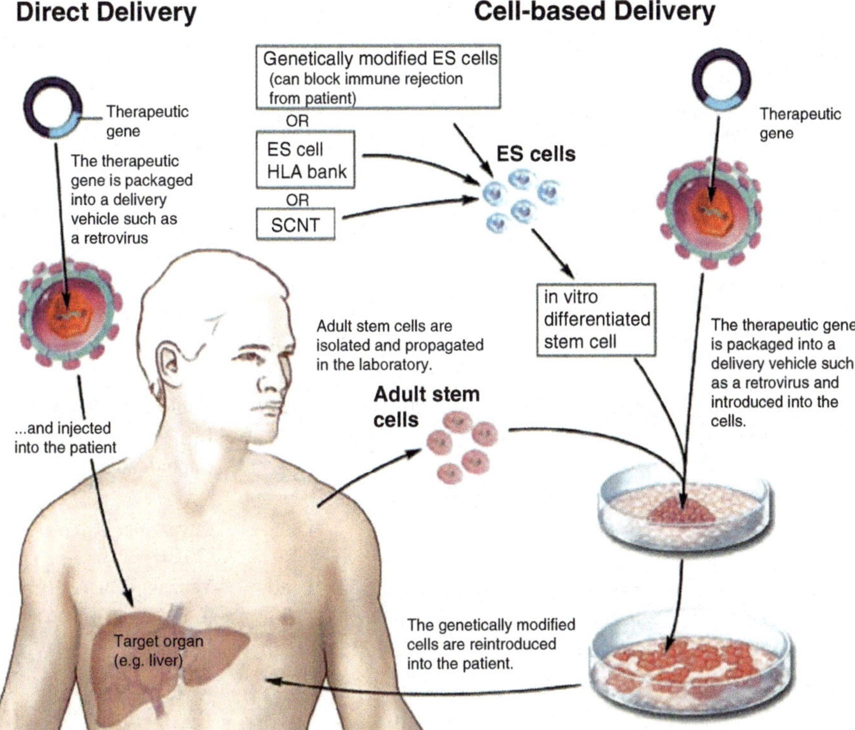

Fig. 3.1 Strategies for delivering therapeutic transgenes into patients (Image used with permission from Terese Winslow LLC)

engraft and populate a particular site specific to each cell type (Fig. 3.1) (Mohit and Rafati 2013). Autologous adult stem cells are preferred to allogeneic stem cells due to the absence of the risk of transplant rejection (Harris 2014).

Clinical trials using adult stem cells and gene therapy have provided several success stories, but these have generally been limited to the use of HSCs. However, other studies that investigate the use of mesenchymal stromal cells (MSCs) are in the pipeline (Myers et al. 2011). Indications explored in clinical trials include SCID, Wiskott-Aldrich syndrome (WAS), β-thalassemia, metachromatic leukodystrophy, and HIV (Watts et al. 2012) (Table 3.1). Cavazzana-Calvo et al. (2000) demonstrated that full correction of the SCID-X1 disease phenotype is possible, given that eight of nine patients that received autologous gene-modified cells were in good health 9 years after treatment (Cavazzana-Calvo et al. 2000; Hacein-Bey-Abina et al. 2011). Full immune reconstitution has also been reported in 30 patients that received autologous gene-modified cells for ADA-SCID (Ferrua et al. 2010). Cavazzana-Calvo and colleagues have also demonstrated that gene therapy can provide transfusion independence for patients with severe β-thalassemia, a common inherited disorder that is not easily treated with gene therapy. This is due to the need

Table 3.1 Clinical trials that have made use of gene therapy and stem cells to date

Stem cell	Disease	Tissue source	NCT number	Vector	Reference
HSC	*Primary immunodeficiencies*				
	X-SCID	Peripheral blood	NCT00028236	Retro	Cavazzana-Calvo et al. (2000)
	X-SCID	Bone marrow	NCT01306019	Lenti	Buckley (2004), Cavazzana-Calvo et al. (2000)
	X-CGD	Peripheral blood	NCT00927134	γ-Retro	Ott et al. (2006), Stein et al. (2010)
	ADA-SCID	Bone marrow	NCT01852071	MMLV	Candotti et al. (2012)
	WAS	Bone marrow	NCT01515462	Retro	Aiuti et al. (2013)
	CGD	Peripheral blood	NCT00394316	MMLV	Kang et al. (2010)
	MLD	Peripheral blood	NCT01560182	Lenti	Biffi et al. (2006), Capotondo et al. (2007)
	Hemoglobinopathies				
	β-Thal	Bone marrow	–	Lenti	Cavazzana-Calvo et al. (2010)
	β-Thal	Peripheral blood	NCT00336362	Lenti	Yannaki et al. (2012)
	β-Thal	Peripheral blood	NCT01639690	Lenti	Boulad et al. (2014)
	Sickle cell anemia	Bone marrow	NCT02247843	Lenti	Romero et al. (2013)
	Other single-gene disorders				
	Gaucher's disease	Bone marrow	NCT00001234	Retro	Fink et al. (1990)
	Fanconi's anemia	Bone marrow	NCT00001399	Retro	Walsh et al. (1994)
	Infectious diseases				
	HIV	Bone marrow	–	MMLV	Bauer et al. (1997), Rossi (2000)
	HIV	Bone marrow	–	MMLV	Kohn et al. (1999)
	HIV	Bone marrow	–	Retro	Kang et al. (2002), Mautino et al. (2001)
	HIV	Bone marrow	–	Retro	Amado et al. (2004), Mitsuyasu et al. (2009)
	HIV	Peripheral blood	NCT01769911	Lenti	DiGiusto et al. (2010)
MSC	Tumors	Bone marrow	NCT01844661	Adeno	Marti et al. (2010)
	Tumors	Bone marrow	NCT02008539	γ-Retro	Niess et al. (2015)
ESC	*Dermatological diseases*				
	Netherton	Skin	NCT01545323	Lenti	Di et al. (2013)
	Junctional epidermolysis bullosa	Skin	–	MML	De Rosa et al. (2014)

HSC hematopoietic stem cell, *MSC* mesenchymal stromal cell, *ESC* epithelial stem cell, *SCID* severe combined immunodeficiency, *WAS* Wiskott-Aldrich syndrome, *β-thal* β-thalassemia, *MLD* metachromatic leukodystrophy, *HIV* human immunodeficiency virus, *CGD* chronic granulomatous disease, *MMLV* murine Moloney leukemia retrovirus for high expression levels of functional forms of the β-hemoglobin chain to achieve a therapeutic effect (Cavazzana-Calvo et al. 2010; Watts et al. 2012). Other therapeutic targets of corrective stem cell gene therapy that have been investigated in animal models but are not yet in clinical trials include Hurler's disease, hemophilia A and B, and alpha-1 antitrypsin deficiency (Vollweiler et al. 2003).

The potential use of MSCs for gene therapy was recognized during the early characterization of these cells (Myers et al. 2011). It was found that MSCs are hypoimmunogenic and immunomodulatory and have the ability to home to damaged tissues to initiate repair processes (Martinez-Quintanilla et al. 2013). MSCs are found in very low numbers in adult tissues but can be isolated from the bone marrow, adipose tissue, and umbilical cord with relative ease. To obtain desired MSC numbers for clinical applications, the isolated MSC population needs to be expanded (Fossett and Khan 2012). MSCs can be specifically induced to be differentiated into osteo-, chondro-, and adipogenic cellular lineages (Choudhery et al. 2013) and have been investigated for use in the treatment of musculoskeletal, vascular, hematological, and neurological diseases and neoplasms (Akram et al. 2012; Fossett and Khan 2012; Harris 2014). Unlike HSCs, MSCs do not differentiate spontaneously when expanded ex vivo; however, factors such as the age and gender of the donor as well as the seeding density and culture conditions can affect the expansion (Fossett and Khan 2012).

The usefulness of MSCs in gene therapy has been demonstrated in animal models. MSCs transduced using the adenoviral delivery system with a modified human bone morphogenetic protein 2 (BMP2) gene under the regulation of the "tet-off" system and transplanted into muscle surrounding the lumbar spine, lead to ectopic bone formation in mice (Martinez-Quintanilla et al. 2013). This demonstrates the usefulness of genetically manipulated MSCs in bone regeneration and repair (Martinez-Quintanilla et al. 2013). Chondrogenesis was induced in rabbit, horse, and pig arthritis models when MSCs transduced with BMP2 and TGFβ were transplanted, resulting in reduced progression of osteoarthritis (Cucchiarini et al. 2005). The ability of MSCs isolated from patients with osteogenesis imperfecta to produce collagen fibrils was restored following genetic disruption of the mutant collagen genes (Chamberlain et al. 2004). MSCs transduced with the antiapoptotic proteins oxygenase-1, angiogenin (Tsubokawa et al. 2010), Bcl2, adrenomedullin (Copland et al. 2008), and integrin-linked kinase (Song et al. 2009) or the angiogenic protein angiopoeitin-1 in combination with the prosurvival protein Akt1 (Shujia et al. 2008) improve heart function in animal models of myocardial infarction (Myers et al. 2011).

Genetically engineered MSCs have been used to treat cancers in rodent xenograft models by modulating the immune system following the expression of IL-2, IL-7, IL-12, and IL-18 or by the expression of TNF, a ligand of the TNF receptors expressed on many tumor types, by inducing cancer-specific apoptosis (Myers et al. 2011). MSCs that deliver the suicide gene herpes simplex virus thymidine kinase (HSV-TK) in combination with ganciclovir treatment have been used to suppress tumor growth and metastasis in mice (Carcinoma 2011).

The first clinical trial that made use of genetically modified MSCs (NCT01844661) was initiated in 2007 (Martı et al. 2010) (Table 3.1). MSCs obtained from the bone marrow of four patients with metastatic neuroblastoma were transduced with an oncolytic adenovirus vector called CELYVER that specifically replicates in cancer cells. The metastatic tumors disappeared within weeks in one patient who was in complete remission 36 months after treatment (Martı et al. 2010). Another clinical trial (NCT02008539) that will make use of genetically modified MSCs and an HSV-TK suicide gene therapy strategy in combination with ganciclovir to suppress cancer metastasis, will be initiated in 2015 (Mátrai et al. 2009).

Epithelial stem cells renew and repair the epidermis. Holoclones, generated from ES cells, are colonies with long-term regeneration potential that can be used to restore large epithelial defects such as skin and ocular burns. Epidermal stem cell holoclones, transduced with a murine leukemia virus (MLV)-based retroviral vector expressing the LAMB3 gene (which codes for laminin 332, previously known as laminin-5), have been applied in clinical trials for the treatment of junctional epidermolysis bullosa, a serious skin disease (De Rosa et al. 2014) (Table 3.1). Functional laminin 332 was observed in newly formed epidermis that was firmly adherent and stable in the absence of blisters, infections, or inflammation. A long-term follow-up indicated the safety and efficacy of using epidermal stem cells in gene therapy for the treatment of skin diseases (De Rosa et al. 2014). Five patients with Netherton skin disease will in 2015 receive epithelial stem cells transduced with lentivirus vectors expressing the serine protease inhibitor Kazal-type 5 gene (SPINK5)(Di et al. 2013).

Other stem cell types that are being considered include ES cells, iPS cells, and multipotent adult progenitor cells (MAPCs) (Kazuki et al. 2010; Narsinh et al. 2009). The recently discovered MAPC population can differentiate into MSCs, endothelial cells, epithelial cells, and hematopoietic cells, which makes it potentially very useful for future gene therapy procedures (Reiser et al. 2006). Undifferentiated ES and iPS cells are said to hold great promise, but safety and ethical hurdles are still to be overcome as they have the potential to form teratocarcinomas. Safety precautions are therefore necessary, and no clinical trials that use ES and iPS cells in gene therapy have yet been approved (Yu et al. 2014).

3.3 Safety

Patient safety following administration of gene therapy products is of utmost importance and should not be compromised to obtain greater therapeutic efficiency (Hadaczek et al. 2010). Given that the preparation of gene therapy products is complex, there are several facets that could possibly compromise patient safety, which are discussed in detail below.

3.3.1 Proof of Concept

Each gene therapy application has specific requirements, including whether a single or multiple genes are to be targeted for being "knocked in" or "knocked out," the need for transient or long-term transgene expression, and the target cell type.

The majority of diseases that are being studied in clinical trials are monogenic disorders, which require the corrected expression of a single gene in target cells (Scaramuzza et al. 2009). The demonstration of effective rescuing of the phenotype is usually straightforward and does not require much optimization. The correct version of the gene with its promoter is amplified from a healthy patient, cloned into a virus vector, and transduced into the target cells (Tsuruta 2013). Regarding the treatment of β-thalassemia and ADA-SCID, the therapeutic genes are functional forms of the β-hemoglobin chain and adenosine deaminase genes, respectively (Von Kalle et al. 2004). The only matter of consideration in a proof-of-concept study is the type of vector that will be used to obtain the safest integration profile (Corrigan-curay et al. 2009).

The gene therapy treatment of some diseases requires the elimination of host genes and/or pathogens, which is more complicated (Bunnell and Morgan 1998). Genes can be eliminated by a range of techniques. The most commonly used is RNA interference (RNAi), which includes short hairpin RNA (shRNA), microRNA (miRNA), and ribozymes to disrupt the posttranscriptional expression of the target gene(s) (Rossi 2009). Although shRNA and miRNA loops are very efficient, several consecutive loops could be required to fully eliminate target gene expression, and this requires significant optimization, especially when multiple genes are targeted (Bunnell and Morgan 1998). With the treatment of HIV, host and virus genes can be silenced with various RNAi techniques. Care should be taken not to use an RNAi technique that nonspecifically targets other genes (Tiscornia et al. 2007). Gene editing techniques, which disrupt target genes at the genomic level, include zinc finger nucleases, transcription activator-like effector nucleases (TALENs) (Bobis-Wozowicz et al. 2014), and the clustered regularly interspaced short palindromic repeat (CRISPR/Cas9) system (Cong et al. 2013). These techniques require lengthy optimization and have the risk of nonspecific disruption of nontarget genes. The advantage is that the constructs can be transiently expressed and do not require insertion into cellular genomes. Gene editing techniques can be used to "knock-in" genes at specific loci; however, the insertion efficiency is low and is only useful in certain settings (Bobis-Wozowicz et al. 2014).

Various methods have been developed to transfer therapeutic gene(s) into target cells, each with its own advantages and disadvantages. Non-viral vectors (e.g., plasmids) used in direct delivery methods are easy to produce, are capable of delivering synthetic compounds such as oligonucleotides or small interfering RNA (siRNA), do not have the risk of vector infection, and reduce inflammatory complications (Seow and Wood 2009). However, the transfection efficiency obtained with non-viral vectors is poor compared to viral vectors, transgene expression is transient, and cell specificity is low (Molas et al. 2003). Replication-deficient viral vectors (e.g., adenovirus, adeno-associated virus (AAV), retrovirus, lentivirus, poxvirus, HSV) are more difficult to produce and run the risk of causing insertional mutagenesis (Blumenthal et al. 2007). However, cellular specificity can be adjusted, transduction efficiency is high, and stable transgene expression can be obtained in the long term. Replication-competent oncolytic vectors (e.g., measles, reovirus, vesicular stomatitis virus, vaccinia) have the advantage of being able to spread to and transduce multiple cells in vivo. However, immune activation as well as strict control over cellular specificity

Table 3.2 Comparison of gene therapy vectors (adapted from Gould and Favorov 2003)

Vector	Transgene capacity	Immunogenicity	Genome integration	Long-term expression	Transfer into dividing (D) and quiescent (Q) cells
Plasmid-naked	Unlimited	Low	No	Only in muscle	D and Q
Plasmid-complex	Unlimited	Low	No	No	D and Q
AAV	4 kb	High	Yes and episomal	Yes	D and Q
HSV	35 kb	High	No	Yes	D and Q
Retrovirus	Up to 8 kb	Low	Yes	Yes	D
Lentivirus	Up to 8 kb	Low	Yes	Yes	D and Q
Bacteria	Unlimited	High	No	No	D and Q

AAV adeno-associated virus, *HSV* herpes simplex virus

is problematic (Seow and Wood 2009). Microbial vectors (e.g., listeria, salmonella, *E. coli*, bacteriophage) can be used to deliver therapeutic genes in vivo; however, they are currently limited to the targeting of cancer cells and face safety and efficacy limitations (Baban et al. 2010) (Table 3.2).

Early clinical trials (Table 3.1) demonstrated that the use of γ-retroviruses for stem cell genetic engineering has major safety issues and does not provide efficient transduction (Doering et al. 2011). The insertion sites of therapeutic genes in two patients that developed leukemia with γ-retroviral vector stem cell gene therapy were found to be near the *LMO2* oncogene, and the integration profile of γ-retrovirus vectors has been found to favor oncogenes. This initiated the search for other virus vector options (Hacein-bey-abina et al. 2008). Clinical trials have been conducted to test the use of viruses constructed from the backbone of the human foamy virus, murine Moloney leukemia retrovirus (MMLV), adenovirus, and lentivirus (Bouard et al. 2009). Retroviral and lentiviral vectors have been further improved. However, lentiviral vectors have the advantage of having a large packaging capacity (10 kb from long terminal repeat (LTR) to LTR), the ability to transduce nondividing cells, stable integration and long-term expression of transgenes, low immunogenicity of the virus particles, and an integration profile that does not favor oncogenes (Naldini et al. 1996) (Table 3.2). The latest self-inactivating (SIN) lentiviral vectors reduce the risk of activating oncogenes or deactivating anti-oncogenes and limit interference of endogenous gene regulation (Baum 2008). The safe and efficient transduction of a wide range of cell types has made lentivirus the vector of choice (Throm et al. 2009). Lentiviral vectors can be derived from a range of viruses, but those derived from HIV are the most widely used. The safety and efficacy of lentiviral vectors are progressively being improved to prevent replication-competent recombination (RCR) (http://lentilab.unique.ch/) and insertional oncogene activation. The polypurine tract/central termination sequence, the woodchuck hepatitis virus post-transcriptional regulatory element, the rev response element, the tat-independent

Table 3.3 Tissue-specific promoters (adapted from Gould and Favorov 2003)

Promoter	Target tissue	Disease
Saliary gland amylase promoter	Salivary gland acinar epithelial cells	Sjogren's syndrome
Kallikrein promoter	Salivary gland ductal epithelial cells	Sjogren's syndrome
Involucrin promoter	Keratinocytes	Scleroderma
Keratin 14 promoter	Basal layer of epidermis	Scleroderma
L-type pyruvate kinase promoter	Liver (hepatocytes)	Diabetes and other autoimmune diseases
Rat insulin promoter	Pancreatic β-islet cells	Diabetes
Collagen II promoter	Joints (chondrocytes)	Rheumatoid arthritis
Human glial fibrillary acidic protein promoter	Brain (astrocytes)	Multiple sclerosis
Neuron-specific enolase promoter	Brain (neurons)	Multiple sclerosis
Interleukin-2 promoter	Activated T cells	All autoimmune diseases
MHC-II specific HLA-DRα promoter	Antigen presenting cells	All autoimmune diseases
Dectin-2 promoter	Langerhans cells	All autoimmune diseases
GATA-1 enhancer	Erythroid cells	All autoimmune diseases

self-inactivating, and the Rous sarcoma virus promoter elements have been incorporated in lentivirus constructs to improve safety and transduction efficiency (Giry-laterrière et al. 2011) (Table 3.3).

Although preclinical proof-of-concept studies may indicate a potentially significant benefit, translation to the clinic may be hindered by low transgene efficiency, risk to patients, and ethical constraints (Sanchez and Silberstein 2013). After proof of concept has been demonstrated, techniques will need to be adapted for clinical purposes (Baoutina et al. 2007). It would be unethical to test experimental medical therapies in humans if clear benefit and safety have not been demonstrated in the preclinical setting. Most experimental procedures are tested either in tissue culture or in animals; however, predicted outcomes are not always representative of actual outcomes in humans (Seok et al. 2013). In order to bridge the gap between tissue culture experiments and patients, mice have been genetically manipulated to be able to receive human tissue without rejection (American Cancer Society 2014). Various genes involved in the immune system of these mice have been functionally compromised to achieve a mouse strain that provides in vivo conditions comparable to those found in humans. The most widely used immunodeficient mouse strain is the NSG mouse (NOD.Cg-*Prkdc^{scid} Il2rg^{tm1Wjl}*/SzJ) (Brehm et al. 2015). Human HSCs engraft well into irradiated newborn NSG mice or after myeloablation of adult mice, which then provide a fully functional human immune system (Ishikawa et al. 2014). These "humanized mice" can be used to investigate cancer, infectious human blood-borne diseases, transplantation of cells and tissues, and even genetically modified stem cells (Shultz et al. 2013). The infection and replication properties of HIV have been investigated in humanized NSG mice and were found to be similar to infection and replication in humans (Singh et al. 2012). It has further been demonstrated that the engraftment of human HSCs, genetically modified with lentivirus constructs to contain genes that prevent HIV replication, reduces HIV viral load and increases human CD4+ T-cell counts in these mice (Kiem et al. 2012).

3.3.2 The Preclinical Testing Program

The Center for Biologics Evaluation and Research (CBER), the Office of Cellular, Tissue and Gene Therapies (OCTGT) of the USA, as well as the European Medicines Agency (EMA) have produced guidelines that should be considered when gene therapy techniques are being developed for clinical application (European Medicines Agency 2008; OCTGT 2013). The selection of a species/model for investigating gene therapy products must be relevant and best suited to the provision of data comparable to the clinical setting. Immunodeficient animals provide information on the potential for adverse immune responses to the ex vivo genetically modified cells, the vector, or the expressed transgene (European Medicines Agency 2008). Dogs and nonhuman primates offer good preclinical models for HSC-related protocols, with nonhuman primates being better due to the fact that they are largely homologous with humans and hence allow for nearly all human cytokines to be functional in this model (Watts et al. 2012). Appropriate animal models allow for accurate analysis of the potential toxicity generated by a vector, transgene, and cell type, as well as the potential risks of the delivery procedure prior to clinical trials (Ciurea and Andersson 2009). The route of administration and the procedures used to administer the product may nullify the therapeutic effect seen in vitro and in animal models. For example, intra-arterial delivery of MSCs avoids the accumulation of the cells in the lungs, compared to intravenous delivery, allowing the cells more time to migrate to the intended tissues (Kean et al. 2013). The delivery procedure of the manipulated stem cells may require bone marrow conditioning which has the risk itself of inducing morbidity and mortality; however, this risk can be mitigated by carefully selecting an appropriate conditioning regimen (Ciurea and Andersson 2009). The in vivo behavior and activity of the transduced cells must be determined, including distribution, localization, trafficking, and persistence (European Medicines Agency 2008). Scaramuzza et al. found that CD34+ cells from a WAS patient, intended for gene therapy, proliferated slower in vitro with reduced cytokine receptor production when compared to healthy donor cells, thereby affecting the efficacy of the gene therapy (Scaramuzza et al. 2009).

The risk of aberrant gene expression caused by the transgene and the integration profile of the vector should be thoroughly investigated and characterized in animal models to prevent treatment-related fatalities. Cavazzana-Calvo et al. managed to determine that dominant clonal expansion is possible due to truncated cellular transcripts of a proto-oncogene that can be spliced to an acceptor site within the vector (Cavazzana-Calvo et al. 2010). Low cell viability, high cell proliferation, and abnormal cytokine production of the engrafted cells can affect the safety and efficacy of the therapy (Gnecchi et al. 2008).

Controlled procedures should be followed during virus stock production and the transduction of HSCs to prevent pathogen contamination. Large-scale production of vector particles must not be adversely affected by the therapeutic genes in the vector construct (Logan et al. 2002). Virus vector production and concentration protocols can cause cellular toxicity; for example, virus concentration with polyethylene glycol 6000 precipitation provides high vector titers with significantly reduced toxicities compared to ultracentrifugation (Kutner et al. 2009). The target gene that is either knocked in or knocked out should not interfere with cellular activities. A siRNA can cause nonspecific gene silencing or cellular toxicity if overexpressed (Tiscornia et al. 2007). Promoters such as the U6 Pol-III promoter that are used to drive strong expression of shRNA can cause toxicity in vivo due to saturation of the endogenous miRNA pathway (Ely et al. 2008). Furthermore, a low transduction rate can be sufficient to provide proof of efficacy in vitro; however, it has been demonstrated that a relatively pure transduced population of cells may be necessary to obtain effective results in mouse models. Myburgh et al. observed that the transplantation of CD34+ cells, in which only 20–30% of positively transduced cells contained an anti-CCR5 miRNA construct, did not reduce HIV viral load in NSG mice. However, a reduction in HIV was observed with a CD34 cell population enriched for positively transduced cells (Myburgh et al. 2015).

3.3.3 Suicide Genes

There is always the risk that stem cells used in gene therapy may behave abnormally and have off-target effects, which in turn may become detrimental to the patient. The history of gene therapy demonstrates the potential for treating patients but also shows that the gene therapy itself may be detrimental and lead to significant morbidity and mortality (Check 2002). Hence it would be beneficial if the transplanted cells could be eliminated in the event of a life-threatening situation. Additionally, the techniques used to eliminate transplanted cells should do so exclusively and not affect the recipient's cells. This can be achieved with the use of "suicide genes" (Zhan et al. 2013). The first clinical trials testing suicide genes were initiated in 2002, and since then many have followed (Table 3.4). The trials listed in Table 3.4 used T cells and CD34+ cells transduced with suicide genes using replication-competent virus vectors to target cancer cells in patients (Zhan et al. 2013). The field of suicide gene therapy has made significant progress (Blumenthal et al. 2007). Cells intended for transplantation for gene therapy can be transduced with a suicide gene, in combination with the therapeutic gene, which can be activated to specifically eliminate the transduced cells in cases of unintended oncogenesis or graft-versus-host disease (GVHD) (Blumenthal et al. 2007; Zhan et al. 2013). The two most widely investigated suicide genes are Caspase-9 and HSV-TK (NCT01204502, NCT01744223).

Table 3.4 Suicide genes tested in clinical trials

NCT number	Suicide gene	Virus vector	Conditions	Start date
NCT00423124	HSVTK	Retro	Hematological malignancies	Jul 2002
NCT00844623	HSVTK	Adeno	Hepatocellular carcinoma	Dec 2002
NCT00415454	HSVTK	Adeno	Pancreatic cancer	Nov 2006
NCT00710892	iCasp9	Retro	Acute lymphoblastic leukemia/non-Hodgkin's lymphoma/myelodysplastic syndrome/chronic myeloid leukemia	Dec 2008
NCT00964756	HSVTK	Adeno	Ovarian cancer	Aug 2009
NCT01086735	HSVTK	Retro	Hematological malignancies	Feb 2010
NCT01204502	HSVTK	Adeno	Haploidentical stem cell transplantation	Jan 2011
NCT01744223	iCasp9	Retro	Acute lymphoblastic leukemia/acute myelogenous leukemia/lymphoma	Mar 2013
NCT01822652	iCasp9	Retro	Neuroblastoma	Aug 2013
NCT01875237	iCasp9	Retro	Leukemia/myeloma/myeloproliferative diseases	Dec 2013

HSVTK human herpes simplex virus thymidine kinase gene, *iCasp9* inducible caspase 9

3.4 Efficacy

The success of a particular gene therapy depends mainly on the ability to achieve a safe level of gene dosage in vivo and an effective therapeutic level of gene-modified stem cells (Watts et al. 2012). The transgene should be stably and not excessively expressed. Stable expression of the transgene is affected by the promoter and vector choice (Taylor et al. 2013). Obtaining an effective therapeutic level of gene-modified stem cells depends on the selective advantage and engraftment capacity of the modified cells.

3.4.1 Promoter Selection

The expression of the therapeutic transgene is differentially driven depending on the nature of the promoter (Johnston et al. 2013). Viral promoters such as CMV are commonly used in in vitro experiments to obtain strong constitutive protein expression. However, viral promoters are prone to long-term inactivation in vivo (Papadakis et al. 2004). Therefore human promoters like the human elongation factor-1 alpha and the phosphoglycerate kinase promoters are preferentially used in vectors for gene therapy (Johnston et al. 2013). Tissue-specific promoters reduce the risk of gene expression in unwanted tissues and thereby increase the safety of the therapy. Table 3.3 lists tissue-specific promoters which can be used to selectively express a transgene in a given tissue type such as hepatocytes, Langerhans cells, and chondrocytes. These are important considerations as the promoter choice influences virus titer and transduction rates, which in turn will impact on the efficacy of the gene therapy (Giry-laterrière et al. 2011).

3.4.2 Transduction Efficiency

Transduction efficiency has a significant impact on the success of a gene therapy (Van Griensven et al. 2005). Only the positively transduced stem cells, which are generally a minor proportion, provide a therapeutic effect. However, the population of untransduced cells will also engraft, but will not contribute to providing the therapeutic effect. In fact, these cells may even inhibit the therapeutic effect of the successfully transduced cells (Myburgh et al. 2015). Within an HSC population, the primitive subpopulation that has long-term engraftment capacity is very small (Van Griensven et al. 2005; Vollweiler et al. 2003). Therefore, even with a high transduction rate, the probability of obtaining a long-term therapeutic effect from the transduced HSCs is low (Watts et al. 2012). This has indeed been demonstrated in clinical trials, where the short-term engraftment in the majority of trials was satisfactory, but long-term engraftment was poor (Watts et al. 2012). This effect may also be important in other stem cell types such as MSCs that contain small primitive stem cell proportions (Myers et al. 2011). Since the transduced cells engraft at extremely low rates (Müller et al. 2008), the number of engrafted cells that contain the transgene is insufficient to effectively convey the efficacy of the gene therapy. Examples where this has been observed include treatments for Fanconi's anemia and Gaucher's disease. The therapeutic benefit of the transgenes was not obtained due to the low number of primitive HSCs that were manipulated (Sidransky et al. 2007). The transduction efficiency of the target cell population is thus important and is affected by parameters that include the period of pre-stimulation, transduction duration, and post-transduction cell culture (Liu et al. 2009).

3.4.3 Engraftment Efficiency

A selective advantage of gene-modified cells and pretransplant conditioning are important for engraftment. Genetic correction of HSCs in the treatment of SCID provides a selective advantage over mutant cells, allowing a small number of manipulated cells to achieve a therapeutic effect without a transplant conditioning regimen (Aiuti and Roncarolo 2009; Fischer et al. 2010). However, in the treatment of Fanconi's anemia and Gaucher's disease, the therapeutic efficacy of the HSC gene therapy was insufficient due to the small numbers of manipulated cells obtained (mentioned in Sect. 3.4.2) that had no selective advantage over non-manipulated cells, and no preparative regimen was provided to overcome this limitation (Liu et al. 1999; Sidransky et al. 2007). Finding a balance between in vivo selection of cells and pretransplant conditioning is difficult and is one of the current limitations in gene therapy (Watts et al. 2012). If the gene-corrected cells have a selective advantage that allows them to persist and function, then no preparative regimen is provided. However, in cases where there is no selective advantage, conditioning should be provided. Increasing the number or proportion of gene-modified cells by in vitro or in vivo selection and/or ex vivo expansion will improve the therapeutic effect of the treatment and provide engraftment with less or no conditioning (Watts et al. 2012).

In order to increase the proportion of transduced cells, and hence select for a purer population, a selection marker can be used. The selection of positively transduced cells in vitro and in vivo has been accomplished with human genes such as $DHFR_{L22Y}$ which provide resistance to the antifolate drug trimethotrexate (TMTX) and $\alpha 1_{Q118R/N129D}$ that provides resistance to ouabain, a selective Na^+/K^+-ATPase inhibitor (Treschow et al. 2007). Chemotherapy resistance genes, such as O^6-benzylguanine (O6BG), bis-chloroethylnitrosourea (BCNU), and cytidine deaminase, have also been used (Momparler et al. 2002). This type of selection can greatly increase engraftment; however, a large initial number of cells is needed due to the removal of the non-transduced portion of the cells (Biffi et al. 2013; Kang et al. 2010). The success of gene therapy is generally affected by the number of stem cells used, and the efficacy can further be affected by donor variables such as age and health status at the time of collection (Harris 2014). Due to the difficulty in expanding HSCs without differentiation, it is important to have a high number of isolated cells for transduction (Biffi et al. 2013; Kang et al. 2010). The transplantation of expanded HSCs is being investigated in non-gene therapy clinical trials (NCT01474681, NCT01816230). The expansion of gene-modified HSCs will improve transplant success by increasing cell numbers as well as safety by providing a window during which transduced cells can be screened for mutagenic integration sites (Watts et al. 2012).

The number of MSCs used in gene therapies varies greatly depending on the application, site of injection, and research group (Lewis and Suzuki 2014). The advantage of MSCs is that they can be expanded in culture to obtain the desired cell number. The time it takes to expand these cells should be taken into account, which could take up to several weeks (Choudhery et al. 2013). Cryopreservation of stem cells in stem cell banks offers the opportunity to preserve freshly isolated or expanded stem cells prior to the onset of disease. Stem cell isolation can be performed at a time of good health and cryopreserved to provide a robust population of stem cells for gene therapy (Harris 2014).

3.5 Gene Therapy Procedures

Providing stem cell gene therapies to patients comes with many safety, ethical, and socioeconomic concerns, all of which affect the outcome of the therapy (Vattemi and Claudio 2009). Manipulated stem cells are considered drugs in the USA and the European Union and are classified as cell-based medicinal products (CBMPs) or advanced therapeutic products (Giancola et al. 2012). The handling of these cells has special requirements concerning equipment, safety standards, and training of personnel (Bosse et al. 1997). Appropriate handling conditions include the use of clean room facilities operated according to current good manufacturing practices (cGMP) and require a quality control system (Giancola et al. 2012). The production protocols and clinical use of the gene-modified cells serve as a basis for the preparation of a risk management plan and the production and distribution of CBMPs. These regulations depend on the relevant national authorities and legislations (Giancola et al. 2012). A cGMP facility is specifically designed as a production facility for the manufacturing of pharmaceutical or cellular products which include the manufacturing space, the raw and finished product, warehouse storage, and support laboratory areas (Giancola et al. 2012). Furthermore, the vectors used (mostly viral vectors) require an accredited vector production platform that can provide high-titer vector samples free of pathogenic contamination and traces of replication-competent virus vectors (Bosse et al. 1997; Spencer et al. 2009). The costs involved in establishing these facilities, together with the maintenance and consumable costs, make gene therapy procedures very expensive and create a degree of uncertainty regarding the financial feasibility of this procedure (Mavilio 2012; Soares et al. 2005). From a cost-benefit perspective, the feasibility of such therapies is measured against the costs of current therapies (some of which are lifelong) and the life expectancy of patients suffering from the disease to be treated (Jackson and Pepper 2013). Gene therapy can, for example, be potentially used to treat HIV as a once-off intervention and thereby allow infected patients to reduce or discontinue highly active antiretroviral treatments (HAART). Although the once-off gene therapy treatment is expensive compared to lifelong HAART treatment, it may be more cost-effective (Bollinger and Stover 1999). The economic impact of HIV on countries with high infection rates could be positively impacted (Bollinger and Stover 1999; Jackson and Pepper 2013). The cost of these gene therapies is also anticipated to decrease as the technology develops (Soares et al. 2005; Tremblay et al. 2013).

The introduction of genetically modified stem cells into patients is associated with an element of risk. In cases where allogeneic stem cells are used, there is a the risk of GVHD, which is regarded as the most lethal complication of HSC transplantation (Zaia and Forman 2013). Since bone marrow ablation is often required to free up a niche for transplanted HSCs, the additional risk of chemotherapy-related mortality is present (Biffi et al. 2013; Kuramoto et al. 2004). The appropriate patient population should therefore be selected based on prognostic determinants, hence limiting the application of certain gene therapies to specific patient populations. Alternative techniques to total body irradiation are being investigated to provide safer procedures for creating space in the bone marrow. These techniques include the use of reduced intensity and combined chemotherapy regimens, anti-c-Kit antibodies (Czechovicz et al. 2008), and the use of G-CSF prior to transplantation (Ugarte and Forsberg 2013).

Close monitoring of patients receiving gene therapy is extremely important. The Committee for Medicinal Products for Human Use at the EMA has provided guidelines for follow-up of patients who have received gene therapy medicinal products (European Medicines Agency 2009). Delayed adverse reactions such as oncogenesis and immunological reactions as well as vector reactivation should be considered. This is due to the long life-span or persistence of modified cells, the bio-distribution, and delayed effects associated with the integrated vector and product expression. Finally, the route of administration also influences bio-distribution and the potential for serious delayed reactions. For example, a change in the route of administration could lead to an increase in the dose of cells that is distributed to tissues not represented in safety studies (European Medicines Agency 2009).

3.6 Way Forward

Significant progress has been made in the field of stem cell gene therapy with isolated successful cases reported to date. Due to an increase in confidence in genetic manipulation techniques, a broader range of disease types are being investigated and included in clinical trials (Myers et al. 2011). Many research groups are able to produce proof-of-concept data on novel gene therapy techniques in cell and animal models but may lack the capability (expertise, facilities, and funding) to get to the clinic (Galis et al., 2015). Fragmented gene therapy research efforts are a significant hurdle to clinical translation, which supports the need for the establishment of an international gene therapy cooperative. Such a consortium would provide the platform for smaller groups to benefit from broader expertise and infrastructure and hence enable more effective translation of gene therapies into the clinical trial stages of development. Following the example of other international consortia such as the Human Genome Project and the ENCODE project, large budgets could be secured from governments and industries to make more funding available for gene therapy research (Tremblay et al. 2013; U.S. Congress 1988). Smaller funded gene therapy consortia such as the Transatlantic Gene Therapy Consortium and the EU Seventh Framework program already exist (Tremblay et al. 2013). These consortia have demonstrated the benefits of sharing of expertise, the increased cost-effectiveness, and the increased probability of therapeutic success (Tremblay et al. 2013).

3.7 Conclusion

Stem cell gene therapies are being investigated for the treatment of many diseases. However, as with any new form of therapy, only by thorough evaluation of these techniques both with regard to efficacy and safety, can success be ensured. Significant progress has been made since the first gene therapy clinical trial was initiated 26 years ago. Gene therapies are currently dominated by the use of HSCs to treat hematological diseases; however, other cell types such as MSCs and epithelial stem cells are expected to be used more commonly in the future. The translation of a potential therapy from the proof-of-concept phase to the clinic is a long journey, and many factors need to be addressed. It is essential to be aware of each of these factors before a subsequent phase of the process is initiated. The future of stem cell gene therapy lies in combining the experience obtained from clinical trials and the advances made in basic research to provide the safest and most efficacious product.

References

Aiuti A, Roncarolo MG (2009) Ten years of gene therapy for primary immune deficiencies. Hematology 23:682–689

Aiuti A, Biasco L, Scaramuzza S, Ferrua F, Cicalese MP, Baricordi C, Dionisio F et al (2013) Lentiviral hematopoietic stem cell gene therapy in patients with Wiskott-Aldrich syndrome. Science (New York, NY) 341(6148):1233151

Akram KM, Samad S, Spiteri M, Forsyth NR (2012) Mesenchymal stem cell therapy and lung diseases. Adv Biochem Eng Biotechnol 130:105–129. doi:10.1007/10_2012_140

Amado RG, Mitsuyasu RT, Rosenblatt JD, Ngok FK, Bakker A, Cole S, Chorn N et al (2004) Anti-human immunodeficiency virus hematopoietic progenitor cell-delivered ribozyme in a phase I study: myeloid and lymphoid reconstitution in human immunodeficiency virus type-1—infected patients. Hum Gene Ther 262:251–262

American Cancer Society (2014) Clinical trials: what you need to know, pp 1–37

Baban CK, Cronin M, O'Hanlon D, O'Sullivan GC, Tangney M (2010) Bacteria as vectors for gene therapy of cancer. Bioeng Bugs 1(6):385–394. doi:10.4161/bbug.1.6.13146

Baoutina A, Alexander IE, Rasko JEJ, Emslie KR (2007) Potential use of gene transfer in athletic performance enhancement. Am Soc Gene Ther 15(10):1751–1766. doi:10.1038/sj.mt.6300278

Bauer BG, Valdez P, Kearns K, Bahner I, Wen SF, Zaia JA, Kohn DB (1997) Inhibition of human immunodeficiency virus-1 (HIV-1) replication after transduction of granulocyte colony-stimulating factor—mobilized CD34+ cells from HIV-1—infected donors using retroviral vectors containing anti—HIV-1 genes, 1. Blood 89(7):2259–2267

Baum C (2008) Insertional mutagenesis in gene therapy and stem cell biology. Curr Opin Hematol 14:337–342

Biffi A, Capotondo A, Fasano S, Carro U, Marchesini S, Azuma H, Malaguti MC et al (2006) Gene therapy of metachromatic leukodystrophy reverses neurological damage and deficits in mice. J Clin Invest 116(11):3070–3082. doi:10.1172/JCI28873.3070

Biffi A, Montini E, Lorioli L, Cesani M, Fumagalli F, Plati T, Baldoli C et al (2013) Lentiviral hematopoietic stem cell gene therapy benefits metachromatic leukodystrophy. Science (New York, NY) 341(6148):1233158. doi:10.1126/science.1233158

Blumenthal M, Skelton D, Pepper KA, Jahn T, Methangkool E, Kohn DB (2007) Effective suicide gene therapy for leukemia in a model of insertional oncogenesis in mice. Mol Ther 15(1):183–192. doi:10.1038/sj.mt.6300015

Bobis-Wozowicz S, Galla M, Alzubi J, Kuehle J, Baum C, Schambach A, Cathomen T (2014) Non-integrating gamma-retroviral vectors as a versatile tool for transient zinc-finger nuclease delivery. Sci Rep 4:4656. doi:10.1038/srep04656

Bollinger L, Stover J (1999) The economic impact of AIDS in South Africa. The Policy Project. 1–16

Bosse R, Singhofger-Wohra M, Rosenthal F, Schulz G (1997) Good manufacturing practice production of human stem cells for somatic cell and gene therapy. Stem Cells 15(Suppl 1):275–280

Bouard D, Alazard-Dany D, Cosset F-L (2009) Viral vectors: from virology to transgene expression. Br J Pharmacol 157(2):153–165. doi:10.1038/bjp.2008.349

Boulad F, Wang X, Qu J, Taylor C, Ferro L, Karponi G, Bartido S et al (2014) Safe mobilization of CD34+ cells in adults with β-thalassemia and validation of effective globin gene transfer for clinical investigation. Blood 123(10):1483–1486. doi:10.1182/blood-2013-06-507178

Brehm MA, Racki WJ, Leif J, Burzenski L, Hosur V, Wetmore A, Gott B et al (2015) Engraftment of human HSCs in nonirradiated newborn NOD-scid IL2rγ null mice is enhanced by transgenic expression of membrane-bound human SCF. Blood 119(12):2778–2789. doi:10.1182/blood-2011-05-353243.The

Buckley RH (2004) Molecular defects in human severe combined immunodeficiency and approaches to immune reconstitution. Annu Rev Immunol 22(2):625–655. doi:10.1146/annurev.immunol.22.012703.104614

Bunnell B, Morgan R (1998) Gene therapy for infectious diseases. Clin Microbiol Rev 11(1):42–56

Burnett JC, Zaia JA, Rossi JJ (2012) Creating genetic resistance to HIV. Curr Opin Immunol 24(5):625–632. doi:10.1016/j.coi.2012.08.013

Candotti F, Shaw KL, Muul L, Carbonaro D, Sokolic R, Choi C, Schurman SH et al (2012) Gene therapy for adenosine deaminase—deficient severe combined immune deficiency: clinical comparison of retroviral vectors and treatment plans plenary paper Gene therapy for adenosine deaminase—deficient severe combined immune deficiency: clinical. Blood 120:3635–3646. doi:10.1182/blood-2012-02-400937

Capotondo A, Cesani M, Pepe S, Fasano S, Gregori S, Tononi L, Venneri MA et al (2007) Safety of arylsulfatase a overexpression for gene therapy of metachromatic leukodystrophy. Hum Gene Ther 18(9):821–836. doi:10.1089/hum.2007.048

Carcinoma H (2011) Selective targeting of genetically engineered mesenchymal stem cells to tumor stroma microenvironments using tissue-specific suicide gene expression suppresses growth of hepatocellular carcinoma. Ann Surg 254(5):767–775. doi:10.1097/SLA.0b013e3182368c4f

Cavazzana-Calvo M, Hacein-bey S, De Saint Basile G, Gross F, Yvon E, Nusbaum P, Selz F et al (2000) Gene therapy of human severe combined immunodeficiency SCID—X1 disease. Science 288:669–672. doi:10.1126/science.288.5466.669

Cavazzana-Calvo M, Payen E, Negre O, Wang G, Cavallesco R, Gillet-legrand B, Caccavelli L (2010) Transfusion independence and HMGA2 activation after gene therapy of human β-thalassaemia. Nature 467(7313):318–322. doi:10.1038/nature09328.Transfusion

Chamberlain JR, Schwarze U, Wang P, Hirata RK, Hankenson KD, Pace JM, Underwood RA et al (2004) Gene targeting in stem cells osteogenesis imperfecta. Science 303:1198–1202

Check E (2002) A tragic setback. Nature 420:116–118

Choudhery MS, Badowski M, Muise A, Harris DT (2013) Comparison of human mesenchymal stem cells derived from adipose and cord tissue. Cytotherapy 15(3):330–343. doi:10.1016/j.jcyt.2012.11.010

Ciurea SO, Andersson BS (2009) Busulfan in hematopoietic stem cell transplantation. Biol Blood Marrow Transplant 15(5):523–536. doi:10.1016/j.bbmt.2008.12.489

Cong L, Ran FA, Cox D, Lin S, Barretto R, Habib N, Hsu PD et al (2013) Multiplex genome engineering using CRISPR/Cas systems. Science (New York, NY) 339(6121):819–823. doi:10.1126/science.1231143

Copland IB, Jolicoeur EM, Gillis M, Cuerquis J, Eliopoulos N, Annabi B, Calderone A et al (2008) Coupling erythropoietin secretion to mesenchymal stromal cells enhances their regenerative properties. Cardiovasc Res 79:405–415. doi:10.1093/cvr/cvn090

Corrigan-curay J, Cohen-haguenauer O, Reilly O, Ross SR, Fan H, Rosenberg N, King N et al (2009) Challenges in vector and trial design using retroviral vectors for long-term gene correction in hematopoietic. Mol Ther 20(6):1084–1094. doi:10.1038/mt.2012.93

Cucchiarini M, Madry H, Ma C, Thurn T, Zurakowski D, Menger MD, Kohn D et al (2005) Improved tissue repair in articular cartilage defects in vivo by rAAV-mediated overexpression of human fibroblast growth factor 2. Mol Ther 12(2):229–238. doi:10.1016/j.ymthe.2005.03.012

Czechovicz A, Kraft D, Weissman I, Bhattacharya D (2008) Efficient transplantation via antibody-based clearance of hematopoietic stem cell niches. Science 318(5854):1296–1299. doi:10.1126/science.1149726.Efficient

De Rosa L, Carulli S, Cocchiarella F, Quaglino D, Enzo E, Franchini E, Giannetti A et al (2014) Long-term stability and safety of transgenic cultured epidermal stem cells in gene therapy of junctional epidermolysis bullosa. Stem cell Rep 2(1):1–8. doi:10.1016/j.stemcr.2013.11.001

Di W, Mellerio JE, Bernadis C, Harper J, Abdul-wahab A, Ghani S, Chan L et al (2013) Phase I study protocol for ex vivo lentiviral gene therapy for the inherited skin disease, netherton syndrome. Hum Gene Ther Clin Dev 190:182–190. doi:10.1089/humc.2013.195

DiGiusto DL, Krishnan A, Li L, Li H, Li S, Rao A, Mi S et al (2010) RNA-based gene therapy for hiv with lentiviral vector—modified CD34+ cells in patients undergoing transplantation for aids-related lymphoma. Gene 43:1–8. doi:10.1126/scitranslmed.3000931

Doering CB, Archer D, Spencer HT (2011) Delivery of nucleic acid therapeutics by genetically engineered hematopoietic stem cells. Adv Drug Deliv Rev 62(12):1204–1212. doi:10.1016/j.addr.2010.09.005.Delivery

Ely A, Naidoo T, Mufamadi S, Crowther C, Arbuthnot P (2008) Expressed anti-HBV primary MicroRNA shuttles inhibit viral replication efficiently. Mol Ther 16(6):1105–1112. doi:10.1038/mt.2008.82

European Medicines Agency (2008) Guideline on the non-clinical studies required before first clinical use of gene therapy medicinal products. Eur Med Agency (November), EMEA/CHMP/GTWP/125459/2006:1–10

European Medicines Agency (2009) Guideline on clinical follow-up gene therapy. Eur Med Agency (October), EMEA/CHMP/GTWP/60436/2007: 1–12

Ferrua F, Brigida I, Aiuti A (2010) Update on gene therapy for adenosine deaminase-deficient severe combined immunodeficiency. Curr Opin Allergy Clin Immunol 10(6):32833. doi:10.1097/ACI.0b013e32833fea85.Update

Fink JK, Correll PH, Perry LK, Brady R, Karlsson S (1990) Correction of glucocerebrosidase deficiency after retroviral-mediated gene transfer into hematopoietic progenitor cells from patients with Gaucher disease. Proc Natl Acad Sci U S A 87:2334–2338

Fischer A, Hacein-Bey-Abina S, Cavazzana-Calvo M (2010) 20 years of gene therapy for SCID. Nat Immunol 11(6):457–460. doi:10.1038/ni0610-457

Fossett E, Khan WS (2012) Optimising human mesenchymal stem cell numbers for clinical application: a literature review. Stem Cells Int 2012:465259. doi:10.1155/2012/465259

Galis ZS, Black JB, Skarlatos SI (2015) Therapeutic approaches. Circ Res 112:1212–1219. doi:10.1161/CIRCRESAHA.113.301100

Giancola R, Bonfini T, Iacone A (2012) Cell therapy: cGMP facilities and manufacturing. Muscles Ligaments Tendons J 2(3):243–247

Giry-laterrière M, Verhoeyen E, Salmon P (2011) Lentiviral vectors. In: Merten O-W, Al-Rubeai M (eds) Viral vectors for gene therapy: methods and protocols, vol 737. Springer, New York, pp 183–209. doi:10.1007/978-1-61779-095-9

Gnecchi M, Zhang Z, Ni A, Dzau VJ (2008) Paracrine mechanisms in adult stem cell signaling and therapy. Circ Res 103(11):1204–1219. doi:10.1161/CIRCRESAHA.108.176826

Gould DJ, Favorov P (2003) Vectors for the treatment of autoimmune disease. Gene Ther 10(10):912–927. doi:10.1038/sj.gt.3302018

GTWP (2008) Guideline on the non-clinical studies required before first clinical use of gene therapy medicinal products. Eur Med Agency (November), EMEA/CHMP/GTWP/125459/2006: 1–10

Hacein-Bey-Abina S, Le Deist F, Carlier F, Bouneaud C, Hue C, De Villartay J-P, Thrasher AJ et al (2002) Sustained correction of X-linked severe combined immunodeficiency by ex vivo gene therapy. N Engl J Med 346(16):1185–1193

Hacein-Bey-Abina S, Von Kalle C, Schmidt M, McCormack MP, Wulffraat N, Leboulch P, Lim A et al (2003) LMO2-associated clonal T cell proliferation in two patients after gene therapy for SCID-X1. Science (New York, NY) 302(5644):415–419. doi:10.1126/science.1088547

Hacein-bey-abina S, Garrigue A, Wang GP, Soulier J, Lim A, Morillon E, Clappier E et al (2008) Insertional oncogenesis in 4 patients after retrovirus-mediated gene therapy of SCID-X1. J Clin Invest 118(9):3132–3142. doi:10.1172/JCI35700.3132

Hacein-Bey-Abina S, Hauer J, Lim A, Picard C, Wang GP, Berry CC, Martinache C et al (2011) Efficacy of gene therapy for X-linked severe combined immunodeficiency. N Engl J Med 363(4):355–364. doi:10.1056/NEJMoa1000164.Efficacy

Hadaczek P, Eberling JL, Pivirotto P, Bringas J, Forsayeth J, Bankiewicz KS (2010) Eight years of clinical improvement in MPTP-lesioned primates after gene therapy with AAV2-hAADC. Mol Ther 18(8):1458–1461. doi:10.1038/mt.2010.106

Harris DT (2014) Stem cell banking for regenerative and personalized medicine. Biomedicine 2:50–79. doi:10.3390/biomedicines2010050

Horn PA, Thomasson BM, Wood BL, Andrews RG, Morris JC, Kiem H (2015) Distinct hematopoietic stem/progenitor cell populations are responsible for repopulating NOD/SCID mice compared with nonhuman primates. Am Soc Hematol 102(13):4329–4336. doi:10.1182/blood-2003-01-0082.Supported

Ishikawa F, Yasukawa M, Lyons B, Yoshida S, Miyamoto T, Watanabe T, Akashi K et al (2014) Development of functional human blood and immune systems in NOD/SCID/IL2 receptor γ chain null mice. Blood 106(5):1565–1573. doi:10.1182/blood-2005-02-0516

Jackson CS, Pepper MS (2013) Opportunities and barriers to establishing a cell therapy programme in South Africa. Stem Cell Res Ther 4:54

Johnston J, Denning G, Doering CB, Spencer HT (2013) Generation of an optimized lentiviral vector encoding a high expression factor VIII transgene for gene therapy of hemophilia A. Gene Ther 20:607–615. doi:10.1038/gt.2012.76

Kang E, De Witte M, Malech H, Morgan RA, Phang S, Carter C, Leitman SF et al (2002) Nonmyeloablative conditioning followed by transplantation of genetically modified HLA-matched peripheral blood progenitor cells for hematologic malignancies in patients with acquired immunodeficiency syndrome. Blood 99(2):698–701

Kang EM, Choi U, Theobald N, Linton G, Priel DAL, Kuhns D, Malech HL (2010) Retrovirus gene therapy for X-linked chronic granulomatous disease can achieve stable long-term correction of oxidase activity in peripheral blood neutrophils. Gene Ther 115(4):783–791. doi:10.1182/blood-2009-05-222760.The

Kazuki Y, Hiratsuka M, Takiguchi M, Osaki M, Kajitani N, Hoshiya H, Hiramatsu K et al (2010) Complete genetic correction of ips cells from Duchenne muscular dystrophy. Mol Ther 18(2):386–393. doi:10.1038/mt.2009.274

Kean TJ, Lin P, Caplan AI, Dennis JE (2013) MSCs: delivery routes and engraftment, cell-targeting strategies, and immune modulation. Stem Cells Int 2013:1–13

Kiem H-P, Jerome KR, Deeks SG, Joseph MM (2012) Hematopoietic stem cell-based gene therapy for HIV disease. Cell Stem Cell 10(2):137–147. doi:10.1016/j.stem.2011.12.015. Hematopoietic

Kohn BDB, Bauer G, Rice CR, Rothschild JC, Carbonaro DA, Valdez P, Hao Q et al (1999) A clinical trial of retroviral-mediated transfer of a rev-responsive element decoy gene into CD34+ cells from the bone marrow of human immunodeficiency virus-1–infected children. Blood 94(1):368–371

Kuramoto K, Follman D, Hematti P, Sellers S, Laukkanen MO, Seggewiss R, Metzger ME et al (2004) The impact of low-dose busulfan on clonal dynamics in nonhuman primates. Gene Ther 104(5):1273–1280. doi:10.1182/blood-2003-08-2935.Reprints

Kutner RH, Zhang X-Y, Reiser J (2009) Production, concentration and titration of pseudotyped HIV-1-based lentiviral vectors. Nat Protoc 4(4):495–505. doi:10.1038/nprot.2009.22

Lewis CM, Suzuki M (2014) Therapeutic applications of mesenchymal stem cells for amyotrophic lateral sclerosis. Stem Cell Res Ther 5(2):32

Liu JM, Kim S, Read EJ, Futaki M, Dokal I, Carter CS, Leitman F et al (1999) Engraftment of hematopoietic progenitor cells transduced with the Fanconi anemia group C gene (FANCC). Hum Gene Ther 10(14):2337–2346. doi:10.1089/10430349950016988.Published

Liu Y, Hangoc G, Campbell TB, Goodman M, Tao W, Pollok K, Srour EF et al (2009) Identification of parameters required for efficient lentiviral vector transduction and engraftment of human cord blood CD34+ SCID repopulating cells NOD. Exp Hematol 36(8):947–956. doi:10.1016/j.exphem.2008.06.005.Identification

Logan AC, Lutzko C, Kohn DB (2002) Advances in lentiviral vector design for gene-modification of hematopoietic stem cells. Curr Opin Biotechnol 13(5):429–436

Marti J, Arriero MM, Garci J, Alemany R, Cascallo M, Madero L, Rami M (2010) Treatment of metastatic neuroblastoma with systemic oncolytic virotherapy delivered by autologous mesenchymal stem cells: an exploratory study. Cancer Gene Ther 17:476–483. doi:10.1038/cgt.2010.4

Martinez-Quintanilla J, Bhere D, Heidari P, He D, Mahmood U, Shah K (2013) Therapeutic efficacy and fate of bimodal engineered stem cells in malignant brain tumors. Stem Cells 31:1706–1714

Mátrai J, Chuah MKL, Vandendriessche T (2009) Recent advances in lentiviral vector development and applications. Mol Ther 18(3):477–490. doi:10.1038/mt.2009.319

Mautino MR, Keiser N, Morgan RA (2001) Inhibition of human immunodeficiency virus type 1 (HIV-1) replication by HIV-1-based lentivirus vectors expressing transdominant rev. J Virol 75(8):3590–3599. doi:10.1128/JVI.75.8.3590

Mavilio F (2012) Gene therapies need new development models. Nature 490:7

Mitsuyasu RT, Merigan TC, Carr A, Zack JA, Mark A, Workman C, Bloch M et al (2009) Phase 2 gene therapy trial of an anti-HIV ribozyme in autologous CD34+ cells. Nat Med 15(3):285–292. doi:10.1038/nm.1932

Mohit E, Rafati S (2013) Biological delivery approaches for gene therapy: strategies to potentiate efficacy and enhance specificity. Mol Immunol 56(4):599–611. doi:10.1016/j.molimm.2013.06.005

Molas M, Bermudez J, Bartrons R, Perales JC (2003) Receptor-mediated gene transfer vectors: progress towards genetic pharmaceuticals. Curr Gene Ther 3:468–485

Momparler RL, Momparler LF, Galipeau J (2002) Human cytidine deaminase as an ex vivo drug selectable marker in gene-modified primary bone marrow stromal cells. Gene Ther 9(7):452–462. doi:10.1038/sj/gt/3301675

Mullen CA, Snitzer K, Culver KW, Morgan RA, Anderson WF, Blaese RM (1996) Molecular analysis of T lymphocyte-directed gene therapy for adenosine deaminase deficiency: long-term in vivo of genes introduced with a retroviral expression vector. Human Gene Ther 7(9):1123–1129

Müller LUW, Milsom MD, Kim M, Schambach A, Schuesler T, Williams DA (2008) Rapid lentiviral transduction preserves the engraftment potential of *Fanca* −/− hematopoietic stem cells. Am Soc Gene Ther 16(6):1154–1160. doi:10.1038/mt.2008.67

Myburgh R, Ivic S, Pepper MS, Gers-Huber G, Li D, Audigé A, Rochat M-A et al (2015) Lentivector knock-down of CCR5 in hematopoietic stem cells confers functional and persistent HIV-1 resistance in humanized mice. J Virol 89(13):1–13. doi:10.1128/JVI.00277-15

Myers TJ, Granero-molto F, Longobardi L, Li T, Yan Y, Spagnoli A (2011) Mesenchymal stem cells at the intersection of cell and gene therapy. Exp Opin Biol Ther 10(12):1663–1679. doi:10.1517/14712598.2010.531257.Mesenchymal

Naldini L, Blomer U, Gallay P, Ory D, Mulligan R, Gage FH, Verma IM et al (1996) In vivo gene delivery and stable transduction of nondividing cells by a lentiviral vector. Science 9(13):263–267

Narsinh KH, Wu JC, Wu CJC (2009) Gene correction in human embryonic and induced pluripotent stem cells: promises and challenges ahead. Mol Ther 18(6):1061–1063. doi:10.1038/mt.2010.92

Neff T, Shotkoski F, Stamatoyannopoulos G (1997) Stem cell gene therapy, position effects and chromatin insulators. Stem Cells 15(Suppl 1):265–271

Niess H, Von Einem JC, Thomas MN, Michl M, Angele MK, Huss R, Günther C et al (2015) Treatment of advanced gastrointestinal tumors with genetically modified autologous mesenchymal stromal cells (TREAT-ME1): study protocol of a phase I/II clinical trial. BMC Cancer 15(237):1–13. doi:10.1186/s12885-015-1241-x

OCTGT (2013) Guidance for industry preclinical assessment of investigational cellular and gene therapy products, pp 1–30

Ott MG, Schmidt M, Schwarzwaelder K, Stein S, Siler U, Koehl U, Glimm H et al (2006) Correction of X-linked chronic granulomatous disease by gene therapy, augmented by insertional activation of MDS1-EVI1, PRDM16 or SETBP1. Nat Med 12(4):401–409. doi:10.1038/nm1393

Papadakis ED, Nicklin SA, Baker AH, White SJ (2004) Promoters and control elements: designing expression cassettes for gene therapy. Curr Gene Ther 40:89–113

Reiser J, Zhang X, Hemenway CS, Mondal D, Russa VFL (2006) Potential of mesenchymal stem cells in gene therapy approaches for inherited and acquired diseases. Exp Opin Biol Ther 5(12):1571–1584

Romero Z, Urbinati F, Geiger S, Cooper AR, Wherley J, Kaufman ML, Hollis RP et al (2013) β-globin gene transfer to human bone marrow for sickle cell disease. J Clin Investig 123(8):3317–3330. doi:10.1172/JCI67930DS1

Rossi JJ (2000) Ribozyme therapy for HIV infection. Adv Drug Deliv Rev 44:71–78

Rossi JJ (2009) NIH public access. Hum Gene Ther 19(4):313–317. doi:10.1089/hum.2008.026. Expression

Sanchez AR, Silberstein LE (2013) Strategies for more rapid translation of cellular therapies for children: a US perspective. Pediatrics 132(2):351–358. doi:10.1542/peds.2012-3383

Scaramuzza S, Biasco L, Ripamonti A, Castiello MC, Loperfido M, Draghici E, Hernandez RJ et al (2009) Preclinical safety and efficacy of human CD34+ cells transduced with lentiviral vector for the treatment of Wiskott-Aldrich syndrome. Mol Ther 21(1):175–184. doi:10.1038/mt.2012.23

Seok J, Warren HS, Cuenca AG, Mindrinos MN, Baker HV, Xu W, Richards DR et al (2013) Genomic responses in mouse models poorly mimic human inflammatory diseases. Proc Natl Acad Sci U S A 110(9):3507–3512. doi:10.1073/pnas.1222878110

Seow Y, Wood MJ (2009) Biological gene delivery vehicles: beyond viral vectors. Mol Ther 17(5):767–777. doi:10.1038/mt.2009.41

Shujia J, Haider HK, Idris NM, Lu G, Ashraf M (2008) Stable therapeutic effects of mesenchymal stem cell-based multiple gene delivery for cardiac repair. Cardiovasc Res 77(3):525–533. doi:10.1093/cvr/cvm077

Shultz LD, Brehm MA, Garcia JV, Greiner DL (2013) Humanized mice for immune system investigation: progress, promise and challenges. Nature reviews. Immunology 12(11):786–798. doi:10.1038/nri3311.Humanized

Sidransky E, LaMarca M, Ginns E (2007) Therapy for Gaucher disease: don't stop thinking about tomorrow. Mol Genet Metab 90:122–125

Singh, M., Singh, P., Gaudray, G., Musumeci, L., Thielen, C., Vaira, D., Vandergeeten, C., et al. (2012). An improved protocol for efficient engraftment in NOD/LTSZ-SCIDIL-2Rc NULL mice allows HIV replication and development of anti-HIV immune responses, 7(6): e38491. doi:10.1371/journal.pone.0038491

Soares C, Cookson C, Beardsley S (2005) The future of stem cells. Sci Am July:1–34

Song S, Chang W, Song B, Song H, Lim S, Him H, Cha M et al (2009) Integrin-linked kinase is required in hypoxic mesenchymal stem cells for strengthening cell adhesion to ischemic myocardium. Stem Cells 27:1358–1365. doi:10.1002/stem.47

Spencer HT, Denning G, Gautney RE, Dropulic B, Roy AJ, Baranyi L, Gangadharan B et al (2009) Lentiviral vector platform for production of bioengineered recombinant coagulation factor VIII. Mol Ther 19(2):302–309. doi:10.1038/mt.2010.239

Stein S, Ott MG, Schultze-strasser S, Jauch A, Burwinkel B, Kinner A, Schmidt M et al (2010) Genomic instability and myelodysplasia with monosomy 7 consequent to EVI1 activation after gene therapy for chronic granulomatous disease. Nat Med 16(2):198 205. doi:10.1038/nm.2088

Taylor P, Kunert R, Casanova E (2013) Recent advances in recombinant protein production. Bioengineered 4(4):258–261. doi:10.4161/bioe.24060

Throm RE, Ouma AA, Zhou S, Chandrasekaran A, Lockey T, Greene M, De Ravin SS et al (2009) Efficient construction of producer cell lines for a SIN lentiviral vector for SCID-X1 gene therapy by concatemeric array transfection. Gene Ther 113(21):5104–5110. doi:10.1182/blood-2008-11-191049.The

Tiscornia G, Singer O, Verma IM (2007) Development of lentiviral vectors expressing siRNA. Spring, (pol III), 23–34

Tremblay J, Xiao X, Aartsma-Rus A, Carlos Barbas A, Blau HM, Bogdanove AJ, Boycott K, Braun S, Breakefield XO, Bueren JA (2013) Translating the genomics revolution: the need for an international gene therapy consortium for monogenic diseases. Mol Ther 21(2):266–268. doi:10.1038/mt.2013.4

Treschow A, Unger C, Aints A, Felldin U, Aschan J, Dilber MS (2007) OuaSelect, a novel ouabain-resistant human marker gene that allows efficient cell selection within 48 h. Gene Ther 14:1564–1572. doi:10.1038/sj.gt.3303015

Tsubokawa T, Yagi K, Nakanishi C, Zuka M, Nohara A, Ino H, Fujino N et al (2010) Impact of anti-apoptotic and anti-oxidative effects of bone marrow mesenchymal stem cells with transient overexpression of heme oxygenase-1 on myocardial ischemia. Am J Physiol Heart Circ Physiol 298:1320–1329. doi:10.1152/ajpheart.01330.2008

Tsuruta T (2013) Recent advances in hematopoietic stem cell gene therapy. INTECH: February: 1–30

U.S. Congress (1988). Mapping our genes—genome projects: how big? How fast? April 1988. Office of Technology Assessment, (April)

Ugarte F, Forsberg EC (2013) Haematopoietic stem cell niches: new insights inspire new questions. EMBO J 32(19):2535–2547. doi:10.1038/emboj.2013.201

Van Griensven J, De Clercq E, Debyser Z (2005) Hematopoietic stem cell-based gene therapy against HIV infection: promises and caveats. AIDS Rev 7:44–55

Vattemi E, Claudio PP (2009) Head & neck oncology cancer. Head Neck Oncol 6:1–6. doi:10.1186/1758-3284-1-3

Vollweiler JL, Zielske SP, Reese JS, Gerson SL (2003) Mini review hematopoietic stem cell gene therapy: progress toward therapeutic targets. Bone Marrow Transplant 37:1–7. doi:10.1038/sj.bmt.1704081

Von Kalle C, Baum C, Williams DA (2004) Lenti in red: progress in gene therapy for human hemoglobinopathies. J Clin Investig 7:889–891. doi:10.1172/JCI200423132.severe

Walsh BCE, Grompe M, Vanin E, Buchwald M, Young NS, Nienhuis AW, Liu JM (1994) Functionally active retrovirus vector for gene therapy in Fanconi anemia group C. Blood 84(2):453–459

Watts K, Adair J, Kiem H (2012) Hematopoietic stem cell expansion and gene therapy. Cytotherapy 13(10):1164–1171. doi:10.3109/14653249.2011.620748.Hematopoietic

Weissman IL, Shizuru JA (2008) The origins of the identification and isolation of hematopoietic stem cells, and their capability to induce donor-specific transplantation tolerance and treat autoimmune diseases. Blood 112(9):3543–3553

Yannaki E, Papayannopoulou T, Jonlin E, Zervou F, Karponi G, Xagorari A, Becker P et al (2012) Hematopoietic stem cell mobilization for gene therapy of adult patients with severe β-thalassemia: results of clinical trials using G-CSF or plerixafor in splenectomized and non-splenectomized subjects. Mol Ther 20(1):230–238. doi:10.1038/mt.2011.195

Yu Y, Wang X, Nyberg SL (2014) Potential and challenges of induced pluripotent stem cells in liver diseases treatment. J Clin Med 3:997–1017. doi:10.3390/jcm3030997

Zaia JA, Forman SJ (2013) Transplantation in HIV-infected subjects: is cure possible? HIV Hematol 2013:389–393. doi:10.1182/asheducation-2013.1.389

Zhan H, Gilmour K, Chan L, Farzaneh F, Mcnicol AM, Xu J, Adams S et al (2013) Production and first-in-man use of T cells engineered to express a HSVTK-CD34 sort-suicide gene. PLoS One 8(10):1–9. doi:10.1371/journal.pone.0077106

Chapter 4
The Safety of Non-Expanded Multipotential Stromal Cell Therapies

Dimitrios Kouroupis, Xiao Nong Wang, Yasser El-Sherbiny, Dennis McGonagle, and Elena Jones

4.1 Introduction

Embryonic stem cells transiently exist in the embryo. Although greatly potent and able to generate many diverse tissues (Desai et al. 2015), they require extensive culturing and pre-differentiation in order to ensure the correct lineage commitment and to prevent tumour development from residual undifferentiated cells. Their culture-induced analogue, so-called induced pluripotent stem cells, offers an opportunity for autologous transplantation without immune rejection but, being derived from patient's own skin cells, similarly requires culturing and pre-differentiation (Takahashi et al. 2007). Adult stem cells, which exist in their native niches in many tissues in the adult body, may be seen as comparatively more restricted in tissues they can generate, but to their advantage, they don't necessarily require culturing prior to transplantation. For example, HSC-based therapies are based on uncultured bone marrow (BM), harvested and processed with minimal manipulation, and bone marrow transplants have been successfully used for the treatment of blood cancers for many decades (Mohty et al. 2015).

Multipotential stromal cells (MSCs) are non-haematopoietic cells first described in the BM by Friedenstein et al. (1974). Since then, culture-expanded MSCs have been commonly generated from BM aspirates and other connective tissues and fluids including subcutaneous adipose tissue (Zuk et al. 2002), synovial tissue (De Bari

D. Kouroupis
Department of Biomedical Research, Foundation for Research and Technology-Hellas,
Institute of Molecular Biology and Biotechnology, University of Ioannina, Ioannina, Greece

X.N. Wang
Institute of Cellular Medicine, University of Newcastle upon Tyne, Newcastle, UK

Y. El-Sherbiny • D. McGonagle • E. Jones (✉)
Leeds Institute of Rheumatic and Musculoskeletal Medicine, University of Leeds, Leeds, St James's University Hospital, Room 5.24 Clinical Sciences Building, Leeds LS9 7TF, UK
e-mail: msjej@leeds.ac.uk

© Springer International Publishing AG 2017
P.V. Pham, A. Rosemann (eds.), *Safety, Ethics and Regulations*, Stem Cells in Clinical Applications, DOI 10.1007/978-3-319-59165-0_4

et al. 2001) and umbilical cord (UC) (Weiss and Troyer 2006; Karahuseyinoglu et al. 2007; Pasquinelli et al. 2007; Ishige et al. 2009) through plastic adherence and in vitro expansion for several passages to achieve stromal cell homogeneity and the removal of contamination of haematopoietic-lineage cells. In 1995 Lazarus et al. have reported the first clinical trial to determine the feasibility of collection, culture expansion and intravenous infusion of BM-derived MSCs (Lazarus et al. 1995). Since then, more than 2000 patients have received autologous or allogeneic MSC-based therapy for the treatment of various musculoskeletal, cardiovascular, neurological, immunological and blood pathologies (Ikebe and Suzuki 2014).

The perceived advantage of cultured MSCs as cell therapy was initially associated with their ease of isolation, high expandability and differentiation capacity towards the bone (Jaiswal et al. 1997), cartilage (Sekiya et al. 2002) and muscle tissues (Westerweel and Verhaar 2008). Subsequently their mechanism of action in some indications was reported to be due to their immunoprivileged and immuno-regulatory properties as well as paracrine production of growth factors and cytokines involved in supporting angiogenesis and inhibiting cell apoptosis (Le Blanc and Ringden 2005; da Silva Meirelles et al. 2009; Salem and Thiemermann 2010; Nguyen et al. 2010; Rehman et al. 2004; Asanuma et al. 2010; Chen et al. 2008; Doorn et al. 2011). It has now become accepted that all these mechanisms may act synergistically resulting in beneficial therapeutic outcome. For example, clinical applications that aim at bone regeneration through direct MSC differentiation are likely to benefit from MSCs' support to neovascularisation (Au et al. 2008) and their immunoregulatory properties (Ringden et al. 2006; Ghannam et al. 2010) that would enhance vascular infiltration and reduce inflammation at the trauma site.

The clinical use of MSCs has been performed in autologous or allogeneic transplant settings. The autologous transplantation of MSCs avoids tissue incompatibility between donor and recipient and thus avoids the risk of immune response-mediated rejections. However, the use of autologous MSCs has some inherent limitations. Firstly, their harvesting involves an additional surgical intervention in often highly compromised and elderly patients. Secondly, ageing is known to affect MSC function (D'Ippolito et al. 1999; Choudhery et al. 2014; Jones et al. 2010). Thirdly, MSCs' reparative capacity may be affected by niche's microenvironment, and MSCs extracted from human osteoarthritic patients may in fact be 'diseased' themselves (Harris et al. 2013). Furthermore, the use of autologous MSCs lacks the potential to become a standardised 'off the shelf' routine treatment to benefit a large number of patients. Because of these limitations, allogeneic MSCs have been utilised in a large number of clinical trials (Syed and Evans 2013). In standard transplantation practice, most solid-organ allogeneic transplantation requires lifelong use of antirejection drug therapy regardless of the fact that the donor and recipient are practically matched for HLA antigens (Malgieri et al. 2010). Remarkably, allogeneic BM MSCs have been used for certain therapeutic indications such as vascular, cardiac, orthopaedic and inflammatory diseases without the use of antirejection drug therapy (Syed and Evans 2013). This is because MSCs possess immunoprivileged and immunosuppressive characteristics (Ghannam et al. 2010; Le Blanc et al. 2003; Melief et al. 2013; Yagi et al. 2010; Wang et al. 2014a). In this chapter, the safety profiles of autologous and allogeneic MSC preparations for diverse clinical indications will be discussed separately.

4.2 The Safety of Current Therapies Based on Culture-Expanded MSCs

4.2.1 Musculoskeletal Tissue Regeneration

The bone is a highly vascularised connective tissue that undergoes continuous remodelling throughout life. Bone regeneration is required primarily in cases of large-bone defects due to trauma, infection or tumour resection or when bone regeneration process is compromised due to underlying disease process. For example, in avascular necrosis (AVN) interruption of blood flow due to bone fracture or joint dislocation leads to cellular death of bone (Hernigou et al. 2009). In the first clinical trial aimed at repairing the bone, autologous BM MSCs were delivered to large long-bone defects using porous calcium phosphate scaffolds (Quarto et al. 2001). Within a few months of post-implantation, all three patients showed abundant callus formation, and no adverse effects were reported. A follow-up of the same three patients has reported good implant integration which was maintained for a minimum of 6 years (Marcacci et al. 2007). Three patients with steroid-induced osteonecrosis of the femoral head were treated with cultured autologous MSCs combined with β-tricalcium phosphate scaffolds (Kawate et al. 2006). At 1 year follow-up, all patients were pain-free and have demonstrated revascularisation at the grafted sites (Kawate et al. 2006).

There is also some evidence that the use of cultured MSCs to treat skeletal diseases such as osteogenesis imperfecta (OI) is safe in the allogeneic settings (Horwitz et al. 2002; Le Blanc et al. 2005). In 2002 allogeneic cultured BM MSCs were infused systemically to treat six children with severe OI. Engrafted MSCs seemed to contribute to new bone formation, and at 18–36 months follow-up, all patients showed increased total body mineral content and accelerated linear growth (Horwitz et al. 2002). In another study, allogeneic foetal liver MSCs were transplanted in utero at 32 weeks of gestation to treat a foetus with severe OI. Transplanted MSCs showed good engraftment and contributed in bone turnover without causing any graft-versus-host disease (GVHD) effect. Importantly, at 2 years follow-up psycho-motor development and growth were normal (Le Blanc et al. 2005).

Injured cartilage shows limited regeneration capacity. In 2004 Wakitani et al. evaluated the effectiveness and safety of autologous cultured BM MSCs combined with collagen I scaffold to repair full-thickness articular cartilage defects. Six months post-implantation, the clinical symptoms have significantly improved in both patients, an effect that remained for at least 4 years later (Wakitani et al. 2004). In another study, autologous cultured BM MSCs were combined with collagen I to repair full-thickness defects in the weight-bearing area of femoral condyle in one patient. This patient returned to normal life 12 months post-implantation (Kuroda et al. 2007). More recently, Kasemkijwattana et al. showed that autologous cultured BM MSCs implanted with collagen I scaffold into cartilage knee defects significantly improved cartilage quality for a mean follow-up of 30 months and no adverse events were reported (Kasemkijwattana et al. 2011).

4.2.2 Myocardial Regeneration

In MSC-based heart regeneration treatments, MSCs are transplanted or directly infused into infarcted myocardium. The observed myocardial regeneration by MSCs is believed to be mediated by their paracrine effects (trophic, immunomodulatory and anti-scarring) rather than through their direct differentiation to cardiomyocytes (da Silva Meirelles et al. 2009).

The first clinical trials utilised autologous or allogeneic cultured BM MSCs infused in patients up to 10 days postmyocardial infarction directly into the ischemic myocardium. One recent clinical trial didn't show any improvement of cardiac function after delivery of cultured BM MSCs (Hare et al. 2012). A meta-analysis of 33 randomised trials revealed heterogeneous results between trials; however, in all cases there were no signs of increased morbidity or mortality (Clifford et al. 2012).

In a dose-escalation study using allogeneic BM MSCs, results showed no major adverse effects, increased left ventricular ejection fraction and remodelling 6 months follow-up (Hare et al. 2009). In a recently published study, allogeneic BM MSC administration also showed no adverse effects after 6 months follow-up (Chullikana et al. 2015).

4.2.3 Neuronal Tissue Regeneration

Similar to the cartilage, the nervous system has limited self-repair capacity upon injury. This happens due to limited ability of neural stem cells to generate functional neurons. In the latest clinical trials, MSC-based therapies have been used to treat neuronal system disorders such as multiple sclerosis (MS), amyotrophic lateral sclerosis (ALS), Parkinson's disease, stroke and spinal cord injury. The leading view on the field is that after transplantation, MSCs exert their action through their paracrine effects by promoting endogenous neuronal growth and reducing apoptosis and inflammation (Joyce et al. 2010).

Two pioneering clinical trials performed in 2010 have demonstrated the safety and the initial immunomodulation effects of autologous BM MSCs transplanted intrathecally to MS (Yamout et al. 2010; Karussis et al. 2010) and ALS (Karussis et al. 2010) patients. In another study, intravenously administered autologous cultured BM MSCs in ten MS patients resulted in vision improvement 10 months post-transplantation with no adverse effects documented (Connick et al. 2012). The reparative capacities of BM MSCs have been also evaluated in Parkinson's disease, which is characterised by loss of dopaminergic cells. Both autologous (Venkataramana et al. 2010) and allogeneic (Venkataramana et al. 2012) MSCs improved Parkinson's disease symptoms with a follow-up 10 to 36 months. Autologous BM MSCs have been also evaluated in clinical trials for stroke, which is characterised by compromised blood flow leading to ischemia and eventually brain tissue damage. Intravenously administered MSCs showed no adverse effects in all five patients with cerebral infarct (Bang et al. 2005). In another study, Honmou et al. intravenously administered autologous BM MSCs in 12 patients 36 to 133 days postischaemic stroke. All patients showed excellent recovery and no adverse effects of MSC treatment over a 12-month follow-up (Honmou et al. 2011).

Traumatic spinal cord injury (SCI) results in necrosis and apoptosis of the neighbouring site neurons to the injury site neurons (Vawda 2013). The safety of MSC therapy has been demonstrated in a clinical trial involving eight SCI patients treated intravenously with autologous cultured adipose-derived MSCs 12 months post-injury. None of the patients developed any adverse effects for a 3-month follow-up (Ra et al. 2011). In another clinical trial, Park et al. applied autologous BM MSCs to ten SCI patients; three out of ten patients showed decreased cavity size, improvement in motor power and beneficial electrophysiological changes for a follow-up of 30 months (Park et al. 2012). Although, allogeneic MSCs have not yet been applied in clinical trials for SCI, preclinical studies in animal models show beneficial outcomes (Torres-Espin et al. 2015).

4.2.4   Prevention of Acute Graft-Versus-Host Disease (GVHD)

The most significant results exploiting immunosuppressive effects of MSCs have been seen in prevention of acute graft-versus-host disease (GVHD) . In 2007, Ball et al. (2007) co-transplanted allogeneic cultured BM MSCs in 14 children undergoing transplantation of HLA-disparate CD34$^+$ HSCs from a relative. All patients transplanted with MSCs showed sustained haematopoietic engraftment without any adverse effects for a follow-up up to 28 months. In an open randomised clinical trial, 1 year later, allogeneic BM MSCs were co-transplanted with HLA-identical sibling-matched HSCs in haematological malignancy patients (Ning et al. 2008). The incidence of grades II–IV acute GVHD in ten patients receiving MSCs was significantly lower than in 15 patients not receiving MSCs (Ning et al. 2008). Phase II clinical study showed that 30 out of 55 GVHD patients had complete response (loss of all symptoms of acute GVHD) to allogeneic MSC infusion therapy (Le Blanc et al. 2008).

To conclude, the longest reported follow-up exists for studies that have used autologous cultured MSCs to treat long-bone defects showing no adverse effects 6–7 years post-implantation (Quarto et al. 2001; Marcacci et al. 2007). The longest monitored study with allogeneic BM MSCs (3 years) has been performed in haematological malignancy patients, in which MSCs were co-transplanted with HLA-identical sibling-matched HSCs (Ning et al. 2008). Although no direct comparison between autologous and allogeneic cultured MSCs have been performed in one single study, there is no evidence to suggest that allogeneic cultured MSCs are less safe than autologous MSCs following their implantation or infusion in vivo.

4.2.5 Current Approaches to Ensure and Improve the Safety of Culture-Expanded MSCs

According to US FDA regulatory and guidance documentation, cultured MSCs are categorised as 'human cells, tissues, or cellular and tissue-based products' (The Code of Federal Regulation title 21 part 1271) and on a risk basis can be 'minimally manipulated' and 'more-than-minimally manipulated'. Processes that alter the biological characteristics of the cells such as ex vivo culture expansion are considered more-than-minimally manipulations. Therefore cultured MSCs are not considered minimally manipulated products, and their in vivo application involves a very rigorous control of any potential safety concerns associated with MSC manufacture.

4.2.5.1 Replacing Foetal Calf Serum (FCS)

The most pertinent safety issue with cultured MSCs is related to the use of animal products, mostly foetal calf serum (FCS) in MSC expansion media, which gives rise to prion exposure risk, toxicological risk (due to persistence of toxic agents) and immunological risk (due to contaminating proteins, peptides or other molecules of animal origin) (Herberts et al. 2011). Although all clinical studies reported in the previous section have utilised FCS-cultured MSCs with no adverse events reported, avoiding FCS in MSC manufacture would ensure their additional safety. Previous studies have used allogeneic human serum (Jung et al. 2008; Shafaei et al. 2011); however obtaining large amounts of human serum to generate clinical relevant numbers of MSCs remains challenging (Shahdadfar et al. 2005; Tateishi et al. 2008). Alternatively, serum-free, chemical-defined media can replace FCS and also provide an enhanced proliferative capacity to MSCs (Agata et al. 2009; Chase et al. 2012). Recent studies showed that serum-free cultured MSCs possess different biological characteristics but do not change their therapeutic potential (Wang et al. 2014b). However, until now there are no documented clinical trials involving serum-free cultured MSCs. Human platelet lysate (hPL) contains a broad range of bioactive factors and is considered a promising replacement for FCS in MSC cultures (Bieback 2013). A pilot clinical study using MSCs expanded in hPL-containing medium has observed no severe adverse events during or after the MSC administration (von Bonin et al. 2009). Another study treating paediatric patients suffering from GVHD with MSCs expanded in hPL reported no acute or late side effects at a median follow-up of 8 months, including some patients receiving up to five MSC infusions (Lucchini et al. 2010).

4.2.5.2 Ensuring the Lack of Malignant Transformation

Another safety concern related to culture-expanded MSCs is that their extensive multiplication in vitro could result in acquisition of genomic abnormalities and chromosomal rearrangements that could theoretically lead to malignant transformation in vivo (Herberts et al. 2011). Some preclinical studies demonstrated MSCs' tumour-supporting capacity when administered via intravenous, intraarterial or peritumoural routes in mouse animal models of glioma (Yang et al. 2009; Hong et al. 2009; Nakamizo et al. 2005). However, up to date clinical evidence shows that transplantation of cultured MSCs in humans is safe without any signs of malignant transformation. Serial magnetic resonance imaging performed in 226 patients received hPL-expanded MSCs for various orthopaedic conditions showed no malignant transformation for a mean follow-up of 11 months (Centeno et al. 2011). Similarly, Tarte et al. showed that although clinical-grade MSC production could result in some chromosomal instability, this had no impact on their transformation potential, both in vitro and in vivo (Tarte et al. 2010). In fact, chromosomal abnormalities in cultured MSCs do not directly correlate with tumourigenesis but appear to be a result of cell senescence process due to extensive passaging (Wang et al. 2013; Tarte et al. 2010). Because this issue is still under debate, a more refined genetic stability testing of the cellular product should include more sensitive evaluations such as fluorescent in situ hybridisation to detect any potential abnormalities in cultured MSCs (Wang et al. 2014b).

4.2.5.3 Developing the Safest Route of Delivery

There are two potential routes to deliver cells in human body, systemic infusion or local infusion. Local infusion can be performed either via intraperitoneal, intramuscular, intracardiac injection or embedded into biomaterials. In the last 10–12 years, the commonest administration route of cultured MSCs has been through their systemic infusion into peripheral blood. In the case of high-dose infusion of MSCs (more than 1–2×10^6 per kg body weight) (Schallmoser et al. 2008; Ikebe and Suzuki 2014), potential adverse effects are considered to be allergic response to culturing compounds, prothrombotic effects and pulmonary or peripheral artery embolisation. In addition, the possibility of acute anti-donor antibody-mediated and/or T cell-mediated immune response post-allogeneic cultured MSC infusion is an additional concern (Griffin et al. 2013). In the first MSC clinical trial, allogeneic BM MSCs were intravenously infused in children with OI. Although, four out of five children had impressive functional improvements, only 1% of infused BM MSCs could be detected in the bone, skin and other tissues (Horwitz et al. 2002). Similar results were obtained in preclinical studies, indicating low levels of engraftment in animal models of diseases including myocardial infarction, Parkinson's disease, spinal cord injury and stroke (Baraniak and McDevitt 2010). Low homing of systemically infused MSCs can be attributed to their entrapment into capillaries of various tissues including lungs (Lee et al. 2009; Schrepfer et al. 2007).

4.2.5.4 Optimizing MSC Dose

As mentioned in the above sections, many patients have already received allogeneic cultured MSCs for a variety of clinical indications, with no documented adverse effects (Ankrum and Karp 2010). Specifically, phase I and II clinical studies of autologous or allogeneic cultured MSCs' administration via infusion and other routes showed no acute complications for doses up to 5×10^6 cells per kg of body mass (Hare et al. 2009; Duijvestein et al. 2010; Hare et al. 2012; Vaes et al. 2012). In 2009 Hare et al. (2009) performed a dose-ranging (0.5, 1.6 and 5×10^6 cells per kg of body mass) safety clinical trial of intravenous administered allogeneic cultured BM MSCs in 53 acute myocardial infarction patients. No adverse effects were observed in all different doses administered. In a recent high dose-ranging (10 and 20×10^6 cells per kg of body mass) safety clinical trial, allogeneic cultured UC MSCs were intratracheally administered in nine preterm infants with bronchopulmonary dysplasia (Chang et al. 2014). High-dose MSC administration didn't show adverse effects different from placebo group. Therefore, present data show that even high doses of autologous or allogeneic MSCs do not trigger adverse effects in vivo, but it remains unknown whether such high doses are necessarily required to achieve the desired therapeutic effect.

4.3 The Safety of Current Therapies Based on Non-Expanded MSCs

4.3.1 The Use of Non-Expanded MSCs in Orthopaedics Applications

The early recognition that BM aspirates contain MSCs (bone progenitor cells) as well as HSCs has prompted initial pioneering studies for the use of BM aspirate concentrates (BMACs) in bone repair applications by Hernigou et al. BMAC preparations can be easily made in the operating theatre room environment by centrifugation and volume reduction, with first commercial devices appearing on the orthopaedic market in the early 2000s. Philippe Hernigou and his team were the first orthopaedic academics to use BMAC preparations for the treatment of AVN, fracture non-union (NU) and rotator cuff repair (Table 4.1). In these innovative studies, no local or systemic complications were reported after an average follow-up of 7–10 years. Additionally, the positive treatment outcome was observed to correlate with the critical number of implanted MSCs (~50,000–60,000 CFU-Fs regardless the site of injection) (Hernigou et al. 2005; Hernigou et al. 2015; Hernigou et al. 2014). An up-to-date analysis of all treated patients by the same team (a total of 1873 patients), with an average follow-up of 12.5 years (range 5–22), has revealed no evidence of increased risks of cancers at both the injected site and systemically (Hernigou et al. 2013a).

Other surgical teams have reported the use of BMAC for the treatment of hip AVN (Kang et al. 2013; Mao et al. 2013; Sen et al. 2012; Gangji et al. 2011) and other orthopaedic conditions (Jäger et al. 2011). With a combined follow-up of approx. 5 years, these reports also indicated no adverse effects following BMAC injection at the local site. In Mao et al. study, BMAC was injected systemically; again no complications were reported (Mao et al. 2013). In one clinical study aimed

Table 4.1 Clinical trials using uncultured autologous BMAC/BMMNC for the treatment of orthopaedic defects

	Cell	Delivery	Patient	Average follow-up	Safety outcome
Hernigou and Beaujean (2002)	BMAC	Local injection	Hip osteonecrosis $N = 116$ (189 hips)	7 years (range 5–11)	Not reported
Hernigou et al. (2005)	BMAC	Local injection	Non-unions of tibia $N = 60$	4 months	No local or systemic complications
Hendrich et al. (2009)	BMAC BMAC+ scaffold	Local injection; Local implantation	AVN 69 Non-union 12 Other bone defects 20	14 months (range 2–24)	No infections, no excessive new bone formation, no induction of tumour formation and no morbidity
Hernigou et al. (2009)	BMAC	Local injection	Hip osteonecrosis $N = 342$ (534 hips)	13 years (rang 8–18)	No complications were encountered
Jäger et al. (2011)	BMAC+ scaffold	Local implantation	Various bone deficiencies $N = 39$	Minimal 6 months Maximal 2.5 years	No report of complications. Reported as 'application of BMAC is a safe procedure'
Gangji et al. (2011)	BMMNC	Local injection	Hip osteonecrosis $N = 19$ (24 hips)	5 years	Reported as 'no serious adverse reactions, with minor side effects'
Sen et al. (2012)	BMMNC	Local injection	Hip osteonecrosis $N = 40$ (51 hips) CD + MBAC ($n = 25$) CD alone ($n = 26$)	24 months	Reported as 'no infection or allergic reaction associated with BMNC instillation' and 'the present study also supports the safety of auto BMNC infusion into the necrotic area'
Hernigou et al. (2012)	BMAC	Local injection	Hip osteonecrosis $N = 62$, bilateral, one hip CD + BMAC ($n = 62$) one hip CD alone ($n = 62$)	17 years (range 15–20)	Not reported
Hernigou et al. (2013b)	BMAC	Local injection	Osteonecrosis 1089 Non-union 523 General orthopaedics 261	12.5 years (range 5–22)	No MRI evidence of tumorigenesis at the reimplant site. No increased risk of systemic cancer either

(continued)

Table 4.1 (continued)

	Cell	Delivery	Patient	Average follow-up	Safety outcome
Mao et al. (2013)	BMMNC	Intraarterial	Hip osteonecrosis N = 62 (78 hips)	5 years	Reported as: 'no complications'
Kang et al. (2013)	BMMNC	Local injection	Hip osteonecrosis $N = 52$ (61 hips)	68 months (range 60–88)	Not reported
Hernigou et al. (2014)	BMAC	Local injection	Rotator cuff repair $N = 45$ Matched ctrl $N = 45$	Minimum 10 years	Not reported
Enea et al. (2015)	BMAC on collagen membrane	Local implantation	Knee cartilage repair $N = 9$	29 months (range 18–40)	Reported as safe
Hernigou et al. (2015)	BMAC	Local injection	Ankle non-unions BMMNC = 86 Standard bone grafting = 86	3–15 years	Diabetes-related complications were lower in BMAC group compared to standard bone-grafting group

BMAC bone marrow aspirate concentrate, *BMMNC* bone marrow mononuclear cells, *CD* core decompression

at repairing the cartilage (autologous BMAC loaded on collagen membrane), the procedure was reported as 'safe' with an average follow-up of 2.5 years (Table 4.1).

Regarding the use of allogeneic uncultured MSCs for orthopaedic applications, we are aware of at least one product, which is marketed as a viable allograft material and consisting of allogeneic (cadaveric) bone devoid of donor immune cells but retaining donor MSCs (Neman et al. 2013). This non-immune cell preservation is achieved through proprietary washing procedures (Gonshor et al. 2011). In our recent study, we showed that the donor cellular component of such material is enriched for MSCs (>30% of all live cells), whereas donor lymphocytes represent less than 6% of total live cells (Baboolal et al. 2014). Remarkably, up to now this material has been implanted for over 10 years, and so far no immune reactions or other material-related complications have been reported (Kerr et al. 2011; Hollawell 2012). It is likely that such a low-level frequency of donor lymphocytes in material's matrix is not sufficient to initiate host immune response following the material's implantation. Additionally, in our study we showed that MSCs present in the material have a strong immunoregulatory capacity (Baboolal et al. 2014), which could further diminish any potential immune response at the implantation site.

4.3.2 The Use of Non-Expanded MSCs in Vascular Disease Applications

Peripheral artery disease (PAD) is noncoronary vascular disease affecting the peripheral arteries, most commonly the limb arteries. In about 1–2% of patients, the disease progresses to critical limb ischaemia (CLI); the terminal stage of PAD, which has a 1-year mortality of approximately 25%; and a limb amputation rate of 30% (Norgren et al. 2007). PAD is an increasing important public health issue because of persistent tobacco usage and an anticipated rise in diabetes prevalence. The common standard treatment for severe PAD and CLI includes traditional surgical bypass and contemporary revascularization procedures such as angioplasty (Fernandez et al. 2011). However up to 30% of patients are not candidate for such interventions due to widespread endothelium dysfunction, excessive operative risk or unfavourable vascular involvement (Bradbury et al. 2010). Moreover, the standard procedures rely primarily on anatomical correction rather than physiological modulation and often partially satisfactory clinical outcomes often achieved in only approximately 50% of the patients with CLI (Sultan and Hynes 2014).

The cell-based therapy is emerging as an alternative therapeutic approach for PAD especially for the patients who are ineligible to standard treatment. Based on currently available literature, the cell-based therapeutic product used in nearly all PAD clinical trials has been uncultured mononuclear cells (MNCs) derived from bone marrow aspirate or peripheral blood, harvested with or without granulocyte colony-stimulating factor mobilization (Botti et al. 2012; Compagna et al. 2015). The main active cellular component for these applications is believed to be BM-derived circulating endothelial progenitors cells (EPCs), which directly contribute to angiogenesis in ischaemic tissue. However BM MSCs could additionally contribute to vessel formation and maturation by providing an essential perivascular support. MSCs may also contribute to vascular repair by producing angiogenesis-supporting growth factors (Koike et al. 2004; Au et al. 2008; Pedersen et al. 2014; Murphy et al. 2013). Numerous preclinical studies have initially explored the possibility of using BM mononuclear cells (MNCs) for the treatment of PAD which have rapidly progressed to clinical safety and efficacy testing following the first therapeutic angiogenesis for patients with limb ischaemia by local intramuscular injection of BMMNCs (Tateishi-Yuyama et al. 2002).

In pioneering studies by Tateishi-Yuyama et al. (2002), local injections of either BMMNCs or BMAC did not result in any ectopic bone formation or interstitial fibrosis (Table 4.2). The safety of intramuscular delivery of autologous BMAC was later confirmed by Bartsch et al. (2007) and Iafrati et al. (2011); however the follow-up period for patients participating in these studies remained fairly short (maximum 15 months). In a multicentre, randomised placebo-controlled trial, the renal function of BMAC-treated patients was documented to be stable, and no inappropriate angiogenesis or osteogenesis was observed (Iafrati et al. 2011). As an alternative route of delivery, intraarterial injection of BMMNC or BMAC was attempted (Table 4.2). The aggregate data from these clinical studies suggested no adverse events with a maximum follow-up period of 57 months (Walter et al. 2011; Cobellis et al. 2008; Teraa et al. 2015; Klepanec et al. 2012; Ruiz-Salmeron et al. 2011).

Table 4.2 Clinical trials using uncultured autologous BMAC/BMMNC for the treatment of peripheral artery disease

	Cell and delivery	Patient and cells	Control	Follow-up	Safety outcome
Tateishi-Yuyama et al. (2002)	IM BMAC vs saline	Unilateral LI $N = 25$	Same patients $N = 25$	6 months	Neither bone formation nor increased interstitial fibrosis was detected
Tateishi-Yuyama et al. (2002)	IM BMAC	Bilateral LI $N = 20$	Same patients $N = 20$	6 months	Neither bone formation nor increased interstitial fibrosis was detected
Huang et al. (2005)	IM G-CSF mobilized PBMNC	CLI $N = 14$	CLI $N = 14$ Prostaglandin i.v.	3–14 months	No adverse effects related to the BMMNC treatment was observed, monitored by ECG, ultrasound cardiogram, liver and kidney function
Bartsch et al. (2007)	IA + IM Drug + BMMNC	Chronic PAD $N = 13$	PAD $N = 12$ drug therapy only	11–15 months	No complications or side effects were observed
Cobellis et al. (2008)	IA Whole BM	Advanced PAD $N = 10$	Advanced PAD $N = 9$ Standard care	12 months	No complications or side effects were observed
Prochazka et al. (2010)	IM BMAC	CLI $N = 42$	CLI $N = 54$ standard care	4 months	Not reported
Burt et al. (2010)	IM CD133$^+$ cells from G-CSF mobilized PBSC	CLI $N = 9$	N/A	12 months	No cardiovascular side effects from stem cell mobilization nor complications from leukapheresis or cell injection. No infections
Walter et al. (2011)	IA BMMNC	CLI $N = 19$	CLI $N = 21$ Placebo	3, 6 and 28 months	Concluded as 'safe and feasible' but 'no benefit' for more advanced patients (Rutherford 6)
Iafrati et al. (2011)	IM BMAC	CLI $N = 34$	CLI $N = 14$ Placebo	3 months	No adverse events related to BMMNC injection. Renal function was not affected
Ruiz-Salmeron et al. (2011)	IA	CLI BMMNC $N = 20$	N/A	12 months	No adverse events. BMMNC infusion did not cause any embolic events
Klepanec et al. (2012)	IM vs IA	Advanced CLI MBAC (IM) $N = 21$	Advanced CLI BMAC (IA) $N = 20$	6 months	Not reported

(continued)

Table 4.2 (continued)

	Cell and delivery	Patient and cells	Control	Follow-up	Safety outcome
Losordo et al. (2012)	IM CD34 enriched BMMNC	CLI *N* = 16	CLI *N* = 12 Placebo	12 months	Two possible treatment-related serious adverse events: one moderate hypotension during mobilization and one severe worsening of CLI after injection. Both required prolonged hospitalisation
Teraa et al. (2015)	IA BMMNC	No-option LI *N* = 81	No-option LI *N* = 79 Placebo	6 months	All-cause mortality, occurrence of malignancy or hospitalisation due to infection was not significantly different between treatment and control groups

IM intramuscular, *IA* intraarterial, *LI* limb ischemia, *PAD* peripheral artery disease, *CLI* critical limb ischemia

BMAC bone marrow aspirate concentrate, *BMMNC* bone marrow mononuclear cells, *PBMNC* peripheral blood mononuclear cells

4.3.3   Current Approaches to Ensure and Improve the Safety of Non-Expanded MSCs

Early-stage clinical studies utilising local or systemic injections of autologous BMMNCs or BMAC provided some pilot evidence of their safety; however any risks for long-term complications or emergence of cancers remain to be monitored. It is notable that current clinical evidence is primarily based on single-centre, case-control studies with small patient numbers and rather varied severity of underlying medical conditions. Large-scale, multicentre, randomised, double-blind trials would therefore be needed to fully establish the safety and efficacy of these procedures in a larger number of patients.

Furthermore, the available published information on the initial cell composition of BMMNC and BMAC products remain very limited. For example, the initial concentration of MSCs in BM aspirates critically depends on the surgical technique and the volume of aspirate collected (Hernigou et al. 2013b; Cuthbert et al. 2012). It is well recognised that large-volume collection of BM aspirate leads to its significant dilution with blood (Muschler et al. 1997). Therefore, if significantly diluted BM sample is processed for concentration, the concentration of MSCs in the final BMAC product would remain low. Based on the consideration, Hernigou et al. have recently proposed a standardised method of BMA collection based on small volumes of marrow collected using small aspiration syringes (Hernigou et al. 2013b).

Additionally, the delivered MSC dose is established only retrospectively because the CFU-F assay normally takes 10–14 days to perform. In our laboratory we have recently developed a rapid (<45 min) flow cytometry-based assay for MSC enumeration in BM aspirates based on $CD45^-CD271^+$ MSC phenotype (Cuthbert et al. 2012). It remains to be established whether such an assay can be further automated and hence applicable to measuring MSCs in BMAC before it is delivered to a patient.

Furthermore, we recently showed that CD271-based strategy can be used not only for MSC enumeration but also for their positive selection (Cuthbert et al. 2015). One preclinical study has indicated that intravenous administration of CD271-selected MSCs resulted in their recruitment to the fracture site in mice (Dreger et al. 2014). Using first-generation clinical-grade selection instrumentation (CliniMACs), we achieved over 150-fold increases in MSC purities from BM aspirates; however contaminating lymphocytes remained at an average frequency of ~20% of total live cells. Additionally, our group suggested that uncultured UC MSCs could be selected based on the $CD45^-CD235\alpha^-CD31^-CD146^+$ phenotype (Kouroupis et al. 2014). Further improvements in these, similar isolation technologies or developing other novel ways of MSC 'capture' (e.g. using biomimetic scaffolds) could make these MSC-enriched isolates potentially suitable for future allogeneic applications. Even in autologous applications, large numbers of contaminating immune cells can be detrimental; if activated, they can release MMPs that degrade matrix, as well as free oxygen radicals and other harmful compounds that may lead to MSC and other cell apoptosis at the injection site (Kaux et al. 2011).

Finally, and similar to interventions based on cultured MSCs, only limited data exist on the most effective route of non-expanded MSC delivery. Local delivery has the least potential for MSC loss; however percutaneous injections, even under X-ray guidance, would still not prevent MSC losses to the neighbouring tissues. For this reason, new approaches include the use of so-called barrier membranes that provide this essential barrier function and additionally prevent the invasion of neighbouring non-osteogenic tissues into the repair site (Dimitriou et al. 2012). Regarding the systemic route of delivery, it may seem more practical for some applications; however most researchers working in the field accept that the majority of infused MSCs may end up trapped in lungs (Schrepfer et al. 2007). However, it is now known that non-expanded MSCs have a different integrin profile compared to culture-expanded MSCs (Qian et al. 2012), and therefore the degree of this entrapment with the use of uncultured MSC products (BMAC and BMMNCs) may in fact be less significant.

4.3.4 Valuable Insights from Non-Expanded Haematopoietic Stem Cell Transplant (HSCT) Field

Haematopoietic stem cell transplant (HSCT) is one of very few proven stem cell therapies. This therapy has experienced a constant development journey for over half a century (Appelbaum 2007). Cell therapies using expanded or non-expanded MSCs are still in their infancy with most clinical applications experimental and investigational. The valuable experience from HSCT could shed light on the development of novel therapies using other stem cell types including non-expanded MSCs. International cooperation has been a key factor for the development of HSCT, including the establishment of CIBMTR (Center for International Blood and Marrow Transplant Research), EBMT (European Group for Blood and Marrow Transplantation), NMDP (National Marrow Donor Program) and WMDA (World Marrow Donor Association). These international organizations have enabled analysis of large cohort multicentre transplant outcomes and provided clear guidelines and standard protocols. Such organizations are currently lacking for the development of non-expanded MSC therapies.

The success of HSC therapy is also a result of multidisciplinary team effort including clinical specialists, scientists and data managers. Encouraging close clinical and scientific collaboration could therefore significantly facilitate the development of non-expanded MSC therapy. One of the key achievements in HSC therapy is an ability to select and infuse desirable cell populations to suit the therapeutic purposes such as positive selection of $CD34^+$ HSCs or depletion of alloreactive lymphocytes to avoid detrimental complications (Gaipa et al. 2003; Henig and Zuckerman 2014). Such procedure has paved the way to advance the isolation of non-expanded MSCs for clinical use. Positive selection of MSCs via CD271 labelling and CliniMACS isolation has marked the first step towards the right direction (Cuthbert et al. 2015).

As mentioned at the beginning of this chapter, bone marrow transplantation (BMT) to treat patients with acute leukaemia has been successful for decades (Appelbaum 2007) and since the BMT procedure is indicated for a variety of neoplastic and non-neoplastic disorders. Like other cell therapies, there are two major types of HSC transplants, allogeneic and autologous.

For allogeneic approaches, the procedure was originally carried out using donor whole BM that was infused following irradiation of the recipient's marrow in order to deplete cancer cells. Infused healthy HSCs are then able to reconstitute the whole haematopoietic system of the recipient. Nowadays a refinement that obviates the need for invasive donor BM procurement is widely used which relies on the peripheral blood (PB) mobilization of donor HSCs using chemotherapeutic agents and growth factors such as granulocyte-colony stimulating factor (G-CSF) or both. Umbilical cord blood-derived HSCs represent another source of allogeneic HSCs for BMT. The BMT procedure may also use blood from an HLA-identical sibling, a matched unrelated donor (MUD), a haploidentical family peripheral blood stem cell (PBSC), BM donor or an HLA-mismatched unrelated donor. The major advantage of allogeneic BMT is that there is no risk of reinfusing cancerous cells with an enhanced therapeutic graft-versus-leukaemia effect.

Autologous HSC transplantation relies on high-dose chemotherapy or chemoradiotherapy to first eliminate patient's cancerous cells. When patient is in remission, the HSC source can then be either mobilized PB or BM. Autologous HSCT is commonly used for lymphomas and myeloma and less commonly for leukaemia. In patients with leukaemia and lymphoma, there are considerable concerns regarding the reintroduction of tumour cells alongside HSCs. Consequently, efforts to remove tumour cells from HSC may be undertaken; however, it remains unclear whether such manipulations improve patients' outcomes such as relapse rates (Kreissman et al. 2013).

Safety concerns and treatment complications following HSCT mostly relate to the need of using irradiation and chemotherapy to destroy patients' cancer pool, which also depletes their immune cells leading to vulnerability to infection. Also, HLA-disparity-mediated GVHD complications are another major safety concern that should be taken into account. HLA-identical sibling BM transplant remains the first choice, but the volume of collection must be controlled according to the donor and recipient weight (no more than 10–20 mL/kg donor body weight). PBSC harvest is being increasingly used because it is an easy procedure (no need of anaesthesia or operating theatre); however it is practically not applied in children due to side effects of G-CSF or problems of venous approach. Plerixafor, an immunostimulant used to mobilize haematopoietic stem cells in cancer patients, has been recently used with and without G-CSF to improve mobilization process (Korbling and Freireich 2011). Regarding matched unrelated HSCT donors (MUD), the choice is based on high-resolution allelic typing for HLA-A, HLA-B, HLA-C, HLA-DRB1, HLA-DQB1 and HLA-DPB1. The choice between BM and PBSC depends on donor choice, centre preference and indication for HSCT. PBHSCT is associated with a greater propensity to GVHD reaction; however it is favourable in malignant disease indications to prevent relapse while in non-oncological indications, e.g. aplastic anaemia, it showed lower survival rates (Eapen et al. 2011). Cord blood, another HSC source, possesses no risk for mothers and donors besides the low possibility of transmitting infections and is available for instant use. In comparison to adult-unrelated donor transplants (who need 10/10 allele level matches with the patients), cord blood transplants have a reduced risk of severe GVHD and permit a mismatched transplantation at least in one HLA locus (Gluckman and Rocha 2009; Cutler and Ballen 2012).

Stem cell donor registries and cord blood banks provide HLA typing on their donors and cord units that are available to search for use in HSC transplantation. HLA-typing methods have developed considerably since the discovery of the first HLA antigens. Initially, serological and cellular techniques were used to determine HLA type. Since the application of DNA-based techniques, the refinement of the methods available for HLA typing has increased significantly, and this technology has lately taken the next step with the progress of next-generation sequencing. Choice of donor source is dependent on the indication for HSCT, its urgency, the age of the patient and the expertise and resources of the centre. Altogether, the multiple choices of HSC sources have increased the options of offering HSCT to almost every patient needing a transplant. The safety against GVHD is often achieved by selectively depleting of immunogenic cellular subpopulations, such as alloreactive T cells, or modulating the balance of cells with immunosuppressive properties such as effector T cells and regulatory T cells (Perez et al. 2011). Still, keeping the balance between the levels of immunosuppression required to control a GVHD and retaining a degree of immunity against infectious organisms remain a major challenge.

A quick look at the state of the art in the field of non-expanded HSC therapies shows that the challenges facing the HSC field are different from those relating to non-expanded MSCs. The use of allogeneic cultured MSCs does not require a similar level of control for histocompatibility, as it is for HSCs; however some believe that to ensure the best safety possible, the same rigorous level of HLA typing should be applied to non-expanded MSC products as it currently exists for HSCs. In an alternative opinion, rigorous HLA typing/matching is a unique feature of HSCT as HSCT is a transplant of a whole donor immune system into the recipient. This has a high risk for the engrafted donor immune system to lead to severe GVHD, a life-threatening complication after allogeneic HSCT. Strict HLA matching is not required for most allogeneic solid organ transplants, and it can significantly reduce the donor availability and increase the overall cost. It remains unclear whether low immunogenicity and the lack of rejection reactions with the use of cultured MSCs are a result of their culturing, whereas non-expanded MSCs may still be mildly immunogenic.

The route of delivery for all HSC transplants remains to be intravenous and does not offer the same challenge as it exists for MSCs that are applied using a broader range of options. The dose of HSCs in HSC transplants is carefully controlled, which should be definitely adapted for MSCs, in which doses remain largely empirical. Finally, for the development of non-expanded MSC therapies, the use of approaches to stimulate and mobilize HSCs into the PB, which is much easier to harvest, should be definitely considered and learnt from. Currently, very little is known about which biological stimuli are responsible for maintaining MSC proliferation, mobilization and migration in vivo. Our recent work has shown that platelets and growth factors released by platelets are likely to control BM MSC pool in vivo, but they remain insufficient to mobilize MSCs into the PB (Tan et al. 2015). A recent clinical trial has however shown that limited MSC mobilization is possible at the local level (Philippart et al. 2014). It can be envisaged that novel agents can be found in the future, allowing non-expanded MSC stimulation and PB mobilization, permitting their easier harvesting for therapy.

Furthermore, over the last 20 years, exploitation of novel donor sources has significantly extended the donor availability for HSCT from HLA-identical siblings to HLA-matched unrelated donors then further to umbilical cord blood and haploidentical donors (Henig and Zuckerman 2014). Exploring and expanding novel sources of MSC are a key factor for the future development of cell therapy using MSCs. For example, previous studies have shown that long bones contain large numbers of MSCs, transcriptionally and functionally similar to BM MSCs (Churchman et al. 2013; Cox et al. 2011; Porter et al. 2009). In addition, cord blood can now be collected from more than one cord so pooling non-expanded MSC products from several matched donors may also become a common practice in the future.

4.4 Conclusions and Future Directions

As seen from previous chapters, the current literature indicates that cellular preparations based on expanded as well as non-expanded MSCs are clinically safe; however longer-term follow-up and closer monitoring of clinical complications are

Table 4.3 Future challenges facing new-generation cell therapies based on expanded and non-expanded MSCs

Future challenge	Expanded MSCs	Non-expanded MSCs
The safest route of delivery	+	+
Indication-specific MSC dose that is both efficacious and safe	+	+
Ensuring MSC retention/accumulation at the repair site	+	+
Gaining full understanding of the mechanisms of action underlying the beneficial effects	+	+
Indication-specific selection of MSC tissue source	+	+
Loss of function during passaging	+	−
An increasing proportion of MSCs becomes senescent during passaging	+	−
Increase in chromosomal abnormalities during passaging	+	−
Hazards linked to culture media ingredients	+	−
Control of MSC purity and contaminating cell types	+	+
MSC activation during positive selection	−	+
Very low numbers of MSCs present in some tissue sources requiring stricter control of purity	−	+

required to assess the risks of cancer or incidence of long-term complications or chronic conditions. Furthermore, much stricter guidelines, particularly more rigorous HLA typing, would be needed when these therapies are used in allogeneic setting. Although the assessment of the efficacy of these therapies is beyond the scope of this chapter, similar very strict guidelines should be also applied to their efficacy assessment, which in our opinion, would require the introduction of standard, universally accepted and highly quantitative outcome measures. Although these and many other challenges remain common for expanded and non-expanded MSCs, there are issues specific for these two groups (Table 4.3).

It should also be noted that a new wave of non-cell-based treatments is now emerging in the regenerative medicine arena. One class of such treatments is based on MSC-derived conditioned media or exosomes. Timmers et al. were the first to show that intravenous and intracoronary MSC-conditioned medium (MSC-CM) administration resulted in functional improvement in a porcine myocardial infarction model (Timmers et al. 2008). The cardioprotective effect was mediated by large complexes (>1000 kDa, 50–200 nm) consisting of plasma membrane proteins and phospholipids named as exosomes (Timmers et al. 2008; Lai et al. 2010). A very recent study has however failed to show that exosomes secreted from MSCs suppress lymphocyte proliferation (Gouveia de Andrade et al. 2015). This demonstrates that despite their advantages as a nonliving, 'off-the-shelf' therapeutic agent and associated lower risks of occlusion in microvasculature or uncontrolled cell growth (Yeo et al. 2013), more preclinical work is needed to establish whether exosome-based therapies have comparable therapeutic efficacy and are safer than cellular therapies based on MSCs.

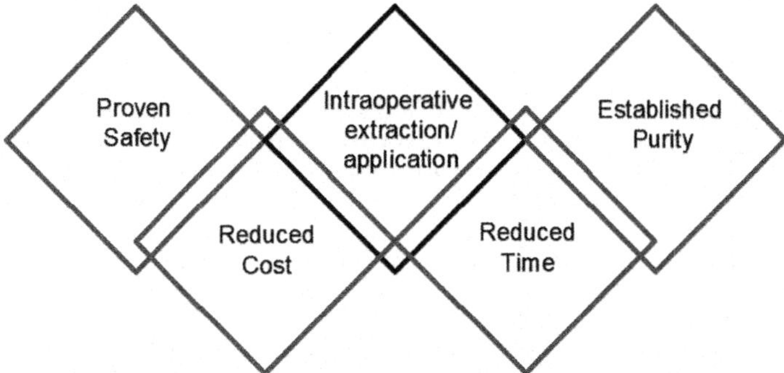

Fig. 4.1 Desired characteristics of cell therapies based on non-expanded MSCs

Another acellular approach for repairing damaged tissue is to induce endogenous MSC homing to the repair site using the controlled release of bioactive molecules, as pioneered by Lee et al. (2010) and elaborated in more detail in previously published review articles (Jones and Sanjurjo-Rodriguez 2014; Eseonu and De Bari 2015). To our best knowledge, these exciting developments have now been extensively tested in a preclinical domain (Chen et al. 2015); however future clinical trials would be needed to assess their safety and efficacy in humans.

In summary, culture-expanded MSCs have been used extensively as cell therapy for a number of traumatic, degenerative and immune-related diseases. Although the efficacy and cost-effectiveness of such interventions in comparison to current standard of care are yet to be determined for many of these diseases, it is now clear that such therapies are safe and non-tumorigenic, with long-term follow-up showing no incidence of serious adverse events. Still, therapies based on expanded MSCs remain expensive and require further improvements in culture media formulations, as well as better potency characterisation, in order to progress from clinical trials to broader clinical practice. Therapies based on mixed cell isolates enriched for non-expanded MSCs, such as BMAC, are comparatively newer technology, and up to date these cellular 'biologics' have been primarily used in orthopaedic and vascular surgery. In preclinical and clinical investigations, MSCs present in such isolates may not be the sole active ingredient, with EPCs and other lineage cells providing an additional benefit to graft survival and subsequent tissue regeneration. In this chapter we have outlined an up-to-date literature pertaining to current safety record of culture-expanded and non-expanded MSCs, highlighting the need to further improve their safety, for example, by optimal dosages and routes of administration. The advantages of non-expanded MSCs in autologous settings would include low cost, low regulatory burden and the need for only one surgery for both MSC harvesting and implantation (Fig. 4.1).

Although there remain many limitations for the use of non-expanded MSCs in the clinic, mostly related to the need for high MSC doses for some indications, low regulatory burden associated with these therapies and their significantly lower cost suggest that these therapies may become economically viable and available to much larger groups of patients compared to culture-expanded MSCs.

References

Agata H, Watanabe N, Ishii Y, Kubo N, Ohshima S, Yamazaki M, Tojo A, Kagami H (2009) Feasibility and efficacy of bone tissue engineering using human bone marrow stromal cells cultivated in serum-free conditions. Biochem Biophys Res Commun 382(2):353–358

Ankrum J, Karp JM (2010) Mesenchymal stem cell therapy: two steps forward, one step back. Trends Mol Med 16(5):203–209

Appelbaum FR (2007) Hematopoietic-cell transplantation at 50. N Engl J Med 357(15):1472–1475

Asanuma H, Meldrum DR, Meldrum KK (2010) Therapeutic applications of mesenchymal stem cells to repair kidney injury. J Urol 184(1):26–33

Au P, Tam J, Fukumura D, Jain RK (2008) Bone marrow-derived mesenchymal stem cells facilitate engineering of long-lasting functional vasculature. Blood 111(9):4551–4558

Baboolal TG, Boxall SA, El-Sherbiny YM, Moseley TA, Cuthbert RJ, Giannoudis PV, McGonagle D, Jones E (2014) Multipotential stromal cell abundance in cellular bone allograft: comparison with fresh age-matched iliac crest bone and bone marrow aspirate. Regen Med 9(5):593–607

Ball LM, Bernardo ME, Roelofs H, Lankester A, Cometa A, Egeler RM, Locatelli F, Fibbe WE (2007) Cotransplantation of ex vivo expanded mesenchymal stem cells accelerates lymphocyte recovery and may reduce the risk of graft failure in haploidentical hematopoietic stem-cell transplantation. Blood 110(7):2764–2767

Bang OY, Lee JS, Lee PH, Lee G (2005) Autologous mesenchymal stem cell transplantation in stroke patients. Ann Neurol 57(6):874–882

Baraniak PR, McDevitt TC (2010) Stem cell paracrine actions and tissue regeneration. Regen Med 5(1):121–143

Bartsch T, Brehm M, Zeus T, Kogler G, Wernet P, Strauer B (2007) Transplantation of autologous mononuclear bone marrow stem cells in patients with peripheral arterial disease (the TAM-PAD study). Clin Res Cardiol 96(12):891–899

Bieback K (2013) Platelet lysate as replacement for fetal bovine serum in mesenchymal stromal cell cultures. Transfus Med Hemother 40(5):326–335

Botti C, Maione C, Coppola A, Sica V, Cobellis G (2012) Autologous bone marrow cell therapy for peripheral arterial disease. Stem Cells Cloning 5:5–14

Bradbury AW, Adam DJ, Bell J, Forbes JF, Fowkes FGR, Gillespie I, Ruckley CV, Raab GM (2010) Bypass versus angioplasty in severe Ischaemia of the leg (BASIL) trial: an intention-to-treat analysis of amputation-free and overall survival in patients randomized to a bypass surgery-first or a balloon angioplasty-first revascularization strategy. J Vasc Surg 51(5 Suppl):S5–S17

Burt RK, Testori A, Oyama Y, Rodriguez HE, Yaung K, Villa M, Bucha JM, Milanetti F, Sheehan J, Rajamannan N, Pearce WH (2010) Autologous peripheral blood CD133+ cell implantation for limb salvage in patients with critical limb ischemia. Bone Marrow Transplant 45(1):111–116

Centeno CJ, Schultz JR, Cheever M, Freeman M, Faulkner S, Robinson B, Hanson R (2011) Safety and complications reporting update on the re-implantation of culture-expanded mesenchymal stem cells using autologous platelet lysate technique. Curr Stem Cell Res Ther 6(4):368–378

Chang YS, Ahn SY, Yoo HS, Sung SI, Choi SJ, Oh WI, Park WS (2014) Mesenchymal stem cells for bronchopulmonary dysplasia: phase 1 dose-escalation clinical trial. J Pediatr 164(5):966–972

Chase LG, Yang S, Zachar V, Yang Z, Lakshmipathy U, Bradford J, Boucher SE, Vemuri MC (2012) Development and characterization of a clinically compliant xeno-free culture medium in good manufacturing practice for human multipotent mesenchymal stem cells. Stem Cells Transl Med 1(10):750–758

Chen L, Tredget EE, Wu PYG, Wu Y (2008) Paracrine factors of mesenchymal stem cells recruit macrophages and endothelial lineage cells and enhance wound healing. PLoS One 3(4):e1886

Chen P, Tao J, Zhu S, Cai Y, Mao Q, Yu D, Dai J, Ouyang H (2015) Radially oriented collagen scaffold with SDF-1 promotes osteochondral repair by facilitating cell homing. Biomaterials 39(0):114–123

Choudhery MS, Badowski M, Muise A, Pierce J, Harris DT (2014) Donor age negatively impacts adipose tissue-derived mesenchymal stem cell expansion and differentiation. J Transl Med 12:8

Chullikana A, Majumdar AS, Gottipamula S, Krishnamurthy S, Kumar AS, Prakash VS, Gupta PK (2015) Randomized, double-blind, phase I/II study of intravenous allogeneic mesenchymal stromal cells in acute myocardial infarction. Cytotherapy 17(3):250–261

Churchman SM, Kouroupis D, Boxall SA, Roshdy T, Tan HB, McGonagle D, Giannoudis PV, Jones EA (2013) Yield optimisation and molecular characterisation of uncultured CD271+ mesenchymal stem cells in the reamer irrigator aspirator waste bag. Eur Cell Mater 13(26):252–262

Clifford DM, Fisher SA, Brunskill SJ, Doree C, Mathur A, Clarke MJ, Watt SM, Martin-Rendon E (2012) Long-term effects of autologous bone marrow stem cell treatment in acute myocardial infarction: factors that may influence outcomes. PLoS One 7(5):e37373

Cobellis G, Silvestroni A, Lillo S, Sica G, Botti C, Maione C, Schiavone V, Rocco S, Brando G, Sica V (2008) Long-term effects of repeated autologous transplantation of bone marrow cells in patients affected by peripheral arterial disease. Bone Marrow Transplant 42(10):667–672

Compagna R, Amato B, Massa S, Amato M, Grande R, Butrico L, de Franciscis S, Serra R (2015) Cell therapy in patients with critical limb ischemia. Stem Cells Int 2015:931420

Connick P, Kolappan M, Crawley C, Webber DJ, Patani R, Michell AW, Du M-Q, Luan S-L, Altmann DR, Thompson AJ, Compston A, Scott MA, Miller DH, Chandran S (2012) Autologous mesenchymal stem cells for the treatment of secondary progressive multiple sclerosis: an open-label phase 2a proof-of-concept study. Lancet Neurol 11(2):150–156

Cox G, McGonagle D, Boxall SA, Buckley CT, Jones E, Giannoudis PV (2011) The use of the reamer-irrigator-aspirator to harvest mesenchymal stem cells. J Bone Joint Surg Br 93-B(4):517–524

Cuthbert R, Boxall SA, Tan HB, Giannoudis PV, McGonagle D, Jones E (2012) Single-platform quality control assay to quantify multipotential stromal cells in bone marrow aspirates prior to bulk manufacture or direct therapeutic use. Cytotherapy 14(4):431–440

Cuthbert RJ, Giannoudis PV, Wang XN, Nicholson L, Pawson D, Lubenko A, Tan HB, Dickinson A, McGonagle D, Jones E (2015) Examining the feasibility of clinical grade CD271+ enrichment of mesenchymal stromal cells for bone regeneration. PLoS One 10(3):e0117855

Cutler C, Ballen KK (2012) Improving outcomes in umbilical cord blood transplantation: state of the art. Blood Rev 26(6):241–246

D'Ippolito G, Schiller PC, Ricordi C, Roos BA, Howard GA (1999) Age-related osteogenic potential of mesenchymal stromal stem cells from human vertebral bone marrow. J Bone Miner Res 14(7):1115–1122

da Silva Meirelles L, Fontes AM, Covas DT, Caplan AI (2009) Mechanisms involved in the therapeutic properties of mesenchymal stem cells. Cytokine Growth Factor Rev 20(5–6):419–427

De Bari C, Dell'Accio F, Tylzanowski P, Luyten FP (2001) Multipotent mesenchymal stem cells from adult human synovial membrane. Arthritis Rheum 44(8):1928–1942

Desai N, Rambhia P, Gishto A (2015) Human embryonic stem cell cultivation: historical perspective and evolution of xeno-free culture systems. Reprod Biol Endocrinol 13(1):9

Dimitriou R, Mataliotakis G, Calori G, Giannoudis P (2012) The role of barrier membranes for guided bone regeneration and restoration of large bone defects: current experimental and clinical evidence. BMC Med 10(1):81

Doorn J, van de Peppel J, van Leeuwen JPTM, Groen N, van Blitterswijk CA, de Boer J (2011) Pro-osteogenic trophic effects by PKA activation in human mesenchymal stromal cells. Biomaterials 32(26):6089–6098

Dreger T, Watson JT, Dvm WA, Molligan J, Achilefu S, Schon LC, Zhang Z (2014) Intravenous application of CD271-selected mesenchymal stem cells during fracture healing. J Orthop Trauma 28(Suppl 1):S15–S19

Duijvestein M, Vos ACW, Roelofs H, Wildenberg ME, Wendrich BB, Verspaget HW, Kooy-Winkelaar EMC, Koning F, Zwaginga JJ, Fidder HH, Verhaar AP, Fibbe WE, van den Brink GR, Hommes DW (2010) Autologous bone marrow-derived mesenchymal stromal cell treatment for refractory luminal Crohn's disease: results of a phase I study. Gut 59(12):1662–1669

Eapen M, Le Rademacher J, Antin JH, Champlin RE, Carreras J, Fay J, Passweg JR, Tolar J, Horowitz MM, Marsh JCW, Deeg HJ (2011) Effect of stem cell source on outcomes after unrelated donor transplantation in severe aplastic anemia. Blood 118(9):2618–2621

Enea D, Cecconi S, Calcagno S, Busilacchi A, Manzotti S, Gigante A (2015) One-step cartilage repair in the knee: collagen-covered microfracture and autologous bone marrow concentrate. A pilot study. Knee 22(1):30–35

Eseonu OI, De Bari C (2015) Homing of mesenchymal stem cells: mechanistic or stochastic? Implications for targeted delivery in arthritis. Rheumatology 54(2):210–218

Fernandez N, McEnaney R, Marone LK, Rhee RY, Leers S, Makaroun M, Chaer RA (2011) Multilevel versus isolated endovascular tibial interventions for critical limb ischemia. J Vasc Surg 54(3):722–729

Friedenstein AJ, Chailakhyan RK, Latsinik NV, Panasyuk AF, K-B IV (1974) Stromal cells responsible for transferring the microenvironment of the hemopoietic tissues. Cloning in vitro and retransplantation in vivo. Transplantation 17(4):331–340

Gaipa G, Dassi M, Perseghin P, Venturi N, Corti P, Bonanomi S, Balduzzi A, Longoni D, Uderzo C, Biondi A, Masera G, Parini R, Bertagnolio B, Uziel G, Peters C, Rovelli A (2003) Allogeneic bone marrow stem cell transplantation following CD34+ immunomagnetic enrichment in patients with inherited metabolic storage diseases. Bone Marrow Transplant 31(10):857–860

Gangji V, De Maertelaer V, Hauzeur J-P (2011) Autologous bone marrow cell implantation in the treatment of non-traumatic osteonecrosis of the femoral head: five year follow-up of a prospective controlled study. Bone 49(5):1005–1009

Ghannam S, Bouffi C, Djouad F, Jorgensen C, Noel D (2010) Immunosuppression by mesenchymal stem cells: mechanisms and clinical applications. Stem Cell Res Ther 1(1):2

Gluckman E, Rocha V (2009) Cord blood transplantation: state of the art. Haematologica 94(4):451–454

Gonshor A, McAllister BS, Wallace SS, Prasad H (2011) Histologic and histomorphometric evaluation of an allograft stem cell-based matrix sinus augmentation procedure. Int J Oral Maxillofac Implants 26(1):123–131

Gouveia de Andrade AV, Bertolino G, Riewaldt J, Bieback K, Karbanova J, Odendahl M, Bornhauser M, Schmitz M, Corbeil D, Tonn T (2015) Extracellular vesicles secreted by bone marrow- and adipose tissue-derived mesenchymal stromal cells fail to suppress lymphocyte proliferation. Stem Cells Dev 24(11):1374–1376

Griffin MD, Elliman SJ, Cahill E, English K, Ceredig R, Ritter T (2013) Concise review: adult mesenchymal stromal cell therapy for inflammatory diseases: how well are we joining the dots? Stem Cells 31(10):2033–2041

Hare JM, Traverse JH, Henry TD, Dib N, Strumpf RK, Schulman SP, Gerstenblith G, DeMaria AN, Denktas AE, Gammon RS, Hermiller JB Jr, Reisman MA, Schaer GL, Sherman W (2009) A randomized, double-blind, placebo-controlled, dose-escalation study of intravenous adult human mesenchymal stem cells (prochymal) after acute myocardial infarction. J Am Coll Cardiol 54(24):2277–2286

Hare JM, Fishman JE, Gerstenblith G et al (2012) Comparison of allogeneic vs autologous bone marrow-derived mesenchymal stem cells delivered by transendocardial injection in patients with ischemic cardiomyopathy: the poseidon randomized trial. JAMA 308(22):2369–2379

Harris Q, Seto J, O'Brien K, Lee PS, Kondo C, Heard BJ, Hart DA, Krawetz RJ (2013) Monocyte chemotactic protein-1 inhibits chondrogenesis of synovial mesenchymal progenitor cells: an in vitro study. Stem Cells 31(10):2253–2265

Hendrich C, Engelmaier F, Waertel G, Krebs R, Jäger M (2009) Safety of autologous bone marrow aspiration concentrate transplantation: initial experiences in 101 patients. Orthop Rev 1:e32

Henig I, Zuckerman T (2014) Hematopoietic stem cell transplantation-50 years of evolution and future perspectives. Rambam Maimonides Med J 5(4):e0028

Herberts C, Kwa M, Hermsen H (2011) Risk factors in the development of stem cell therapy. J Transl Med 9(1):29

Hernigou P, Beaujean F (2002) Treatment of osteonecrosis with autologous bone marrow grafting. Clin Orthop Relat Res 405:14–23

Hernigou P, Poignard A, Beaujean F, Rouard H (2005) Percutaneous autologous bone-marrow grafting for nonunionsinfluence of the number and concentration of progenitor cells. J Bone Joint Surg 87(7):1430–1437

Hernigou P, Poignard A, Zilber S, Rouard H (2009) Cell therapy of hip osteonecrosis with autologous bone marrow grafting. Indian J Orthop 43(1):40–45

Hernigou P, Poignard A, Lachaniette CHF (2012) Treatment of corticosteroid hip osteonecrosis with stem cells. J Bone Joint Surg Br 94-B(Suppl XXXVII):68

Hernigou P, Homma Y, Flouzat-Lachaniette C-H, Poignard A, Chevallier N, Rouard H (2013a) Cancer risk is not increased in patients treated for orthopaedic diseases with autologous bone marrow cell concentrate. J Bone Joint Surg 95(24):2215–2221

Hernigou P, Homma Y, Flouzat Lachaniette C, Poignard A, Allain J, Chevallier N, Rouard H (2013b) Benefits of small volume and small syringe for bone marrow aspirations of mesenchymal stem cells. Int Orthop 37(11):2279–2287

Hernigou P, Flouzat Lachaniette C, Delambre J, Zilber S, Duffiet P, Chevallier N, Rouard H (2014) Biologic augmentation of rotator cuff repair with mesenchymal stem cells during arthroscopy improves healing and prevents further tears: a case-controlled study. Int Orthop 38(9):1811–1818

Hernigou P, Guissou I, Homma Y, Poignard A, Chevallier N, Rouard H, Flouzat-Lachaniette CH (2015) Percutaneous injection of bone marrow mesenchymal stem cells for ankle non-unions decreases complications in patients with diabetes. Int Orthop 39(8):1639–1643

Hollawell SM (2012) Allograft cellular bone matrix as an alternative to autograft in hindfoot and ankle fusion procedures. J Foot Ankle Surg 51(2):222–225

Hong X, Miller C, Savant-Bhonsale S, Kalkanis SN (2009) Antitumor treatment using interleukin-12-secreting marrow stromal cells in an invasive glioma model. Neurosurgery 64(6):1139–1147

Honmou O, Houkin K, Matsunaga T, Niitsu Y, Ishiai S, Onodera R, Waxman SG, Kocsis JD (2011) Intravenous administration of auto serum-expanded autologous mesenchymal stem cells in stroke. Brain 134(6):1790–1807

Horwitz EM, Gordon PL, Koo WKK, Marx JC, Neel MD, McNall RY, Muul L, Hofmann T (2002) Isolated allogeneic bone marrow-derived mesenchymal cells engraft and stimulate growth in children with osteogenesis imperfecta: implications for cell therapy of bone. Proc Natl Acad Sci U S A 99(13):8932–8937

Huang P, Li S, Han M, Xiao Z, Yang R, Han ZC (2005) Autologous transplantation of granulocyte colony-stimulating factor-mobilized peripheral blood mononuclear cells improves critical limb ischemia in diabetes. Diabetes Care 28(9):2155–2160

Iafrati MD, Hallett JW, Geils G, Pearl G, Lumsden A, Peden E, Bandyk D, Vijayaraghava KS, Radhakrishnan R, Ascher E, Hingorani A, Roddy S (2011) Early results and lessons learned from a multicenter, randomized, double-blind trial of bone marrow aspirate concentrate in critical limb ischemia. J Vasc Surg 54(6):1650–1658

Ikebe C, Suzuki K (2014) Mesenchymal stem cells for regenerative therapy: optimization of cell preparation protocols. Biomed Res Int 2014:11

Ishige I, Nagamura-Inoue T, Honda M, Harnprasopwat R, Kido M, Sugimoto M, Nakauchi H, Tojo A (2009) Comparison of mesenchymal stem cells derived from arterial, venous, and Wharton's jelly explants of human umbilical cord. Int J Hematol 90(2):261–269

Jäger M, Herten M, Fochtmann U, Fischer J, Hernigou P, Zilkens C, Hendrich C, Krauspe R (2011) Bridging the gap: bone marrow aspiration concentrate reduces autologous bone grafting in osseous defects. J Orthop Res 29(2):173–180

Jaiswal N, Haynesworth SE, Caplan AI, Bruder SP (1997) Osteogenic differentiation of purified, culture-expanded human mesenchymal stem cells in vitro. J Cell Biochem 64(2):295–312

Jones E, Sanjurjo-Rodriguez C (2014) Stem cell therapy for bone and cartilage defects: can culture expansion be avoided? J Stem Cell Res Ther 4(1):e118

Jones E, Churchman SM, English A, Buch MH, Horner EA, Burgoyne CH, Reece R, Kinsey S, Emery P, McGonagle D, Ponchel F (2010) Mesenchymal stem cells in rheumatoid synovium: enumeration and functional assessment in relation to synovial inflammation level. Ann Rheum Dis 69(2):450–457

Joyce N, Annett G, Wirthlin L, Olson S, Bauer G, Nolta JA (2010) Mesenchymal stem cells for the treatment of neurodegenerative disease. Regen Med 5(6):933–946

Jung J, Moon N, Ahn J-Y, Oh E-J, Kim M, Cho C-S, Shin J-C, Oh I-H (2008) Mesenchymal stromal cells expanded in human allogenic cord blood serum display higher self-renewal and enhanced osteogenic potential. Stem Cells Dev 18(4):559–571

Kang JS, Moon KH, Kim B-S, Kwon DG, Shin SH, Shin BK, Ryu D-J (2013) Clinical results of auto-iliac cancellous bone grafts combined with implantation of autologous bone marrow cells for osteonecrosis of the femoral head: a minimum 5-year follow-up. Yonsei Med J 54(2):510–515

Karahuseyinoglu S, Cinar O, Kilic E, Kara F, Akay GG, Demiralp DÖ, Tukun A, Uckan D, Can A (2007) Biology of stem cells in human umbilical cord stroma: in situ and in vitro surveys. Stem Cells 25(2):319–331

Karussis D, Karageorgiou C, Vaknin-Dembinsky A et al (2010) Safety and immunological effects of mesenchymal stem cell transplantation in patients with multiple sclerosis and amyotrophic lateral sclerosis. Arch Neurol 67(10):1187–1194

Kasemkijwattana CHS, Kesprayura S, Rungsinaporn V, Chaipinyo K, Chansiri K (2011) Autologous bone marrow mesenchymal stem cells implantation for cartilage defects: two cases report. J Med Assoc Thail 94:395–400

Kaux JF, Le Goff C, Renouf J, Peters P, Lutteri L, Gothot A, Crielaard JM (2011) Comparison of the platelet concentrations obtained in platelet-rich plasma (PRP) between the GPStm II and GPStm III systems. Pathol Biol (Paris) 59(5):275–277

Kawate K, Yajima H, Ohgushi H, Kotobuki N, Sugimoto K, Ohmura T, Kobata Y, Shigematsu K, Kawamura K, Tamai K, Takakura Y (2006) Tissue-engineered approach for the treatment of steroid-induced osteonecrosis of the femoral head: transplantation of autologous mesenchymal stem cells cultured with beta-tricalcium phosphate ceramics and free vascularized fibula. Artif Organs 30(12):960–962

Kerr EJ, Jawahar A, Wooten T, Kay S, Cavanaugh DA, Nunley PD (2011) The use of osteo-conductive stem-cells allograft in lumbar interbody fusion procedures: an alternative to recombinant human bone morphogenetic protein. J Surg Orthop Adv 20(3):193–197

Klepanec A, Mistrik M, Altaner C, Valachovicova M, Olejarova I, Slysko R, Balazs T, Urlandova T, Hladikova D, Liska B, Tomka J, Vulev I, Madaric J (2012) No difference in intra-arterial and intramuscular delivery of autologous bone marrow cells in patients with advanced critical limb ischemia. Cell Transplant 21(9):1909–1918

Koike N, Fukumura D, Gralla O, Au P, Schechner JS, Jain RK (2004) Tissue engineering: creation of long-lasting blood vessels. Nature 428(6979):138–139

Korbling M, Freireich EJ (2011) Twenty-five years of peripheral blood stem cell transplantation. Blood 117(24):6411–6416

Kouroupis D, Churchman SM, McGonagle D, Jones EA (2014) The assessment of CD146-based cell sorting and telomere length analysis for establishing the identity of mesenchymal stem cells in human umbilical cord. F1000Res 3:126

Kreissman SG, Seeger RC, Matthay KK, London WB, Sposto R, Grupp SA, Haas-Kogan DA, LaQuaglia MP, Yu AL, Diller L, Buxton A, Park JR, Cohn SL, Maris JM, Reynolds CP, Villablanca JG (2013) Purged versus non-purged peripheral blood stem-cell transplantation for high-risk neuroblastoma (COG A3973): a randomised phase 3 trial. Lancet Oncol 14(10):999–1008

Kuroda R, Ishida K, Matsumoto T, Akisue T, Fujioka H, Mizuno K, Ohgushi H, Wakitani S, Kurosaka M (2007) Treatment of a full-thickness articular cartilage defect in the femoral condyle of an athlete with autologous bone-marrow stromal cells. Osteoarthr Cartil 15(2):226–231

Lai RC, Arslan F, Lee MM, Sze NSK, Choo A, Chen TS, Salto-Tellez M, Timmers L, Lee CN, El Oakley RM, Pasterkamp G, de Kleijn DPV, Lim SK (2010) Exosome secreted by MSC reduces myocardial ischemia/reperfusion injury. Stem Cell Res 4(3):214–222

Lazarus HM, Haynesworth SL, Gerson SL, Rosenthal NS, Caplan AI (1995) Ex vivo expansion and subsequent infusion of human bone marrow-derived stromal progenitor cells (mesenchymal progenitor cells): implications for therapeutic use. Bone Marrow Transplant 16(4):557–564

Le Blanc K, Ringden O (2005) Immunobiology of human mesenchymal stem cells and future use in hematopoietic stem cell transplantation. Biol Blood Marrow Transplant 11(5):321–334

Le Blanc K, Tammik L, Sundberg B, Haynesworth SE, Ringden O (2003) Mesenchymal stem cells inhibit and stimulate mixed lymphocyte cultures and mitogenic responses independently of the major histocompatibility complex. Scand J Immunol 57(1):11–20

Le Blanc K, Gotherstrom C, Ringden O, Hassan M, McMahon R, Horwitz E, Anneren G, Axelsson O, Nunn J, Ewald U, Norden-Lindeberg S, Jansson M, Dalton A, Astrom E, Westgren M (2005) Fetal mesenchymal stem-cell engraftment in bone after in utero transplantation in a patient with severe osteogenesis imperfecta. Transplantation 79(11):1607–1614

Le Blanc K, Frassoni F, Ball L, Locatelli F, Roelofs H, Lewis I, Lanino E, Sundberg B, Bernardo ME, Remberger M, Dini G, Egeler RM, Bacigalupo A, Fibbe W, Ringden O (2008) Mesenchymal stem cells for treatment of steroid-resistant, severe, acute graft-versus-host disease: a phase II study. Lancet 371(9624):1579–1586

Lee CH, Cook JL, Mendelson A, Moioli EK, Yao H, Mao JJ (2010) Regeneration of the articular surface of the rabbit synovial joint by cell homing: a proof of concept study. Lancet 376(9739):440–448

Lee RH, Pulin AA, Seo MJ, Kota DJ, Ylostalo J, Larson BL, Semprun-Prieto L, Delafontaine P, Prockop DJ (2009) Intravenous hMSCs improve myocardial infarction in mice because cells embolized in lung are activated to secrete the anti-inflammatory protein TSG-6. Cell Stem Cell 5(1):54–63

Losordo DW, Kibbe MR, Mendelsohn F, Marston W, Driver VR, Sharafuddin M, Teodorescu V, Wiechmann BN, Thompson C, Kraiss L, Carman T, Dohad S, Huang P, Junge CE, Story K, Weistroffer T, Thorne TM, Millay M, Runyon JP, Schainfeld R, Autologous CD34+ Cell Therapy for Critical Limb Ischemia Investigators (2012) A randomized, controlled pilot study of autologous CD34+ cell therapy for critical limb ischemia. Circ Cardiovasc Interv 5(6):821–830

Lucchini G, Introna M, Dander E, Rovelli A, Balduzzi A, Bonanomi S, Salvade A, Capelli C, Belotti D, Gaipa G, Perseghin P, Vinci P, Lanino E, Chiusolo P, Orofino MG, Marktel S, Golay J, Rambaldi A, Biondi A, D'Amico G, Biagi E (2010) Platelet-lysate-expanded mesenchymal stromal cells as a salvage therapy for severe resistant graft-versus-host disease in a pediatric population. Biol Blood Marrow Transplant 16(9):1293–1301

Malgieri A, Kantzari E, Patrizi MP, Gambardella S (2010) Bone marrow and umbilical cord blood human mesenchymal stem cells: state of the art. Int J Clin Exp Med 3(4):248–269

Mao Q, Jin H, Liao F, Xiao L, Chen D, Tong P (2013) The efficacy of targeted intraarterial delivery of concentrated autologous bone marrow containing mononuclear cells in the treatment of osteonecrosis of the femoral head: a five year follow-up study. Bone 57(2):509–516

Marcacci M, Kon E, Moukhachev V, Lavroukov A, Kutepov S, Quarto R, Mastrogiacomo M, Cancedda R (2007) Stem cells associated with macroporous bioceramics for long bone repair: 6-to 7-year outcome of a pilot clinical study. Tissue Eng 13(5):947–955

Melief SM, Zwaginga JJ, Fibbe WE, Roelofs H (2013) Adipose tissue-derived multipotent stromal cells have a higher immunomodulatory capacity than their bone marrow-derived counterparts. Stem Cells Transl Med 2(6):455–463

Mohty M, Richardson PG, McCarthy PL, Attal M (2015) Consolidation and maintenance therapy for multiple myeloma after autologous transplantation: where do we stand? Bone Marrow Transplant 50(8):1024–1029

Murphy MB, Moncivais K, Caplan AI (2013) Mesenchymal stem cells: environmentally responsive therapeutics for regenerative medicine. Exp Mol Med 45:e54

Muschler GF, Boehm C, Easley K (1997) Aspiration to obtain osteoblast progenitor cells from human bone marrow: the influence of aspiration volume. J Bone Joint Surg Am 79A(11):1699–1709

Nakamizo A, Marini F, Amano T, Khan A, Studeny M, Gumin J, Chen J, Hentschel S, Vecil G, Dembinski J, Andreeff M, Lang FF (2005) Human bone marrow-derived mesenchymal stem cells in the treatment of gliomas. Cancer Res 65(8):3307–3318

Neman J, Duenas V, Kowolik CM, Hambrecht AC, Chen MY, Jandial R (2013) Lineage mapping and characterization of the native progenitor population in cellular allograft. Spine J 13(2):162–174

Nguyen B-K, Maltais S, Perrault L, Tanguay J-F, Tardif J-C, Stevens L-M, Borie M, Harel F, Mansour S, Noiseux N (2010) Improved function and myocardial repair of infarcted heart by intracoronary injection of mesenchymal stem cell-derived growth factors. J Cardiovasc Transl Res 3(5):547–558

Ning H, Yang F, Jiang M, Hu L, Feng K, Zhang J, Yu Z, Li B, Xu C, Li Y, Wang J, Hu J, Lou X, Chen H (2008) The correlation between cotransplantation of mesenchymal stem cells and higher recurrence rate in hematologic malignancy patients: outcome of a pilot clinical study. Leukemia 22(3):593–599

Norgren L, Hiatt WR, Dormandy JA, Nehler MR, Harris KA, Fowkes FGR (2007) Inter-society consensus for the management of peripheral arterial disease (TASC II). J Vasc Surg 45(1 Suppl):S5–S67

Park JH, Kim DY, Sung IY, Choi GH, Jeon MH, Kim KK, Jeon SR (2012) Long-term results of spinal cord injury therapy using mesenchymal stem cells derived from bone marrow in humans. Neurosurgery 70(5):1238–1247

Pasquinelli G, Tazzari P, Ricci F, Vaselli C, Buzzi M, Conte R, Orrico C, Foroni L, Stella A, Alviano F, Bagnara GP, Lucarelli E (2007) Ultrastructural characteristics of human mesenchymal stromal (stem) cells derived from bone marrow and term placenta. Ultrastruct Pathol 31(1):23–31

Pedersen T, Blois A, Xue Y, Xing Z, Sun Y, Finne-Wistrand A, Lorens J, Fristad I, Leknes K, Mustafa K (2014) Mesenchymal stem cells induce endothelial cell quiescence and promote capillary formation. Stem Cell Res Ther 5(1):23

Perez L, Anasetti C, Pidala J (2011) Have we improved in preventing and treating acute graft-versus-host disease? Curr Opin Hematol 18(6):408–413

Philippart P, Meuleman N, Stamatopoulos B, Najar M, Pieters K, De Bruyn C, Bron D, Lagneaux L (2014) In vivo production of mesenchymal stromal cells after injection of autologous platelet-rich plasma activated by recombinant human soluble tissue factor in the bone marrow of healthy volunteers. Tissue Eng A 20(1-2):160–170

Porter RM, Liu F, Pilapil C, Betz OB, Vrahas MS, Harris MB, Evans CH (2009) Osteogenic potential of reamer irrigator aspirator (RIA) aspirate collected from patients undergoing hip arthroplasty. J Orthop Res 27(1):42–49

Prochazka V, Gumulec J, Jaluvka F, Salounova D, Jonszta T, Czerny D, Krajca J, Urbanec R, Klement P, Martinek J, Klement GL (2010) Cell therapy, a new standard in management of chronic critical limb ischemia and foot ulcer. Cell Transplant 19(11):1413–1424

Qian H, Le Blanc K, Sigvardsson M (2012) Primary mesenchymal stem and progenitor cells from bone marrow lack expression of CD44 protein. J Biol Chem 287(31):25795–25807

Quarto R, Mastrogiacomo M, Cancedda R, Kutepov SM, Mukhachev V, Lavroukov A, Kon E, Marcacci M (2001) Repair of large bone defects with the use of autologous bone marrow stromal cells. N Engl J Med 344(5):385–386

Ra JC, Shin IS, Kim SH, Kang SK, Kang BC, Lee HY, Kim YJ, Jo JY, Yoon EJ, Choi HJ, Kwon E (2011) Safety of intravenous infusion of human adipose tissue-derived mesenchymal stem cells in animals and humans. Stem Cells Dev 20(8):1297–1308

Rehman J, Traktuev D, Li J, Merfeld-Clauss S, Temm-Grove CJ, Bovenkerk JE, Pell CL, Johnstone BH, Considine RV, March KL (2004) Secretion of angiogenic and antiapoptotic factors by human adipose stromal cells. Circulation 109(10):1292–1298

Ringden O, Uzunel M, Rasmusson I, Remberger M, Sundberg B, Lonnies H, Marschall H-U, Dlugosz A, Szakos A, Hassan Z, Omazic B, Aschan J, Barkholt L, Le Blanc K (2006) Mesenchymal stem cells for treatment of therapy-resistant graft-versus-host disease. Transplantation 81(10):1390–1397

Ruiz-Salmeron R, de la Cuesta-Diaz A, Constantino-Bermejo M, Perez-Camacho I, Marcos-Sanchez F, Hmadcha A, Soria B (2011) Angiographic demonstration of neoangiogenesis after intra-arterial infusion of autologous bone marrow mononuclear cells in diabetic patients with critical limb ischemia. Cell Transplant 20(10):1629–1639

Salem HK, Thiemermann C (2010) Mesenchymal stromal cells: current understanding and clinical status. Stem Cells 28(3):585–596

Schallmoser K, Rohde E, Reinisch A, Bartmann C, Thaler D, Drexler C, Obenauf AC, Lanzer G, Linkesch W, Strunk D (2008) Rapid large-scale expansion of functional mesenchymal stem cells from unmanipulated bone marrow without animal serum. Tissue Eng Part C Methods 14(3):185–196

Schrepfer S, Deuse T, Reichenspurner H, Fischbein MP, Robbins RC, Pelletier MP (2007) Stem cell transplantation: the lung barrier. Transplant Proc 39(2):573–576

Sekiya I, Vuoristo JT, Larson BL, Prockop DJ (2002) In vitro cartilage formation by human adult stem cells from bone marrow stroma defines the sequence of cellular and molecular events during chondrogenesis. Proc Natl Acad Sci U S A 99(7):4397–4402

Sen RK, Tripathy SK, Aggarwal S, Marwaha N, Sharma RR, Khandelwal N (2012) Early results of core decompression and autologous bone marrow mononuclear cells instillation in femoral head osteonecrosis: a randomized control study. J Arthroplast 27(5):679–686

Shafaei H, Esmaeili A, Mardani M, Razavi S, Hashemibeni B, Nasr-Esfahani MH, Shiran MB, Esfandiari E (2011) Effects of human placental serum on proliferation and morphology of human adipose tissue-derived stem cells. Bone Marrow Transplant 46(11):1464–1471

Shahdadfar A, Frønsdal K, Haug T, Reinholt FP, Brinchmann JE (2005) In vitro expansion of human mesenchymal stem cells: choice of serum is a determinant of cell proliferation, differentiation, gene expression, and transcriptome stability. Stem Cells 23(9):1357–1366

Sultan S, Hynes N (2014) Contemporary management of critical lower limb ischemia in TASC D lesions with subintimal angioplasty in femoro-popliteal lesions, tibial angioplasty and sequential compression biomechanical device for infra-inguinal arterial occlusion. Experience and quality of life outcome learned over 25 years. J Cardiovasc Surg 55(6):813–825

Syed BA, Evans JB (2013) Stem cell therapy market. Nat Rev Drug Discov 12(3):185–186

Takahashi K, Tanabe K, Ohnuki M, Narita M, Ichisaka T, Tomoda K, Yamanaka S (2007) Induction of pluripotent stem cells from adult human fibroblasts by defined factors. Cell 131(5):861–872

Tan HB, Giannoudis PV, Boxall SA, McGonagle D, Jones E (2015) The systemic influence of platelet-derived growth factors on bone marrow mesenchymal stem cells in fracture patients. BMC Med 13:6

Tarte K, Gaillard J, Lataillade JJ, Fouillard L, Becker M, Mossafa H, Tchirkov A, Rouard H, Henry C, Splingard M, Dulong J, Monnier D, Gourmelon P, Gorin NC, Sensebe L, Soc Francaise Greffe M (2010) Clinical-grade production of human mesenchymal stromal cells: occurrence of aneuploidy without transformation. Blood 115(8):1549–1553

Tateishi K, Ando W, Higuchi C, Hart DA, Hashimoto J, Nakata K, Yoshikawa H, Nakamura N (2008) Comparison of human serum with fetal bovine serum for expansion and differentiation of human synovial msc: potential feasibility for clinical applications. Cell Transplant 17(5):549–557

Tateishi-Yuyama E, Matsubara H, Murohara T, Ikeda U, Shintani S, Masaki H, Amano K, Kishimoto Y, Yoshimoto K, Akashi H, Shimada K, Iwasaka T, Imaizumi T (2002) Therapeutic angiogenesis for patients with limb ischaemia by autologous transplantation of bone-marrow cells: a pilot study and a randomised controlled trial. Lancet 360(9331):427–435

Teraa M, Sprengers RW, Schutgens REG, Slaper-Cortenbach ICM, van der Graaf Y, Algra A, van der Tweel I, Doevendans PA, Mali WPTM, Moll FL, Verhaar MC (2015) Effect of repetitive intra-arterial infusion of bone marrow mononuclear cells in patients with no-option limb ischemia: the randomized, double-blind, placebo-controlled rejuvenating endothelial progenitor cells via transcutaneous intra-arterial supplementation (JUVENTAS) trial. Circulation 131(10):851–860

Timmers L, Lim SK, Arslan F, Armstrong JS, Hoefer IE, Doevendans PA, Piek JJ, El Oakley RM, Choo A, Lee CN, Pasterkamp G, de Kleijn DPV (2008) Reduction of myocardial infarct size by human mesenchymal stem cell conditioned medium. Stem Cell Res 1(2):129–137

Torres-Espin A, Redondo-Castro E, Hernandez J, Navarro X (2015) Immunosuppression of allogenic mesenchymal stem cells transplantation after spinal cord injury improves graft survival and beneficial outcomes. J Neurotrauma 32(6):367–380

Vaes B, Van't Hof W, Deans R, Pinxteren J (2012) Application of MultiStem(®) allogeneic cells for immunomodulatory therapy: clinical progress and pre-clinical challenges in prophylaxis for graft versus host disease. Front Immunol 3:345

Vawda RFM (2013) Mesenchymal cells in the treatment of spinal cord injury: current & future perspectives. Curr Stem Cell Res Ther 8(1):25–38

Venkataramana NK, Kumar SKV, Balaraju S, Radhakrishnan RC, Bansal A, Dixit A, Rao DK, Das M, Jan M, Gupta PK, Totey SM (2010) Open-labeled study of unilateral autologous bone-marrow-derived mesenchymal stem cell transplantation in Parkinson's disease. Transl Res 155(2):62–70

Venkataramana NK, Pal R, Rao SAV, Naik AL, Jan M, Nair R, Sanjeev CC, Kamble RB, Murthy DP, Chaitanya K (2012) Bilateral transplantation of allogenic adult human bone marrow-derived mesenchymal stem cells into the subventricular zone of parkinson's disease: a pilot clinical study. Stem Cells Int 2012:931902

von Bonin M, Stolzel F, Goedecke A, Richter K, Wuschek N, Holig K, Platzbecker U, Illmer T, Schaich M, Schetelig J, Kiani A, Ordemann R, Ehninger G, Schmitz M, Bornhauser M (2009) Treatment of refractory acute GVHD with third-party MSC expanded in platelet lysate-containing medium. Bone Marrow Transplant 43(3):245–251

Wakitani S, Mitsuoka T, Nakamura N, Toritsuka Y, Nakamura Y, Horibe S (2004) Autologous bone marrow stromal cell transplantation for repair of full-thickness articular cartilage defects in human patellae: two case reports. Cell Transplant 13(5):595–600

Walter DH, Krankenberg H, Balzer JO, Kalka C, Baumgartner I, Schluter M, Tonn T, Seeger F, Dimmeler S, Lindhoff-Last E, Zeiher AM, PROVASA Investigators (2011) Intraarterial administration of bone marrow mononuclear cells in patients with critical limb ischemia: a randomized-start, placebo-controlled pilot trial (PROVASA). Circ Cardiovasc Interv 4(1):26–37

Wang Y, Zhang Z, Chi Y, Zhang Q, Xu F, Yang Z, Meng L, Yang S, Yan S, Mao A, Zhang J, Yang Y, Wang S, Cui J, Liang L, Ji Y, Han ZB, Fang X, Han ZC (2013) Long-term cultured mesenchymal stem cells frequently develop genomic mutations but do not undergo malignant transformation. Cell Death Dis 4:e950

Wang Y, Chen X, Cao W, Shi Y (2014a) Plasticity of mesenchymal stem cells in immunomodulation: pathological and therapeutic implications. Nat Immunol 15(11):1009–1016

Wang Y, Wu H, Yang Z, Chi Y, Meng L, Mao A, Yan S, Hu S, Zhang J, Zhang Y, Yu W, Ma Y, Li T, Cheng Y, Wang Y, Wang S, Liu J, Han J, Li C, Liu L, Xu J, Han ZB, Han Z (2014b) Human mesenchymal stem cells possess different biological characteristics but do not change their therapeutic potential when cultured in serum free medium. Stem Cell Res Ther 5(6):132

Weiss M, Troyer D (2006) Stem cells in the umbilical cord. Stem Cell Rev Rep 2(2):155–162

Westerweel PE, Verhaar MC (2008) Directing myogenic mesenchymal stem cell differentiation. Circ Res 103(6):560–561

Yagi H, Soto-Gutierrez A, Parekkadan B, Kitagawa Y, Tompkins RG, Kobayashi N, Yarmush ML (2010) Mesenchymal stem cells: mechanisms of immunomodulation and homing. Cell Transplant 19(6):667–679

Yamout B, Hourani R, Salti H, Barada W, El-Hajj T, Al-Kutoubi A, Herlopian A, Baz EK, Mahfouz R, Khalil-Hamdan R, Kreidieh NMA, El-Sabban M, Bazarbachi A (2010) Bone marrow mesenchymal stem cell transplantation in patients with multiple sclerosis: a pilot study. J Neuroimmunol 227(1-2):185–189

Yang B, Wu X, Mao Y, Bao W, Gao L, Zhou P, Xie R, Zhou L, Zhu J (2009) Dual-targeted anti-tumor effects against brainstem glioma by intravenous delivery of tumor necrosis factor-related, apoptosis-inducing, ligand-engineered human mesenchymal stem cells. Neurosurgery 65(3):610–624

Yeo RWY, Lai RC, Tan KH, Lim SK (2013) Exosome: a novel and safer therapeutic refinement of Mesenchymal stem cell. Exosomes Microvesicles 1:7

Zuk PA, Zhu M, Ashjian P, De Ugarte DA, Huang JI, Mizuno H, Alfonso ZC, Fraser JK, Benhaim P, Hedrick MH (2002) Human adipose tissue is a source of multipotent stem cells. Mol Biol Cell 13(12):4279–4295

Part II
Ethics and Regulations

Chapter 5
A Normative Case for the Patentability of Human Embryonic Stem Cell Technologies

Fikile M. Mnisi

5.1 Introduction

World Trade Organization (WTO) et al. (2013) document on 'Promoting Access to Medical Technologies and Innovation' reports that "the past decade has seen considerable growth in the use of patents for medical technologies in terms of the volume of the patent filings, the geographical base of activity (with notable rise in patents from certain emerging economies) and the diversity of private and public entities seeking patents. This same period has also been marked by an intense debate on the role of patent system regarding innovation in and access to, medical products", this further including research and scientific progress and growth. Patents have at the same time been criticised for its failure based on the same grounds and notions that the same system was meant for (with grounds such as occurrence of inventive genius independent of the promise reward of patent protection), blockage of rapid technological development in the future prospect, patent creation of monopolies which are anticompetitive and neglect of concerns of the developing world (Anand 2011).

In the light of this failure created by default through the patent system and in order to mitigate this negative effect thus, 'promoting the development of new medicine and impacting on prices was recognized in the Doha Declaration. In addition, there have been debates as to whether the patent system provides sufficient and appropriate incentive to ensure the development of new products in certain areas' (WTO et al. 2013). Against this background, South Africa's (SA) patent system is currently under review with a reformulated legal and policy framework included in the Government Gazette of 2013 (No. 36816), the draft National Intellectual Property Policy. This draft policy states that SA is currently granting weak patents as there is no patent examination system. On the other hand, this is detrimental to

F.M. Mnisi (✉)
Steve Biko Centre for Bioethics, Witwatersrand University, Johannesburg, South Africa
e-mail: mnisimf@gmail.com

© Springer International Publishing AG 2017
P.V. Pham, A. Rosemann (eds.), *Safety, Ethics and Regulations*, Stem Cells in Clinical Applications, DOI 10.1007/978-3-319-59165-0_5

the SA's pro-public health objectives (article a (ii)). Due to this, SA is now working on incorporating the flexibilities mandated by the TRIPS Agreement (after the Doha Declaration). These flexibilities will allow for stronger patentability grounds in order to promote access to public health. Moreover, based on the TRIPS Agreement, the draft policy has numerous recommendations including: Commitment to the protection of data in terms of Article 39 of the TRIPS Agreement (1995), (Article a (vii)), Bolar provision before patent expiration (Article ii) and generic medicine and companies (Article iii).

In order to tackle, analyse and discuss the above-mentioned recommendations, I will discuss the SA patent system to analyse what is regarded as a patentable subject matter and what examination system is used to examine the patentability criteria (novelty, inventive step, industrial applicable and *ordre public* and morality) as well as compulsory licensing agreements. I will only analyse this system in terms of biotechnology inventions, by only focusing and using human embryonic stem cell (hESC) technology in order to invoke the need for ethical standard within the patent system which will maintain ethical responsibility in commercial exploitation of the patents, granting of patents that will not be contrary to public morality, that will not limit access to affordable stem cell healthcare and therapies and furthermore will be able to sustain research and scientific progress in this field of biotechnology. Therefore, for this normative analysis I will draw an ethical principle by Hans Jonas based on his work concerning 'new' modern technology on 'Ethics of Responsibility' which may be applied and integrated within the *South African Patent Act of* 1978 (No. 58, amendment Act No. 58 of 2002). I am using the SA patent system for hESC patents in order to invoke and use as a tool to argue that hESC inventions should be patentable subject matters and that these patents should not be prohibited without ethically and morally justifiable reasons. Thus, the legal and regulatory framework needs to facilitate an environment for hESC patents that will ensure that any patents that are either granted or revoked are based on solid ethical justification and not on one's desires or moral view of the human embryo itself as this may often lead to arbitrary and unjust decisions.

5.2   South African Patent System

Human embryonic stem cell technology inventions' patent eligibility is determined by the *SA Patent Act of* 1978 (No. 58, amendment Act No. 58 of 2002) as with any other biotechnology inventions. Currently, in SA there has not been any stem cell technology (human stem cell) or invention that have been granted patents or any court cases thereof unlike in the USA and European Union (EU) countries. This could be due to this technology being at its early stage and thus there have not been many private and/or academic researchers who are working on this technology. Furthermore it may also be due to the legal framework of stem cell technologies more especially hESC in the country which seems to have lacunas that need to be addressed and resolved. However, progress within the legal framework is noticeable (by this I mean the amendments made with the regulations which supplement the *National Health Act of 2003* and this drafted National Intellectual Property Policy), without any only promising but showings some progress within the legal framework which will thereby bring growth and future developments within this field of biotechnology. However, with that said there is still a need of clear and coherent policies in general within the SA legal regime of hESC and more so policies pertaining to stem cell technology and ethical guidelines for these patents. Lack of these public policies further hinders research progress and access to 'possible' stem cell therapies. Furthermore, hESC patents have been opposed by some professional guidelines (General Ethical Guidelines for Biotechnology Research Guidelines: Health Professional Council of South Africa (HPCSA) and Guideline on Ethics for Medical Research: Reproductive Biology and Genetic Research (MRC Booklet 2)), and both guidelines suggest that hESC patents be prohibited; whether this opposition is based on the morality clause found in the Patent Act or on the moral status of the hESC or not is unclear, and hence this seemingly negative attitude towards hESC technology has by default slowed down or hampered hESC research progress and development of therapies in SA. Patents were generally created not only as an economic incentive for the patent holder(s) but as well as for scientific progress and growth. Therefore, for this to happen there is a need for certain issues to be addressed, dealt with as well as resolutions taken and implemented to the patent system in order to create an inducive environment for biotechnology patents as a whole.

SA allows for an exclusive right over an invention of 20 years for which during this period the patent holder has exclusive rights to exploit and/or work his/her patent. The same basic patent criteria that are found with other national and international patent system are also found in SA, and these include novelty ('new'), inventive step (non-obvious) and should be capable of being used or applied in a trade or industry or agriculture (useful) and morality clause (*ordre public* and morality—which is found in certain countries). Furthermore, the invention must also be sufficiently disclosed in order for that patent to be granted. For an invention to be eligible for patentability, thus it must pass all the patent criteria including that of patent disclosure. Thus, the current SA patent system is based on a depository registration system unlike with other countries where their patent system is based on an examination system. The depository system is when, 'the Registrar "examines", in the prescribed manner, every application for a patent and every complete specification accompanying such application. If there is a compliance with the requirements of the Patent Act, the Registrar accepts the patent application for registration. Meaning that SA has no quality examiners within the South African Patent Office' according to Wen and Matsaneng (2013). Therefore, the current registration system may become a hurdle for this country in identifying and granting strong and ethically acceptable patents especially those of biotechnology as many of these patents may have contentious issues. Therefore, an examination system may be an ideal system and needs to be implemented within patent system to solve problems of possibly granting 'weak' patents in the future.

According to WTO et al. (2013) within the examination system, thus, 'the patent offices and courts interpret and apply national patentability requirements on a case-by-case bases within the applicable legal framework…with many patent officers providing the patent guidelines'. These examination guidelines will help to assess and test the patent criteria, and this will help grant 'strong' or 'solid' patents. This is a very crucial point for which SA can follow and integrate as part of the system especially since SA is currently granting weak patent, and in order for SA to grant 'solid' patent, it is vital that they incorporate within the system an examination system instead of the registration system. It is also important when it comes to biotechnology patents as these patents are complex and require a lot of analysis, assessment and testing to prevent patent infringements, to limit and/or prevent patent litigations in the future as well as make certain that issues concern the morality clause is well addressed and dealt with before granting that particular patent. Therefore in order for this system to be integrated, thus, an examination board will have to be developed and this board must have a certain guideline to assess and test patents. Being able to assess and test novelty or an inventive step when it comes to biotechnology inventions such as hESC may be challenging, and whether a claimed invention is new or a discovery or human ingenuity, this requires those professionals who specialise in the subject matter and not only individuals who are trained and specialise in patent law. Therefore, an in-depth analysis of the patent application is needed before any patents may be granted and/or invoked for that matter. More so, with hESC patents, this technology has many contentious ethical issues which require proper examination to be able to either grant or revoke these patents.

South Africa may experience some problems if it has to rely on international policies with some points or guidance within these international policies not applicable or may not work well within a South African context, taking into consideration the cultural differences (even if minor). Thus, the Government Gazette (No. 36816 of 2013) reports that, 'S.A approaches to IP matters is fragmented and not informed by national policies'. Therefore, 'this being a problem as international obligations attracted even if their cost of implementation outweigh the benefits'. This is a problem that SA has to address, come up with resolutions and implement IP policies that will be SA based and of course, whatever resolutions implemented within these policies should also not be outside the scope of international obligations—as with this drafted policy, albeit with some lacuna. We need to establish policies that will be both locally and internationally applicable for SA to have a strong patent system. However, these policies will ensure that there will be benefits for the country at large and does not place a strain into SA economy through obligations that may be costly. There is always a compromise that can be made through public policies that will accommodate SA's current situation but not compromising international obligations and standard and in turn also ensuring 'strong' patents. These policies may also help when it comes to the patent examination system as well as other parts of the patent system that I will not be dealing with in this paper, which may be equally important. What needs to be kept in mind is that whatever it may be, the system must foster research and scientific progress and access to public health by establishing a process where patents do not affect or limit availability of healthcare in society.

It is therefore clear that the SA patent system still has a vacuum and requires some work. This is going to be of great concern when hESC patents start to pick up in this country as these patents are already viewed as being against *ordre public* and morality, i.e. immoral and unethical by some other countries. Moreover, such patents may block or slow down research progress and technological (or innovation) advancement due to patent pools, infringements and 'high' licencing fees or royalty prices as we are seeing with the WARF patents amongst others. Therefore, IP policies and examination system that will address, assess and test for such patent is advisable and required especially with hESC patent which has many ethically controversial issues to deal with, such as patenting of human life or human body and this being unethical and immoral.

5.3   Patentable Subject Matter

'Patents are only available for patentability subject matter, generally defined as "invention" in patent law. In absence of an internationally agreed definition of patentable subject matter, national law define the requirement either positively or through a negative list of excluded subject matter'—or both (WTO et al. 2013). TRIPS Agreement defines requirement for patent eligibility based on a list of certain criteria and the list includes (Anand 2011) novel (new), inventive step (non-obvious) and industrial applicable, (Aymé et al. 2008) excluded from patentability are inventions for which commercial exploitation of patents is contrary to *ordre public* and morality (i.e. morality clause) and (Correa 2005) (a) diagnostic, therapeutic and surgical methods for the treatment of human or animals, (b) plants and animals other than micro-organism and essential biological process for the production of plants or animals other than non-biological and microbiological process. However, members shall provide for the protection of plant varieties either by patent or by an effective *sui generis* system or by any combination thereof. The provision of this paragraph shall be reviewed 4 years after the date of entry into force of the WTO Agreement.

Based on the *SA Patent Act of* 1978 (No. 58, amendment Act No. 58 of 2002) apart from TRIPS Agreement list also include within the Act are listed in Section 25 (2) title 'Anything which consist of'—discovery; scientific theory; mathematical method; a literacy, dramatic, musical or artistic work or any other aesthetic creation; a scheme, rule or method for performing a mental act, playing a game or doing business; and or a program for a computer—these shall not be regarded as patentable subject matter according to the Act. Therefore, according to the above list the Act does not have a specific exclusion to hESC inventions as being unpatentable subject matter. Therefore, hESC inventions will have to go through the same patents criteria as with any other patents. It is not clear at this stage whether patentability of stem cells will be influenced by the Professional Guidelines or by EU decision or not or SA patent system will decide on including hESC inventions as patentable subject matter as long as they meet all the patent criteria. One of the criteria for patent exclusion, i.e. *ordre public* and morality, is seen to be applied in order to exclude hESC from being patent eligible. This exclusion has been related to ethical concern of the invention or rather the entity than it is with the other criteria, albeit these other criteria may have an influence that may lead to being offensive to public's morality. However, even if this morality clause gets applied there is no clear indication of what normative principle(s) will be applied to determine these morality grounds that may be used to prohibit hESC patents or patentability of their subject matters. In EU hESC is unpatentable subject matter based on the *ordre public* and morality clause and within the interpretation of *ordre public* and morality thus Art. 6 (2) of the Biopatent Directive has a list of what is unpatentable based on this morality clause and to guide patent officers and courts (Zachariades 2013). Amongst the list was the exclusion of hESC patent—by excluding the 'use of human embryo for industrial and commercial purposes'—based on the morality clause. This clause was later used in *Brüstle v. Greenpeace* case, where the German Federal Supreme Court declined the patent application based on the morality clause.

On the contrary, unlike with the EU's and SA's patent system, the US patent law does not include *ordre public* and morality clause as a criterion of excluding an invention from patentability. 'In the U.S., laws of nature, natural phenomena and abstract ideas are not patent- eligible subject matter' (Zacheriades 2013). However, cases such as those of *Diamond v. Chakrabay*; *Association for Molecular Pathology et al. Pettioners v. Myriad Genetic, Inc., et al.*; and *Mayo Collaborative Services v. Prometheus Laboratories Inc.*, have prompt the court's decisions and made the court to realise that 'rules against patents on naturally occurring things has limits' and, clearly there is a need to make certain distinction (especially with naturally occurring matters and/or 'human body' or cells) on what may be patentable under these status and under what circumstances. In addition, US patent law does 'recognise inventions involving hESC as patent—eligible subject matter' (Zachariades 2013). Of course, the first hESC patent was held by a US company named Wisconsin Alumni Research Foundation (WARF). Thus, the US law has come a long way in distinguishing what is patent eligible based on naturally occurring matter (with inclusion to 'human body' and/or cells) or not by distinguishing what is a mere discovery and what is a human ingenuity (for which patent eligibility will be granted upon).

SA on the one hand has not yet dealt with such cases and matters as with international laws relating to biotechnology patents (hESC) where Section 25 is used as a determining criterion for what may be regarded as a patentable subject matter as well as on the morality clause. It is not yet known if SA patent, legal and regulatory system will follow either the EU's direction or USA's concerning hESC inventions or not or will it make its own decision regarding this matter as to whether hESC inventions are patent eligible or not based on the country's cultural point of view concerning the moral stand of the human embryo. However, because the law does include within its law the exclusion of patentability on the ground based on the morality clause, thus there is a need for some ethical guidelines and rules as to what this clause will be based upon and to help patent examiner or officer or courts to be able to assess and test this criterion in addition to court cases—or decision of the Court case law— (nationally and/or internationally) and/or following international laws and policies. However, SA needs to make its own decision regarding the grounds or guidelines for determining grounds for the morality clause which will be based on SA sociocultural views. Morality is subjective from nation to nation even though there may be similarities; however, every nation has some notable differences in what they consider to be morally and ethically acceptable based on its society's cultural differences. Therefore, exclusion of inventions based on the morality clause is vital for granting of ethically acceptable strong patents which will not be contrary to *ordre public* and morality and moreover that will not limit access to available therapies at affordable prices. This will also not hinder research progress, development of viable therapies and facilitating an environment for economic growth as well through this technology. Ethical decisions need to be made which will not be arbitrary and/or prohibit certain technologies or research or inventions from patentability based on 'by popular demand', but these decisions will be based on sound ethical and justifiable reasons, for which the morality clause will be based upon. However, even if an invention is found to be patent eligible based on this criterion, it must still be tested and assessed for the other patent criteria before a patent can be granted for that invention; these other standard patent criteria are crucial in determining patentability of a subject matter, and they may further influence and affect the morality clause. Therefore, examination of these criteria is of vital importance and required to be implemented within the patent system in order to determine which subject matter is patent eligible or not. This will further indicate how that patent may or may not be offensive to public morality in the future once the patent is exploited commercially.

5.3.1   Novelty ('New Invention')

For an invention to be identified as being novel according to the patent law—what does that mean? As stipulated in the Act, Section 25 (5): 'an invention is deemed to be new if it does not form part of the state of the art immediately before the prior date of that invention. This is an invention whose application has not previously been disclosed before or prior to application'. Therefore, this criterion is very strict and may be jeopardised by any disclosure or leakage of knowledge to the application (Soini et al. 2008). It is important that prior to applying for a patent relating to any research invention being patented, thus there are no disclosure of the invention made publicly, only after or through patent application and filing is the information concerning the invention disclosed to the public. Whether this includes disclosure through publication or oral presentation or not before patent application is then determined by the national law—the national law will have to determine what constitutes as public disclosure before granting a patent when assessing novelty of the document (WTO et al. 2013).

5.3.2 Inventive Step ('Non-Obviousness')

WTO et al. (2013) reports that 'patent law, in general, defines only the basic concept of what constitute an inventive step and leaves interpretation to patent offices and supervisory courts'…These criteria is understood and stipulated to mean that… 'an inventive step is not obvious to a person skilled in the art, having regard to any matter which forms, immediately before the priority date of the invention…' according to Section 25 (10). Therefore, the person skilled in that art (meaning a person who is working in that technical area related to the field) must not find that invented step to be obvious prior to the patent application, once disclosed to the public. Thus, the invention or invented step must not be obvious to an average person who is working in that field before the time of invention. Being able to test for 'non-obviousness' is imperative as it has been noted that patents of DNA (deoxyribonucleic acid) sequences or of genetics and as well as those of hESC.

Not being able to examine and notice obviousness by the patent officer and courts may lead to granting of the patents that may lead to patents or 'working' of these patents which may be offensive to public's morality in the future or may slow down or halt scientific progress. Soini et al. (2008) state that 'HUGO (human genome organisation) has expressed a concern that, *the patenting of partial and uncharacterised cDNA (complimentary DNA) will reward those who make routine discoveries but paralyse those who determine biological function or application. Such an outcome would impede the development of diagnostic and therapeutic'.* The same concern has been noticed with the three WARF hESC patents where it was later found that these patents within the scope of their patent application included what is regarded as 'routine' or an obvious step to an individual who is trained and skilled as a cell biologist. Being able to test and examine this criterion is of paramount importance for the development of therapies, progress of research and economic growth. This is a good reasons for SA patent system to implement an examination system and an examination guideline, to help guide patent officers on how to assess and test this criterion since this may not only require legally trained individual but may also requires a technically trained individual who is qualified in this particular field of work.

5.3.3 Industrial Applicability or Utility

Industrial applicability (or utility) means that the invention can be made or used in any industry, including agricultural, or that it has a specific credible and substantial utility (WTO et al. 2013). For an invention to be regarded as patent eligible, thus it must fulfil not only the above-mentioned criteria but must be eligible for commercial exploitation. Industrial applicability or utility is a vital criterion especially when it comes to the field such as biotechnology. Thus, even though a patent is used as incentive if utility is not defined and fulfilled thus it will be difficult to understand when assessing the patent application as to how the patent can be used, or worked or exploited commercially. Moreover, this criterion is important in determining whether or not exploitation of the patent may lead offenses that may be contrary to *ordre public* and morality or not. The Brüstle patent was declined based on hESC commercial and industrial application which is regarded as being offensive to *ordre public* and morality and hence falling under the unpatentable subject matter. Therefore, one can see how important this criterion is because some inventions may be deemed as unpatentable subject matter based on this. Albeit, I don't agree with the decision regarding the Brüstle case taken by the European Court of Justice as I do not see (a moral justifiable reason) as to how exploitation of hESC patent commercially is immoral or unethical. However, this is an example of how utility as a criterion may cause *ordre public* and morality or a biological entity be considered to be morally offensive.

Moreover, in the field of biotechnology, thus utility is important since patents are also commercial incentives for companies who invest in these research projects. Industrial applicability examination will further assist in not granting patents with inventions that may not benefit society (health benefit that may stem from hESC technology) or scientific benefits or may be deemed to be 'obvious' to a person skilled in that state of art (such as the WARF patents or so it is claimed and contested) but patents that will be able to benefit society at large. However, some of these patents may not show utility or how they may be commercially exploited in the future to benefit and that is why being able to examine this particular criterion is important. As already mentioned, patenting of hESC is regarded as being immoral and unethical by some as with any other patents on the 'human body'; therefore in order to mitigate the issues brought by these patents thus unless otherwise for other special reason, hESC patent should be granted if and when they show industrial applicability (they are not just an invention but an innovation) amongst other patent criteria. This will help to benefit the public health and society and also further research and development progress in this field. In other words, hESC inventions should not only prove to be inventions but should extend to innovations (that is, they must show to be marketable) therefore industrial applicability and how they can be utilised commercially. Innovations are the ones that will benefit public healthcare at large as well as the country. Therefore, examination or assessing or testing of these criteria is just as crucial as the above and thus the examination board must be able to recognise patentable inventions that have an innovation potential and healthcare benefits. I submit that this will not only strengthen the patent system but science and technology as well as the economy. This is particularly important when it comes to biotechnology patent (i.e. hESC) as they already are regarded by some as being ethically controversial and immoral so having an examination board may mitigate some of these ethical issues attitudes towards these patents.

5.3.4 Disclosure

'Sufficient disclosure of an invention is required in order to grant a patent' (WTO et al. 2013). In addition, TRIPS Agreement also state that the applicant shall disclose the invention in a manner that is clear and complete in order to be carried out by a person skilled in that state of the art and indicate how the invention is to be carried. Furthermore, information regarding foreign applications and grants must be provided as well. *SA Patent Act of* 1978 (No. 58, Amendment no. 58 of 2002) Section 32 describes the manner in which an applicant shall disclose a patent. Depending on what type of specification the patent application falls under, i.e. provisional or complete. Where a provisional specification of the application shall fairly describe the invention, whereas, a complete specification of the application shall have certain particular requirements which are also found within (Article 29) of the TRIPS Agreement (1995). However, *SA Patent Act of* 1978 also requires description of an abstract, drawing and illustration and a complete specification claims.

'One of the fundamental questions raised with respect to "disclosure" of the patent is the extent a patentee need to disclose his/her invention in order to contribute to the promotion of innovation and to the transfer and dissemination of the technology to a mutual advantage of producers and users of technological knowledge' (WTO et al. 2013). To answer this question, thus patent disclosure has to be sufficient enough in order to 'work' the patent and the technical information disclosed should provide examiners with enough information for them to be able to assess and examine the patent without any confusion or missing information before revoking or granting a patent or even qualifying their invention as either patentable subject matter or not. This is very important, as it will help assess and test the other patent criteria that will in turn allow patent office and courts to grant stronger patents. More so, hESC inventions, which will require clear and sufficient information regarding the invention, especially if issues of *ordre public* and morality and whether the claimed invention or process is regarded as patentable subject matter or not; are to be examined, analysed and from which decisions of granting a patent are based upon. It will further make it easier for the patent examiners to test the other patent criteria and make a better decision concerning granting or revoking of patents. What and how information is disclosed is very crucial. Moreover, what may be important is how these criteria are tested and what tests are applied in examining patent eligibility.

5.4 Examination of Patentable Subject Matter or Patent Criteria

Being able to define and test for a patentable subject matter, especially those that are defined in patent law as being non-patentable subject matter such as abstract ideas, scientific theories and/or natural occurring phenomena, amongst others, is important. Therefore, patent law will have to avoid overprotection to accomodate felxibility in cases of 'non-patentable subject matters' by seeting out rules that will bring unorthodox inventions and discoveries within the scope of patentability, rules such as those noted within Section 25 (2) and 25 (2) (4) of the *SA Patent Act of* 1978 (No. 58, amendment no. 58 of 2002). However, there are no set-up guidelines on how to actually test for a subject matter, which are non-patentable inventions even though section 34 of the Patents Act provides for this. The line for what may be patentable and unpatentable may be blurred at times especially when it comes to the field of biotechnology. Some of the patented inventions are claimed to be immoral or 'obvious' to a person skilled in that state of the art, or inventions are non-novel but mere discoveries and therefore should be considered as unpatentable subject matters. This has been mostly seen within the gene sequences field as well as in the field of hESC amongst others. Where claims are made that these biological entities occur in nature and therefore are unpatentable subject matters but mere discoveries or form part of scientific theories and/or methods. It must be pointed out that, patents which are granted on subject matters that have not been subjected to strict examination could easily be challenged on the grounds of invalidity and this is what SA may have to face in the future.

The question is how do patent officers and courts determine or assess or test patent eligibility of an invention when it comes to biotechnological or biological inventions; those that may occur naturally or through natural phenomena and/or found in the human body, and/or their process may or may not involve methods that may be considered to be 'obvious'. How do the patent officers and courts determine and test all this? It has been found that courts do rule a patent valid which may be considered as unpatentable subject matter by some or regarded as scientific discoveries and not novel invention, such as the *Laboratory Corporation of American Holding, DBA LAB CORP, Petitioner v. Metabolite Laboratories, Inc.*, et al. and the *Bilski v. Kappos* cases, which have been amongst the few court cases that patents were considered to be granted unfairly. The *Bilski v. Kappos* case was based on a claimed invention which was designed for the business world and whether this claimed invention was patent eligible or not. The patent was claimed for a procedure to be used to instruct buyers and sellers on how to protect themselves against the risk of price fluctuations in a discrete section of the economy. Thus, three arguments are advanced for the proposition that the claimed invention is outside the scope of the patent law as, firstly, it is not tied to a machine and does not transform an article; secondly, it involves a method of conducting business; and, thirdly, it is merely an abstract idea, as the Supreme Court of the United State (2010) reports. Thus, the so-called machine-or-transformation test was applied in determining the patentability of the invented claim by the court, with this test being considered to be the sole test for testing patent eligibility of claimed invention and process under the US patent law. However, Tysver (2013) reports that 'this machine—or- transformation test may be used to determine patent eligibility of an invention, however the Supreme Court has also recognised that this test is not the sole test that can or should be used to determine a patentable subject matter', albeit this may have been the case with regard to the *Bilski v. Kappos* case. He further states that 'the majority portion of this opinion indicates clearly that the machine – or- transformation test is not the sole test for determining patent eligibility for process, but instead "a useful and important clue, an investigative tool for determining whether some claims invention are process under Section 101'". Even though this test may be important and eligible in order to determine the patentability of some invention and/or process, however, this may not always work and be challenging when it comes to other fields such as business methods (as seen with the *Bilski v. Kappos*) as well as in biotechnology inventions (i.e. hESC).

This machine-or-transformation test is important in determining which invention is eligible as a patentable subject matter. Because most patent laws find abstract ideas, discoveries, scientific theories, business methods, naturally occurring phenomena, mathematical methods, etc. to be unpatentable subject matters, this test may help determine and test 'process' that may be patent eligible, i.e. a 'process' that is new, useful or non-obvious and industrially applicable. Although this test may help determine these patentable 'process' or inventions, thus, special attention may need to be considered when it comes to other fields such as hESC for which their claimed inventions and process may raise special issues that need more scrutiny and may not be as straightforward or obvious as with other fields. For instance, concerning the field of biotechnology claimed inventions of seemingly naturally occurring phenomena may be applied for and, for which the machine-or-transformation test may not be applicable as such. An example to illustrate the above point is the *Diamond v. Chakrabarty* case, where the claimed invention was a naturally occurring phenomena with the patent being granted based on human 'ingenuity' by the court. The court decision in this case opened doors for patents to be granted based on human ingenuity in cases where the claimed invention is not a mere 'discovery' and where the machine-or-transformation test is inapplicable. Granting of patents based on human ingenuity is important when it comes to bio-technology inventions, which are usually based on naturally occurring phenomena or scientific theories or discoveries. As a result of the machine-or-transformation test that only tests if a claimed process or invention is patent eligible and based on; '(Anand 2011) the invention being tied to a particular machine or apparatus-(this may not always be the case with biological invention such as those of stem cell) or (Aymé et al. 2008) if it transforms a particular article into a different state or thing-that is, it causes charge or the "process" must bring change somehow into a thing for which this too may be challenging to determine with some biological process or inventions *prima facia*', (Supreme Court of the United State 2010). Therefore more of a reason why the machine-or-transformation test is not regarded as the sole test for determining patent eligibility and other reasons or factors (such as human ingenuity) are taken into consideration in 'special' cases such as biotechnology inventions and/or 'process', which may be incorporated within the patent system. Since, 'human ingenuity and creativity are acknowledged and rewarded' as stipulated by Section 2(2)(d) of the Intellectual Property Rights from Publicly Financed Research and Development Act (No. 51 of 2008). However, even if an invention or 'process' is claimed to be patent eligible in terms of human ingenuity, thus it must still be prone to qualify under the other patent criteria as well, such as novelty. Moreover, it must further prove to indicate that the claimed invention and/or process are novel through documentation and illustration and by providing sufficient information concerning that particular invention and/or process. All this will help examine if the technical part of the 'process' is indeed an inventive step. And it does have trade or industrial or agricultural application or usefulness as well, amongst being not against or contrary to *ordre public* and morality.

In addition, Correa (2007) published a paper on 'Guidelines for the examination of pharmaceutical patents: developing a public health perspective', and makes some particular recommendations regarding patents in this field. Some of these recommendations may also be applied to patents of hESC research and therapies seeing that both fields are closely related and may use similar starting materials and procedures and so forth. I will only mention some of these recommendation that may be specifically applicable for hESC technologies. In one of his recommendations regarding formulations and composition (stem cell lines or products from stem cell may be used in creation of formulations or in composition for new therapies or medicines), these should not be deemed as patentable subject matter because nothing new would be invented. I do agree with the author, as with many formulations and compositions there is usually nothing new that is invented but simple additions or replacement or admixture of active ingredients or just working on different dosage of the ingredients in the formula. Moreover, the synergistic effect that may also be caused by the combination 'may be deemed as discoveries and not "inventions", since the synergy may take place in the body and found through clinical trials' (Correa 2007). I agree with the author (Correa 2007) when he says that these may be cases for which patentability can be granted in special cases based on human ingenuity and if the claimed invention or formulation or composition shows to be novel, non-obvious and useful and/or solve an important problem that may lead to be advantageous to public healthcare or science.

Regarding the analogy process, where the starting material or end result may be novel, thus this process may be deemed as 'non-novel or obvious process' regardless of whether the starting material, intermediaries and/or end product is novel or inventive; these should be considered as unpatentable subject matters (Correa 2007). Additionally, this process may only be deemed patent eligible if it indicates to be novel and the process is an inventive step and/or has a new and non-obvious end result. This may also be so when it comes to hESC analogy process, in which either the process itself or end result is shown through the patent application as being new and non-obvious, apart from the other criteria. Lastly, in cases of patents that include methods of treatments; these patents include claims of cure, relief of pain, surgical methods, prophylaxis and diagnosis. Correa (2007) states that 'these claims do not cover the product per se, but the way in which it is used in order to obtain certain effect. Thus, Article 27.2 of the TRIP Agreement (1995) explicitly allows members to exclude therapeutic, diagnostic and surgical methods from patent protection and many countries do follow this approach'. Including SA as Section 25 (11) of the Act stipulates that *...An invention of a method of treatment of the human or animal body by surgery or therapy or of diagnosis practised on the human or animal body shall be deemed not to be capable of being used or applied in trade or industry or agriculture.* This is prevention of patentability of 'methods of treatment' that may lead to being offensive to *ordre public* and morality, and thereby being protected and excluded from patentability. These methods will ensure that justice is maintained in medical treatments and benefits from such are without limits and may be accessible to the public healthcare sector without exacerbation of the costs. Additionally, that scientific and research progress as well as technological advancement are maintained as well.

However, there may be special cases when it comes to hESC methods of treatment, i.e. therapies or medical devices, which may indicate that the method is novel and non-obvious and/or can solve a problem and shows tremendous advantage to public healthcare sector. Therapeutic and medical devices need to be separated from so-called methods of treatments as such, and maybe there is also a need of definitions or distinction between the terms used for clarity. Many biotechnology claimed treatments may include therapies and medical devices which are novel and non-obvious and not necessary methods of treatments and these should be deemed as patent eligible. Otherwise if this distinction is not made or not understood by the patent officer or court (since SA does not have an examination system and guideline), it will be difficult to understand what may be deemed patentable or unpatentable subject matter in terms of these patents. In addition to improving national patent system, Correa (2007) recommends a certain mechanism to enhance the examinations of pharmaceutical (and biotechnology) patents from a public healthcare perspective. These mechanisms include pre- and post-grant examinations, which are mentioned in the WTO et al. (2013) documentation as well. The pre-grant opposition mechanism is applied to help patent examiners in order to improve the process of patent analysis undertaken, 'as the third parties may bring to their attention precedents that may not have been identified, and lead to granting of more solid patents procedure'. Since, 'filing a pre-grant opposition or observation requires capacity to monitor published patent applications and the skills necessary to make the search and analysis of precedent that may be opposed' (Correa 2007). This can be done by enhancing the technical knowledge, understanding patent laws and claims, interpretation and drafting.

Whereas, post-grant opposition or observation or procedures have been used in EPO but rarely in the USA. Even though, this procedure may enhance the quality of patents granted as it 'may generally be completed at a lower cost and in a shorter time than the court procedure'. This procedure may be particularly valuable for developing countries such as SA, to improve and strengthen the patent system by ensuring that 'strong' or 'solid' patents are being granted. Because this requires less time and has lower costs therefore this would be more feasable in developing countries in order to cut cost but still produce 'strong' and 'solid' patents. When a 'system (SA patent system) lacks pre-grants and post-grants oppositions phases that provides an opportunity to a Patent Office to receive submission from third parties pertinent to the patent application that could assist in the decision to grant or reject a patent application…In this case poor quality or questionable patents could be easily granted' (Wen and Matsaneng 2013). This is another area that needs to be addressed and possibly amended within SA patent system. The last examination procedure that may be adopted for biotechnology patents as well regard examination rules and procedure for pharmaceutical patents. Particular or special examination rules, procedures and guidelines are required for contentious issues such as patents of hESC inventions, which may require specialised individuals or experts in the field and other fields apart from the patent lawyers. Patent officers and courts (or ultimately examination board) may adopt and use these examination rules and procedure to evaluate and assess patent claims. Within these rules and procedures, there must be definitions of necessary terms such as novelty, inventive step, compulsory licence, etc. and different tests that can be applied to determining what may be patent eligible in terms of hESC inventions, process and therapies or products or medical devices as well as the other patent criteria. What is vital to be disclosed concerning these patents and/or what type of documents is required to obtain all the necessary information concerning the claimed invention or process or product? Of course there is a need of a second and independent type of patent examination system and board specifically for hESC patents and that will be for ethical examinations to evaluate and assess specifically the morality clause. This specific examination board will also require to have its own ethical rules, procedure and guidelines that concern *ordre public* and morality clause and what will be the specific activities that may lead or will be regarded as being against *ordre public* and morality concerning human embryonic stem cell inventions, processes, products or therapies or medical devices. I will further elaborate on this Ethical Examination Review Board, as I will refer to this board from now forth. In the next section I will discuss the patent exemption and limitations which are designed to limit patent infringement and litigations and may be used as a factor to reduce patent 'immorality' (patents being contrary to *ordre public* and morality through blockage and slowing down of research and technological progress and advancement and by so doing this indirectly limits access to beneficial healthcare with the public healthcare sector).

5.5   Patent Exemptions

Recommendations to incorporate TRIPS Agreement after the Doha Declaration by the Government Gazette on existence of patent flexibility to protect public healthcare sector, with the Government Gazette (No. 36816 of 2013)—in order to implement this patent flexibility into the patent system—stating that 'to mitigate patent infringement thus the system should allow upon patent expiry date information disclosure for the use by the public and to also implement the Bolar provision for generic medicine'. The system is now open to the Bolar provision in incorporating patent flexibility and to maintain access to public healthcare. Implementation of Bolar provision as one of the patent exceptions is in accordance with TRIPS Agreement's Article 30 (1995). South Africa has made a recommendation that this provision be used without the generic companies resorting to stock piling and competing with the owner of the patent before the expiration date. Thus, it is still maintaining the rights (exclusive rights) of the patent holder and for others to make use of the patent without any infringement and in accessing affordable medicine, therapies and medicine devices by the public healthcare sector.

Bolar exemption entitles third parties to make use of the patent invention during the patent term without consent of the patent holder for the purpose of developing information to obtain marketing approval. It also it favours the market entry by competitors immediately after the patent expiration and is specifically designed as a means to obtaining and being able to access affordable generic medicines. One can obviously see why SA would actually recommend this exception amongst the other exceptions, with cases such as the HIV-AIDS which took place a couple of years back so that the public healthcare sector is able to gain access to ARV for those infected. So this is, obvious why such an exception will be attractive to SA as it may help the government to combat similar saga as that one with any other urgent or emergency public healthcare issues which may arise. Thus, the Bolar exception should be used for public healthcare emergencies or urgent cases and/or in cases where the patent information may be viable for acquiring or performing scientific studies that may hold vital and tremendously advantageous information regarding public healthcare concerns and crises, which may not be necessarily connected to that particular patent before its expiration date. However, clarification and guidelines on how this may be applied without having issues concerning patent infringements may be of use to fields such as hESC technology.

Apart from this Bolar exemption, other countries have also implemented within their patent system the so-called experiment/research exemption to limit patent infringement, improve research and scientific progress as well as provide early and access to affordable healthcare. WTO et al. (2013) state that 'research exception allows others to use the patent invention for research purpose during the life of the patent'. Moreover, 'this permits certain experiments activities that were deemed necessary to support important objectives, in line with Article 7 (TRIPS Agreement 1995)…The research exception permits scientific research "on" and "with" patented subject matter that could result in better products or processes. Bearing in mind that this could potentially undermine the economic value of the patent' according to Misati and Adachi (2010). Majority of countries that have the experimental/research exception within their legislation and made provision for it through the patent system may be invoked by any party wishing to do so. However, in few cases there has been references made to certain categories of beneficiaries or to activities done in a certain circle (Correa 2005) concerning the experimental or research exemption. This could be something to consider regarding hESC patents especially with their contentious ethical issues regarding the morality not only of the patents but of the human embryo as well as claims of patent infringement which may proceed from these inventions and their patents in the future. This will also permit hESC research and therapies to be patentable or to mitigate the issues of their patent eligibility which may also be exacerbated by patent infringements, litigations and licencing fees and agreement from these patents apart from other legal and ethical issues. This experimental/research exemption can be applicable to 'academic environment' for particular scientific research purposes or experimentations or experimental purposes only, for non- commercial purpose and/or to 'private' companies or researches and for the purpose of industrial use for generic medicine. In this way the provision or exemption can be used parallel to the Bolar provision as they interlink. Against that, the experiment/research exemption should form part of the SA patent system as one of the patent limitations and exemption in order to mitigate patent infringement and to make a provision for patent. Otherwise, this exemption will have to be considered for hESC patents, and this may need to be incorporated as such (to be used and applied only for hESC patents). Both these limitations and exemption can serve for that purpose and be applied hand in hand in facilitating SA patent flexibility and access to affordable therapies and medicines for the public healthcare sector. Moreover, there are some research and/or experiments that may be viable in the future although seemingly not so in the current. So being able to conduct a research exemption on the patent legal system that may be used for such research or experiment may be off importance for those type of research work. For example, some diseases may be predictable based on lifestyle or lifestyle changes, for instance, diabetes, stroke and heart disease where at some point (in the past) not so common amongst 'black' South Africans but currently they are on the rise as of many 'black' South African lifestyle and choices have changed. Such diseases will need medical and therapuetic interventions, some of which may be unorthodox, and this will result in a demand for therapies and medicine which may be used to alleviate these sicknesses and diseases. Now, the requirement for therapies may also include SA's 'black' market or gene pool, which may not have been studied before as this was seen to be unnecessary at that time. Furthermore, looking at the sudden Ebola epidemic that, for a while, was seen to be under control and has suddenly spiralled, thus, an experiment/research exception can therefore be included and created for such cases. These are cases where they may not be importance or emergencies at that moment but could become so in the future therefore research studies would be important when development of their therapies are needed. Moreover, this will assist in promoting research and technological progress which will enable the scientific and medical grow, develop and progress and enables SA to compete on an international level by also indirectly addressing some of the ethical issues (benefit-sharing and justice) brought by these patents.

Although, SA's patent system seems to be heading towards a 'better' direction regarding the patent system in order to mitigate patent infringements and promote access to medicine and therapies through the Bolar provision. However, even though the system has to facilitate access to medicine and therapies at affordable prices, the same system must also be able to facilitate and show flexibility in order to produce therapies of good quality and for a provision as well as the information disclosed may be applied to help improve and provide a way for scientific growth and for the development and improvement of therapies and the technology. Therefore, the experimental/research exemption can create such an environment whereas this may not be the case even if the Bolar provision is adopted. I am not opposing the adoption of the provision but I want to submit that considerations to include the experiment/research exemption be made together with the Bolar exemption. Experimental/research exemption and the Bolar provision are crucial more so when it comes to hESC patent that may be prone (possibly in the future in SA) to many legal and ethical issues and are so complex to deal with. Some of these patents have been seen to block others from making use of what may be 'obvious' to a skilled person (such as the WARF patents), while some have 'high' licencing fees to a point that they either block or slow down research progress, or others use them and infringe the patent's rights (such as the Myriad breast and ovarian cancer, BRAC-1 and BRAC-2 patents). Such cases indicate the importance of patent limitation and exception within the patent system especially limitations regarding research and access to urgent healthcare and therapies required by public healthcare sector. This not only addresses and deals with some issues regarding patent infringement litigations but also issues that concern *ordre public* and morality clause and, licencing fees and agreements from such patents (just in case such limitations may need to be enforced to the patent licensee and/or licensor in case of an emergency such as that of Ebola virus). Other patent limitations may also be implemented within the patent system besides the experiment/research and Bolar exceptions, and these include; licencing, compulsory licencing and licencing agreement and fees.

5.6   Compulsory and Licensing Fees and Agreements

Compulsory licence is when the government allows a third party to make use of the patented product or process without consent from the patent owner, and this is one of the patent flexibility provisions mentioned by the TRIP Agreement after the Doha Declaration. 'Compulsory licenses serve a purpose in keeping patentees from exerting extreme monopoly rights' (Aymé et al. 2008). In addition, it is used in situations where an official or court forces the patent holder to grant a licence to third party (Soini et al. 2008). Laws grant compulsory licences usually under certain conditions to third parties or to the government for their use. Thus, the national laws can set grounds upon which compulsory licences can be granted based upon. As WTO et al. (2013) report, these compulsory licences are 'not being limited to emergency or other urgent situations as it sometimes mistakenly believed' to be the case; hence, the grounds that are set by the national laws may prove or indicate to be so. These grounds can be grouped in the following manner:

(a) *Non-working or insufficient working*—this is applicable where a patentee fails to work a patent in its jurisdiction or where such working by the patentee is insufficient.
(b) *Anti-competitive practice*—this allows granting of a compulsory licence, in order to remedy an anti-competitive practice engaged in by the patentee.
(c) *Public interest*—this is a compulsory licence granted based on public interest without further defining the terms. Although others (meaning laws) do not explicitly mention these specific grounds or any other grounds for that matter, such as in cases of national emergencies or extreme urgency.
(d) *Dependent and blocking patents*—this is when the law makes a provision for requesting compulsory licence where a patent (second or dependent patent) cannot be exploited without infringing another (first or 'blocking patent'). Of course requirement within this ground is that the second invention is an important

technical advancement of considerable economic significance, and the first holder of the patent shall have the right through a cross-licence in order to use the second patent.

(e) Lastly, *government use*—this is when the law explicitly entitles the government or third party authorized by government to use a patented invention without the patent holder authorisation. Ground for this kind of compulsory licence may vary but are typically related to public policy objectives such as national security or health (WTO et al. 2013). *SA patent Act of* 1978 does make provision for compulsory licencing in Section 55 and 56 and has set some grounds for this provision or limitation. These grounds do fall under the different compulsory licencing grounds that are mentioned above. This will ensure that the patent system does provide room and facilitate for access healthcare by public healthcare sector and promotion of research and scientific advancement as well as economic benefit from 'working' the patent.

Another patent issue of interest which may also slow down or block third parties from making use of the patent and thereby limiting access to healthcare is patent licencing fees and agreements. Licensing agreement is when two or more contractual owners of a product or patented product or process give permission to a third party to use that patented product or process. These licenses 'are frequently used to allow other companies with specialised research or development expertise required to produce a complex pharmaceutical (this also including those from the biotechnology industry) under mutual agreed terms' (WTO et al. 2013). Through these patent licences, the patent owner can either sell or licence or transfer ownership to a third party to make use of the patent invention. There are reports of licensing problem caused by licence fees which may be viewed as 'high' and because the terms for patent licencing agreement and negotiations are often subject to confidentiality requirements, it makes it easier for patent holders to manipulate the market and licensees (Soini et al. 2008). Furthermore, Soini et al. (2008) mention all the different types of licencing policies that nations can adopt and these include:

(a) *Exclusivity licence*—this is when the licensee gets exclusive rights to use the invention and associated Intellectual Property Rights (IPR), even the licensor has no rights to use the invention itself or to license it further. The National Academy of Science (NAS) does not recommend this when it comes to the field of genetics and some could be extended to other field within biotechnology.

(b) *Sole or semi-exclusive licence*—within these policies both the licensee and licensor can use the invention; however the licensor is not allowed to grant a licence further.

(c) *Non-exclusive licence*—this policy usually belongs to an active commercialisation and dissemination policy of the company. A licensor may accord a right to use the invention to several parties but not retain the rights to use it and license it further. Thus, the OECD wants to promote non-exclusivity licensing policy in foundational genetics inventions so they are broadly accessible.

(d) Lastly, *Licensing in diagnostic*—in the field of diagnostics, four licensing approaches have been patented and these are (1) free access, (2) patent holder who offers licence to perform testing, (3) exclusive licencing policy and (4) the BiOS (the Biological Innovation for Open Society) licence, whereby the licensee instead of paying royalties he/she must agree to share the improvement on the patented invention.

Based on the above, these licensing policies may be adopted to implement licencing agreements for hESC technology in which both the licensee and licensor can have mutual benefits from 'hESC licencing fees and agreements'. In addition, certain requirement may also need to be put in place for these agreements to work within the system and offer those mutual benefit. Therefore, there may be a need for also including a Material Transfer Agreement (MTA); this will be used for sharing of research materials, as well as a Benefit Sharing Agreement (BSA)—for how benefits may be shared between the licensor and licensee on a case-by-case basis. Of course both the MTA and BSA will depend largely on the type of licencing policy it is and as well as the agreement that was reached. Both the MTA and BSA should be integrated as part of licensing agreement for hESC patents and this forming part of a system that will create a morally acceptable patent system for hESC technology. A system that not only seems as though it is only concerned with one party (the licensor) but also concerned with the progress of the nation's research and healthcare improvements and progress and rights of third parties and its society. An example of a licencing agreement policy that integrated a MTA as part of their licensing agreement is the WARF patents for hESC lines. Thus, WARF adopted a dual licensing policy in order to make the cells available to other parties; within this agreement academic researchers could sign a MTA and obtain two vials of cells for an upfront one-time fee, while researchers who are in the industry would have to pay a significant upfront fee as well as an annual maintenance fee (Jain and George 2007). Albeit, some biotechnology project, i.e. HUGO, does not approve of upfront payments when it comes to genetic sequences. But this may be adopted with hESC technology or debated as to which of the two approaches is actually feasible and ethically acceptable. Therefore, concerning benefit-sharing thus the licence agreement could include requirements such as maintenance of scientific and technological advancement through training, sharing of skills and knowledge, etc. to other researchers or academic researchers. Apart from the MTA agreement, thus, WARF also added the following requirements in the licensing agreement, 'licensee had to agree to share the cells with others, not mingle them to make clones or some form of chimera or attempt to grow them into embryo and, lastly licensee had to certify every year that they were complying with the original agreement' (Jian and George 2007). These requirements are actually based on what is legally permissible and what may also not be legally allowed. Therefore, whatever the requirements are in the licence agreement, these must not only be legally but also ethically acceptable too.

In addition, Aymé et al. (2008) recommend that, the type of licencing agreement should be developed to effectively support and in accordance to those that have already been issued and developed by international organisation, and this must be done by national policymakers. I agree with them and in addition to that policies should not only be developed according to international standard but be locally applicable as well so they can work and be applicable not only nationally but also internationally. Livencing and royalty fee issues need to be addressed as this is often an issue regarding patents in general, provide access to affordable healthcare and facilitate an inducive environment for research and technological progress and development. The guidelines should not violate but promote the licensee and licensor's rights in the negotiations as well as the society's claimed rights, as society is mostly affected by consequences of these agreements and fees, it is said that licensing fees should be within reasonable prices especially for healthcare, but the question is what is 'reasonable'? (WTO et al. 2013). How can a national or international body or bodies be able to determine what is regarded as 'unaffordable or high' or 'affordable' royalties and licencing fees for hESC licence and royalty fees? This will have to be determined before any decision relating to licence fees and royalties are placed. Therefore, such matter must be addressed and some resolution be made or provision as to how this can be dealt with. No matter how this licensing fee is agreed upon or based on what criterion, both parties must at least agree on that fee. However the fee should not be 'high' or block others from that invention or making use of the patents as this was the case with the Myriad BRAC-1 and BRAC-2 patents. Some licences may be free or at a certain price or based on social or voluntary purposes; no matter what or how they are, the patent system mandate is to make sure that there is access to affordable healthcare without eliminating the incentive and rewards for investors or biotechnology companies, respectively.

Another type of agreement which may also serve as a type of agreement in addition to a Licensing Agreement is the Technology Transfer Agreement (TTA) also known as the Tech Agreement. This type of agreement is 'distinguished from the other agreement in that it is an agreement under which a technology supplier undertakes to transfer technical process in return for a financial consideration to produce or develop a certain product, install or operate machines or devices, or provide services. This therefore involves legal processes and acts that may be set forth in the technology transfer agreement' as Licensing Executive Society—Arab Countries (2007). TTA has different types of agreements as Licensing Executive Society—Arab Countries (2007) mentions such as:

1. [Joint Venture Agreement: These are the Agreements under which a simple joint venture is established between two or more partners whereby one of the partners provides the technology.
2. Invention Exploitation Agreement: These are the Agreements that allow exploitation of an invention; however it does not provide for technical support and knowledge transfer (Know-How) nor does it grant a license to use trademarks or the right to sell machines or other technologies.

3. Knowledge Agreement: The main characteristic of these Agreements is that they do not address an invention but only technical information and expertise. The property value of these Agreements relies on the confidentiality of the information and the expertise. Thus, when the technology owner wishes to grant a licensee, he also sets strict confidential clause to the extent of including unfair conditions.
4. Non-disclosure Agreement: These are Agreements under which the licensee is bound to maintain confidentiality of information and to use it for a certain purpose. This type of Agreement may also be embedded with the system as well as within hESC patent system. This will help promote and facilitate an environment for technical growth].

Human embryonic stem cell technology with its ethical controversies and issues and patents being viewed as being immoral will require first that the patent system be resolved to accommodate this technology and its claimed inventions in the future. Secondly, that an examination system or at least examination board be incorporated with this system or at least for hESC patents (which will be complex and will require to be examined especially concerning what may be regarded as a patentable subject matter within this field—there is a thin line between what may be regarded as human ingenuity and what is non-obvious and obvious). Thirdly, that appropriate limitations and patent exemption are also integrated within the system. I do stress that not only the Bolar provision should be included but also the experimentation/research exemption (more so for non-commercial purpose) should be incorporated within the system. This is particularly important in order to try and mitigate and avoid patent infringements and litigations that may stem from these patents. Compulsory license, licensing agreements and fee must be addressed and where necessary some form of generic licencing agreement and guidelines be developed to assist with licencing agreements. Moreover where possible, BSA and MTA concerning licencing agreement between a licensee and licensor must be developed or a generic one for research purposes used. However, whatever the examination guidelines and tests as well as licencing guidelines are, these must all be supported by the Laws and Regulations for hESC research and therapies and not in contradiction with the Laws and Regulations; there is definitely a need of coherence within the legal and regulatory framework. Equally important is also developing ethically and morally acceptable policies, guidelines and test which will make certain that granted and/or revoked patents are within or based on *ordre public* and morality clause and that the patent system grants 'strong' patents. SA patent system must not only be legally acceptable but ethically acceptable through developing an ethically acceptable system. Thus, an ethically acceptable system gives society confidence in its governance as well as hESC patents and their use (or exploitation thereof). Therefore, ethical acceptable patents policies and laws are what is needed in order to be develop a system that will grant 'strong' patent and patents that will not be contrary to *ordre public* and morality.

5.7   Ethics of Responsibility and the Patent System

Based on the morality clause as stated in the *SA Patent Act of* 1978, SA is liable for ensuring and maintaining ethical conducts within its patent system. The recognition and inclusion of this clause, *ordre public* and morality, in the Act is one of the ways which the patent system can address. This morality clause is for society's protection from matters that may violate society's dignity, freedom, respect and democracy through working of patents. Additionally, this clause may be used to also provide protection for any claimed invention and the exploitation of their patents from offending the society's cultural view. It is important to understand what the term *ordre public* means before moving ahead as this will help in understanding and the application of the moral clause appropriately. The Old concept explains *ordre public* as 'protection of important public interests such as security, peace and democracy' (Soini *et al.* 2008). According to that definition, for an invention (either process or working of the patent) to be contraty to *ordre public* and morality, this action will have to violate society's (at large); 1. security - this could be health security or job but not limited to these as well, 2. peace - if such action may lead or lead to social disharmony and, 3. democracy - violates society's claimed constitutional rights such as liberty, dignity and respect. With that said, according to the *SA Patent Act of 1978* (No. 51 of 1978) this morality clause can be used in analysing and revoke an application or patent based on the industrial applicability of the patent as well as the exploitation (commercial) of the patent, as stipulated; section 25 (4) (a) exploitation of an invention by a patent may be based on the manufacturing process of that patent in that state of art. As well as, section 36 (1) (b) that the use of the invention which the application relates would be generally expected to encourage offensive or moral behaviour. In the field of human embryonic we have seen a case where it was patented based on the manufacturing process (its industrial applicability) but later a claimed was filed against these patents as they were claimed to voilate freedom of research amongst other things. This case which appeared in the US court was between the Consumer Watchdog and Wisconsin Alumni Research Foundation (WARF), i.e. *Consumer Watchdog v. Wisconsin Alumni Research Foundation*. In this lawsuit the Consumer Watchdog and the Public Patent Foundation filed a brief with the Court of Appeals for the Federal Circuit. In their brief they argue that patent claims on hESC held by the WARF are invalid under the Myriad decision because they are 'products of nature'. Moreover, 'an appeal was filed in an interparty re-examination of certain patents, re-examination of patent 7.029,913 ('the "913" patent') that was issued on the 18th of April 2006 to Professor James Thomson and now assigned to WARF. The Consumer Watchdog argued that three claims of the patent for the in vitro cultures of human embryonic stem cell lines had been obvious in light of the prior art, with any person of ordinary skill in the art of deriving and maintaining embryonic stem cell lines for any mammal' being able to find the process used to derive these hESC lines, at least, obvious to try (Greely 2013).

This case brings two issues concerning hESC patents under the *ordre public* and morality clause. Firstly, the subject matter being patented or invention being patented, i.e. what may be regarded as patent eligible, is under question as to what exactly is regarded as 'natural', seeing that those things that are regarded as 'natural' may not be patented or are patentable subject matters. Therefore, does it mean that human embryo is not regarded as a 'natural' entity? Clarity on what is regarded as a 'natural' entity that may not be a patentable subject matter needs to be defined more and how patent officers and courts may test for patentable subject matter of naturally occurring subject matters such as those from biological materials of human beings, otherwise this may result in patents or commercialisation thereof being offensive to *ordre public* and morality. And secondly, issues seen with this case is the issue on how the inventive step in itself is defined and determined, i.e. how obvious that process is in that particular state of the art (in this case stem cell or biological), which may result in patents being issued without foreseeing that these patents and/or their commercial exploitation may cause offenses in *ordre public* and morality in the future as is seen now with the WARF patents. WARF's hESC patents are now blocking downstream research and development work and it seems like everything concerning processing of hESC line is being covered under their patents. When evaluating this issue, one realises how these patents can be and/or are against public's morality since they are blocking research and therapeutic development (possible of course) which may be beneficial to the public. Causing insecurity and violating society's claimed democrtatic rights, even though this may not be so now but this it may in the future. Moreover, such issues also tend to increase therapeutic costs from these patents, albeit patents do not affect market prices. However, it does not seem to be translated in that manner, and this has resulted in a few opulent individuals or countries being able to have access to these therapies. How scientific knowledge as well as therapies are distributed and how the public may be able to have access to the technology is of paramount importance as this is the main reason (in my opinion) which may be against the public's morality, and this needs to be prevented by policies.

Policies must be written to define set of ground rules and/ or guidelined that can be followed such as those seen within the TRIPS Agreement document. However, we will need guidelines that will not only be based on the types of inventions but also on the types of activities that are/or may be considered offensive to *ordre public* and morality or lead to such an act or conclusion. Ethical ground that may lead to being offensive to *ordre public* and morality should be based on specific types of activities as it is actually the activities (or action) that result in being contrary to *ordre public* and morality and not an invention or patent (these are just things or means to act in a certain manner that may be regarded as being contrary to *ordre public* and morality but are not in themselves). Such a policy will facilitate a structure or foundation for an ethical and legal regime. Since, it is only the action of the moral agent that may be regarded as being offensive to *ordre public* and morality and not the entity or invention or patent. Thereby, it should be in the government's interest to protect its public from being morally violated through the actions of the moral agent (i.e. patent holder) and how he or she 'works' the patent. But the current international policies as well as laws seem to only mention inventions that are contrary to *ordre public* and morality. This puzzles me as inventions connot be against public's morality but rather human (moral agent's) actions. More of a reason why SA should have its own set of grounds that stipulate what activities may lead to the morality exclusion of what may be regarded as patents eligibility or not.

Another issue of concern is one arising from the working of patents, commercialising of patents, as well as ownership rights of the patents. This is mostly so with hESC technology as this field has been mired in many ethical controversies and issues, regarding the use of human body and furthermore ownership issues (i.e. patenting issues). It is crucial that matters relating to commercialization and onwership of hESC patents be addressed before this technology start growing and expanding. Therefore, 'the patent system therefore provides a useful mechanism which can be applied to address some of these ethical and social concerns of hESC technology not because patents are necessarily the cause of concern but because the system for granting them provides a practical way to regulate compliance with ethical and social values' (Gold and Caulfield 2002). I agree with the authors that the patent system can be used as a tool in addressing these ethical issues an it should be used (somehow) as a tool to address morality issues and this could be a beneficial factor within the legal framework. The patent system can accomplish this by setting up ethical standards which will not only reflect international social culture or standard but importantly local socio-cultural standards too. This will aid in bringing back public's confidence towards science and technology and motivate international collaborations and investments. However, in order to set up appropriate, cultural and ethical standard—to set as grounds for the morality clause, this will require a group of specialists or experts in different fields including Bioethicist/Ethicist, different individual in science (biotechnologist/stem cell biologist), Bioentrepenuer, Bio-economist, etc., to develop and make decision regarding the ethical standards and rules, procedures and guidelines. This group of individual will be able to write and develop ethical guidelines and rule concerning patents and what activities are required to exclude certain inventions and patents from being patent eligible based on the morality clause. Furthermore, they will also be able to test and analyse not only the ethical issues but also the technical steps in patents (for inventive step), issues regarding licencing fees and agreement, additional documentation such as the MTA and BSA when required as well as issues concerning justice from exploitation of these patents. Even though this may seem as though it is unnecessary but it is, unless justice is part of the patent system (by making certain that justice is maintained and facilitated through procedural and distributive justice from exploitation or 'working' of these patents) this may lead to granting or 'working' or exploitation of patents which are offensive to *ordre public* and morality and cause public disorder if justice is not maintained or facilitated through the patent policies and laws and regulations, respectively. Therefore, it is only a qualified team of expert with individuals from different fields that are well knowledgeable and involved in this matter which will be able to understand, evaluate and make ethically sound and justifiable decisions concerning hESC patents and ethical issues pertain from these patents. Only those who are in bioethics or ethics may be able to determine what ethical measures are required in hESC patents. Because 'the legal regime is generally held as neutral, that is patent officers do not have to pay attention to the patents' consequences, as they are not trained to do so more especially with morality issues' (Soini et al. 2008). Hence, more reason for an Ethics Review Board for hESC patents that will be aware of the ethical and moral consequences that may emanate from these patents. This will lessen granting of hESC patents that will be contrary to *ordre public* and morality and thus grant ethically acceptable hESC patents. This may mean that SA patent system may have to integrate within its system a separate group of individuals who will be responsible for handling ethical issues amongst others regarding hESC patents. This group of specialised individual could form part of the 'Ethics Examination Review Board', or this could just be for patent in the field of hESC technology and not necessary with all patents. Their duties will be to develop ethical guidelines for the criterion regarding *ordre public* and morality, that is set up specific activities and not inventions that may be contrary to *ordre public* and morality if an invention is granted a patent or by commercial exploitation of a patent. This is important as other patent criteria may lead to granting of patents which may lead to actions that may be contrary to *ordre public* and morality, and having guidelines that specify on what ground this clause can me breached may help mitigate such matters.

Ethics review board 'should not prevent patent holders from enforcing the patent until the board reaches the final decision and the decision of the board would be subject to be reviewed by courts. Thus, the ethics board will be there to suspend the power to enforce the patent, but not to revoke it' (Gold and Caulfield 2002). Once such guidelines are drafted and implemented they must be used and possible should be enforced through legislations in order to ensure that they are followed and used as they should be. This will ensure that the ethical concern found within the patent application is addressed appropriately before granting or revoking of patents. 'This mechanism will separate the patent granting process from the patent enforcement process, meaning that an ethical review board will not slow down the patent examination and granting procedure' (Gold and Caulfield 2002). Furthermore, it will ensure that 'solid' patents are granted and will take off pressure from the Examination Board or patent officers or courts and make sure that no bias and arbitrary decision for granting or revoking patents were taken and hence an ethical responsible decision were taken and applied within the patent system. Ethical responsibility which is needed in Laws and Regulations including those in patent laws and regulatory framework, so that ethical conduct forms part of governance. For this possibility I then propose that Ethics of Responsibility be incorporated into patent laws and regulatory framework, especially regarding hESC patents.

5.7.1 Ethics of Responsibility

Ethics of Responsibility deals and address ethical conducts concerning 'new modern' technology such as those from biotechnology, albeit is not so 'new'. This ethical principle and theory was originally introduced by Hans Jonas a German philosopher, who reports that, this 'new modern' technologies may affect others and this may not be in the same manner as with the previous technologies and hence this requires that a moral agent act in a responsible manner. Moreover, that the power of this new technology forces upon ethics (traditional ethics) a new dimension of responsibility which was never dreamed about before (Jonas 1984). Therefore, according to Jonas Ethics of Responsibility has to have an objective as well as a subjective side, where the one has to deal with or to do with reasoning and the other deals or has to do with emotions (1984). Therefore, an individual (moral agent) must exhibit and show feelings of responsibility towards others, feelings of responsibility that are important in order for the moral agent to make use of this modern technology in an ethically responsible and moral way. Especially since ethical guidelines or rules alone do not make an individual to act in an ethical responsible manner. Even though, they contribute in making sure that a moral agent does act in a morally acceptable way and can make use of this modern technology in an ethically responsible manner.

Ethics of Responsibility does however assist in making an individual to be accountable for his/her own decision no matter what the consequences are and to be able to give a morally justifiable reason for his or her actions. This principle leaves room for both 'success' and 'failure' from decision made and/or laws, rules, procedures and guidelines developed in order to act (or not act). It stresses on the fact that decision should be based not only on the 'success' or 'benefits' and 'failures' or 'risk' from that particular usage (or lack thereof) of modern technology. But says we ought to be responsible towards the usage and application of that technology to others (i.e. society—present or future) and be able to have a morally justifiable reason for our action. Against that background, it is therefore important to understand that good governance comes with taking responsibility towards society and ensuring that good ethical conduct is maintained and rights are not violated by Laws and Regulations. In order for this to happen therefore the government can apply Ethics of Responsibility when drafting and implementing policies, Laws and Regulation and by having feelings of responsibility towards its society, public healthcare issues, research scientist and investors. Jonas (1984) states in his work that 'the government's responsibility is first and foremost of men for men (and of course women for women) and this is the archetype of all responsibility'. Additionally, that 'these responsibilities that the government has encompass the total being for their object, that is, all aspects of them, from naked existence to the highest interest'. Therefore, this would be referred and known as the, referred as 'Total Responsibility'. The second responsibility is referred as 'Responsibility of Continuity', 'which follows from the total nature of responsibility for which in a tautological sense that its exercise dare not stop'. And, the last responsibility is referred to as 'Responsibility of the Future', 'with future responsibility being concerned with the future (the future of the present generation- the proximity, as well as future generation) and ensures that technological progress is maintained throughout'. Thus the government never stops taking responsibility and he/she has to look into a larger span of things extending from the past to the present to the future life of his/her community's history and future. This sums up not only the past but present as well as future generations and that of individuals within a democratic society. All in all Hans Jonas expresses that, the progress that the technology needs to bring should ultimately result in the 'supreme good' over 'supreme evil'.

Therefore, it is the government's responsibility to make sure that society's claims of liberty and dignity are not violated through hESC technology. The government can accomplish this through natioanl policies and legislations which addresses such matters and uses the principle of Ethics of Responsibility within its legal regime. If this approach is applied in the context of hESC patents, there will be no need for prohibiting patentability of its inventions as well as exploitation of hESC patents as the legal framework will provide and facilitate an environment in which moral agents (who are involved in hESC patents—whether by applying or exploiting or foreseeing there application and granting process) feel and take ethically responsible decision concerning granting or revoking and or working of hESC patents. What the legal framework need to enforce is ethical responsibilities towards 'working' or exploitation of hESC patents in order to ensure that scientists and investors or companies are able to make ethically and morally justifiable decisions concerning the invention and 'working' of hESC patents towards society. Ethics of Responsibility will ensure that society's interests is maintained and not only the interests of the scientists, universities, investors and the government. But that claimed liberty and dignity are not voileted by working of hESC patents.

Therefore, when the patent policies and legislations are being drafted and developed the system should not focus on the morality of the inventions and patents (through traditional ethics only—although still required). However, there is a need to be able to integrate or draft and develop policies and legislation that will bring and incorporate a sense of ethical responsibility towards society by the government, patent holder and so forth. This will set out ground in which patent holders will be able to make use of the patent in an ethical responsible way for not the present generation but for future generation and be able to provide a morally justifiable reason for their action no matter what the consequence are. The same goes for the policymakers and law makers, that whatever laws and regulations and guidelines are drafted and implemented they too will be able to take responsibility of them, and whatever results come from these laws, regulations, policies and guidelines, they will be able to have morally justifiable reason for them. This approach will help mitigate ethical and moral concerns related to hESC patents. In addition, it will assist in ethical conduct that can be adhered to by all parties involved, as well as assist i becoming ethically responsible towards society in working of hESC patents. When the moral agents is aware of their ethical responsibility towards society there are slim chances of violating their rights, or as I would assume. Whether through access of affordable healthcare or scientific or research progress, technology development and advancement but ethical responsibility will be conducted either way through the enforcement by the patent law and patents holders being aware of what should and shouldn't be done. This is particularly important when it comes to hESC technology.

Especially because, there are also many ethical issues and controversies this technology is mired with because hESC source is from the human embryo, which is entangled within the ethical issues regarding the moral status of the human embryo for hESC research, and therefore ethical issues regarding this issues have been debated by various academic scholars with no resolutions on the matter. With many people regarding patenting and commercial exploitation of hESC patents being contrary to *ordre public* and morality based on their sense of morality or rather the moral status of the embryo. This has raised cultural issues about the relationship between the human body and human life as well as ownership of these patents. Therefore, addressing, resolving and coming with better resolutions for addressing and dealing with the patent's morality clause is important as this clause may be applied unfairly in order to prohibit hESC inventions from being patentable subject matters which may benefit public healthcare and for which this on its very own may end being contrary to *ordre public* and morality and may cause public's disharmany and injustice. As important therapies will not be developed (which may have been if research and patenting of this technology was permitted) or allowed for industrial application based on this clause as the current state within the E.U. legislation. This may also be interpreted as violating the patients' dignity, respect and freedom of choice on different stem cell therapies and healthcare therapies by prohibiting inventions that may benefit society, if not now but in the future or future generations.

Ordre public and morality deal with the nations' cultural ways or morality, and this varies from society to society (or nation to nation); hence, SA must be able to set its own ethical standard regarding this criterion, and hence this will also be able to incorporate Ethics of Responsibility within the patent system. SA government has a responsibility towards its own society by ensuring that they are able to create a regime that will provide granting of 'strong' patent, that these patents will benefit society, provide access to affordable healthcare therapies and medicine to the public healthcare and that there is scientific progress as well as technological development and progress. Thus, Ethics of Responsibility is not only geared towards the present generation but also for future generations no matter what consequences it bring as long as these consequences are morally justifiable in this present generation as well as future generations. Not to contradict myself, there is a need to evaluate and make decisions based on the risk vs. benefit ratio. However, responsibility of the results and how the decisions were made is very crucial as sometimes some decisions are made based on fear or doubt. Therefore, Ethics of Responsibility will ensure that these decisions are taken for the overall benefit of society as well as proximity to future generation, and this would have been the best decision to be taken no matter what the consequences may be or appear to be.

5.8　Conclusion

Human embryonic stem cell technology can open great opportunities and benefits in science as well as in providing prominent healthcare therapies, medicine and medicinal devices that will benefit societies, especially the public healthcare sector. However, this technology will of course not come cheap, especially with patent issues not being resolved which may also exacerbate prices and affordability issues concerning them. Patent system can be a tool that may be used not only to address issues that have been surrounding hESC technology especially those of ethics, as it makes provision for patent exclusion by the morality clause. Human ESC is legally patentable in SA based on the standards of the patent criteria: novelty, inventive step, industrial applicability and disclosure. There is no evidence suggesting that any hESC technology is unpatentable subject matter based on *ordre public* or morality or any of the other criteria for that matter. Therefore, despite people's desirability thus unless hESC patents show on good ethical grounds to be immoral or unethical and or exploitation of their patents are in the like manner then only then can they be (their inventions) classified as unpatentable subject matters. Ethics is not based on popular demand but based on what is 'good' and 'bad' and not on how majority of people feel. It is therefore important to make ethical decisions that are not arbitory or based on how the majority of people feel. But we should make ethical decisions that will be for the best interest of society as well as encourage growth and development.

Currently, South African patent system is not providing ground for ethical responsible conduct regarding patents from developments of hESC. Lack of definition of certain terms within the Act, the use of a registrar instead of an examination system and this does not facilitate growth, flexibility and progress in this field. Limitations and provision of patent exception to mitigate patent infringement, lack of licencing agreements and fees as well as policies. This will result in stagnation of patents and innovations and will slow down scientific and healthcare progress in the country. Unless patents are examined and patent are criteria properly assessed before granted or revoked, patent system will only result in granting and/or revoking unfairly. Implementation of such a system is important for hESC patents seeing that hESC has been mired in ethical contraversies therefore, careful assessment is important. Moreover, cases of patent infringement as well as unfair dispute relating to licencing agreements and fees when it comes to concerning biotechnology patents raises the importance of addressing these issues. In order for this to be realised it will take specialised individuals and/or experts in the field of ethics or bioethics can deal and address these ethical issues (i.e. morality issue) stemming from hESC patents as well as those of legal and technical matters.

Human embryonic stem cell technology poses and raises special issues that need to be addressed specifically. Bioethics is able to breach a gap and find a middle ground that will be both legal as well as ethical in terms of policies and legislation. Therefore, the incorporation of an Ethics Examination Review Board to the SA patent regime is needed. This board will be able to develop such policies as well as ethical guidelines for the patent system that will promote ethical conduct by individuals. This will promote ethical responsibility, and if ethical responsibility is part of the system, this will not only promote access to healthcare therapies, research and scientific progress, technological improvements and advancements, but it will also bring society's confidence concerning hESC research and patents from this field. A lot is still required concerning development of hESC technology and its patents, and SA is still at the stage where it can incorporate this principle within its patent system which will result in an ethically responsible manner for granting and/or 'working' of hESC patents.

5.9   Recommendations

I recommend that the SA patent system moves away from being a registration system and implement an examination system, with the Examination Board developing guideline for examining and assessing the patent criteria. This may include tests such as the machine-or-transformation test but can also make provision for human ingenuity in cases of natural and 'human body' inventions and patents such as those from hESC patents. Apart from this, the patent system can also incorporates within its policies and legislation an experimental/research exemption, that will facilitate for an environment which will bring research progress, growth. Additionally will also contribute to development of generic medicine after patent expirations as well as accomodate research that will be beneficiary in some 'emergency' cases. What is equally important concerning hESC patents are licensing agreement and fees. Therefore, development of licensing policies as well as a generic licensing agreement from hESC inventions and patents will assist in mitigating issues that are brought about by these patents. In addition, MTA and BSA agreement will also help with regard to licensing agreements between the licensee and licensor in the research community. These can be regarded as a MTA-licensing document and BSA-licensing document for hESC patents. Therefore, submission of a MTA (as an example of a generic MTA that may be adopted with the patent licencing agreement is from the University of Witwatersrand, Material Transfer Agreement for Human Biological Materials, http://www.witshealth.co.za) and BSA agreement for hESC patent during applications or during licensing agreement will be needed, this can be assessed by the Ethics Review Examination Board as well to be able to determine if the agreement is not contrary to *ordre public* and morality and promotes harmony, justice and scientific progress. The BSA document in incorporation with the licensing agreement should stipulate clearly how benefits will be shared between the licensee and licensor as well as society, whether mandatory or non-mandatory benefits. An example of such document can be seen within the South Africa's Bioprospecting Access and Benefit-sharing Regulatory Framework: Guideline for Providers, Users and Regulators (2012), ultimately an international documents such as those created by HUGO just to name a few may also be reviewed.

In order to have a system that will promote patent flexibility as well as access to affordable healthcare for the public healthcare sector, I submit that these recommendations amongst other measures may be incorporated within the patent legal framework. It may help bring feelings and actions of responsibility between stakeholder's towards society and growth and development with hESC area. Additionally, I recommend that SA adds to its policy a Techno Transfer Agreement which will work in cases of experiments and research studies from patented hESC processes or products. Apart from the above, a very important element to integrate within the patent system is that which will address and deal with ethical issues stemming from hESC patents. Therefore, an Ethics Review Patent Examination Board (preferable for hESC) should be integrated as part of the system this board will work hánd in hand with the Examination Board. It will also have certain duties such as to directly address issues pertaining to the *ordre public* and morality clause regarding hESC patents besides having to draft the rules and guidelines for what may be considered as being offensive to morality by these inventions and 'working' of these patents. The government can have this Board working with all patents or have it just for Biotechnology patents; however there is a need for such a Board for hESC patents. Appropriate definitions for certain terms needs to be incorporate, used and adopted within the Act, regulations and guidelines and these definitions be clearly defined must to assist applications as well as assessments and specific type of activities that will be consider for actions or hESC inventions or exploitation of hESC patents that will be considered as contrary to *ordre public* and morality, thereby being excluded from being a patentable subject matter.

Acknowledgements I would like to thank my supervisor Dr. Kevin Behrens from the Steve Biko Centre for Bioethics for his critical analysis, input and as well as advices and guidance where needed. I would also like to show my gratitude to Advocate Sher Mohammend from Section 27 (NPO), for his critical analysis and input with regard to the legal matters and the law. His legal advice and input has been well appreciated and helped with legal understanding of the patent law.

References

Anand N (2011) Accommodating long term scientific progress: patent prospects in the pharmaceutical industry. J Intellect Prop Rights 16:17–22

Aymé S, Matthijs G, Soini S (2008) Patenting and licensing in genetic testing. Eur J Hum Genet 16:405–411

Correa C (2007) Guidelines for the examination of pharmaceutical patents: developing a public perspective. www.iprsonline.org. Accessed 24 June 2015

Correa CM (2005) The international dimension of research exception. www.iprsonline.org. Accessed 24 June 2015

Gold ER, Caulfield TA (2002) The moral toolbooth: a method that makes use of the patent system to address ethical concerns in. Biotechnology 359:2268–2270

Government Gazette (2013) No 36816, vol 579. http://www.greengazette.co.za. Accessed 25 June 2014

Greely HT (2013) Assessing ESCROs: Yesterday and tomorrow. American Journal of Bioethics 13(1):44–52

Jain S, George G (2007) Technology transfer research offices as institutional entrepreneurs: a case of Wisconsin Aluminium Research Foundation & huam embryonic stem cells. Ind Corp Chang 16(4):545–567

Jonas H (1984) The imperative of responsibility: in search of an ethics for the technological age. The University of Chicago Press, Chicago

Licensing Executive Society-Arab Countries (2007) Guidelines for licensing & technology transfer agreements. http://www.tag-publication.com. Accessed 19 Nov 2014

Misati E, Adachi K (2010) The research and experimentation exceptions in Patent Law: jurisdictional variations and the WIPO Development Agenda. UNCTAD-ICTSD Project on IPRs & Sustainable Development. Policy Brief Number 7, pp 1–8. http://www.unctad.org/en/Docs/iprs_in20102_en.pdf. Accessed 19 May 2015

Patent Act No 57 of 1978 (Patent Amendment Act No 58 of 2002). http://www.cipc.co.za. Accessed 10 Mar 2014

Salter B, Salter C (2013) Bioethical ambition, political opportunity and the European governance of patenting: the case of human embryonic stem cell science. Soc Sci Med 19:286–292

Soini S, Aymé S, Matthijs G (2008) Patenting and licensing in genetic testing: ethical, legal and social issues. Eur J Hum Gene 16:11–50

Supreme Court of the United State (2010) Bilski ET AL. v Kappos, Under Secretary of Commerce for Intellectual Property and Director. Patent and trademark office No. 08–964

TRIPS Agreement (1995) Agreement on trade-related aspect of intellectual property rights. https://www.wto.org. Accessed 10 Mar 2014

Tysver DA (2013) Analysis and summary of Bilski. http://www.bitlaw.com/source/cases/patent/BilskvKappos.html. Accessed 25 June 2014

Wen Y, Matsaneng T (2013) Patents, pharmaceuticals & competition: benefiting from an effective patent examination system. http://www.compcom.co.za. Accessed 19 May 2015

World Trade Organization, World Intellectual Property Organization, World Health Organization (2013) Promoting access to medical technologies and innovation—intersections between public health, intellectual property and trade. https://www.wto.org. Accessed 11 Mar 2014

Zachariades NA (2013) Stem cells: intellectual property issues in regenerative medicine. Stem Cells Dev 22(S1):59–62

Chapter 6
Ethical Considerations in Stem Cell Research on Neurologic and Orthopedic Conditions

John D. Banja

6.1   Background

Stem cells are biological products derived from various sources including aborted fetal brains, umbilical cord blood, spinal cord tissue, hematopoietic cells, immortalized stem cell lines (derived from human teratocarcinoma), adipose tissue, autologous cells (derived from bone marrow and adipose tissue), and human embryos created in laboratories or left over in clinics performing in vitro fertilization (Bliss et al. 2007; Bahney and Miclau 2012; Schmitt et al. 2012). Their interest to contemporary clinical research cannot be exaggerated in light of their combined "pluripotent" and longevity properties, i.e., their ability to change from undifferentiated to differentiated cells and proliferate in the three germ layers, i.e., the ectoderm (regulating skin and the nervous system), the mesoderm (affecting bone and muscle formation), and the endoderm (affecting organs like the lungs and digestive system). The therapeutic and regenerative possibilities stem cells represent for repairing damaged or nonfunctional tissue are breathtaking. Clinicians anticipate that the future of stem cell technologies will improve or restore function in potentially all major organ systems, while in 2013, stem cell transplantation research was already in Phase III trials for Crohn's disease, bone marrow transplant procedures, congestive heart failure, limb and myocardial ischemia, liver failure, leg ulcers, leukemia and lymphoma, retinitis pigmentosa, and cartilage disorders and defects (Sargent 2013; Pharmaceutical Research and Manufacturers of America 2013).

Although not anywhere as advanced, stem cell research for orthopedic conditions such as for bone fracture and cartilage and tendon repair have been underway for some time (Schmitt et al. 2012). Unlike human embryonic stem cells, the cells

J.D. Banja (✉)
Department of Rehabilitation Medicine, Emory University, Atlanta, GA 30322, USA

Center for Ethics, Emory University, Atlanta, GA 30322, USA
e-mail: jbanja@emory.edu

© Springer International Publishing AG 2017
P.V. Pham, A. Rosemann (eds.), *Safety, Ethics and Regulations*, Stem Cells in Clinical Applications, DOI 10.1007/978-3-319-59165-0_6

used in these studies are adult, *multipotent* cells that have limited ability to differentiate into preferred cell types in contrast to pluripotent cells (Bahney and Miclau 2012). Found in the bone marrow, hair follicles, adipose tissue, and the intestinal villus crypt, these cells nevertheless show a remarkable affinity for differentiating into cartilage, adipose tissue, and bone cells (thus their interest in orthopedics). Mesenchymal stem cells harvested from bone marrow have especially shown excellent osteogenic differentiation in animal studies, are known to home in on the kinds of injured tissue typical in fractures, and exhibit paracrine-based mechanisms of tissue healing marked by a reduction of inflammation without immune rejection (Schmitt et al. 2012).

Presently, no stem cell trials aimed at improving neurological or orthopedic functioning have received FDA approval as the challenges of these kinds of research are daunting (Centeno and Bashir 2015). Not only must researchers solve the problems of delivering and engrafting the cells to the correct anatomical site, but the cells must be "coaxed" to differentiate, proliferate, and, most importantly, result in the organism exhibiting a meaningful and empirically measureable structural improvement or functional change. As discussed below, the entire research trajectory—from the moment of securing stem cells for implantation, through the translational and first-in-human process, and then proceeding through clinical trials culminating in product launch—is fraught with unpredictability and ethical challenges. What follows is a brief overview and discussion of some of the more salient ones.

6.2   Stem Cells and Moral Status

Anyone even modestly familiar with ethical dilemmas surrounding stem cell research is probably aware of the most familiar one: strident moral objections to the creation of human embryos from which stem cells are derived and used. Opponents of stem cell research using human embryos understand the human embryo to have "moral status" upon its creation and argue that it must be accorded fundamental human rights from the moment of its conception. As such, the removal of the embryo's stem cells at the blastocyst stage (around the third to fifth day of embryogenesis) resulting in embryonic death constitutes, to them, a frank homicide (Robertson 2010). Indeed, the National Institutes of Health will not fund stem cell research wherein the cells are derived from human embryos created specifically for research purposes (National Institutes of Health 2009). NIH-funded researchers using human embryonic cells must secure these cells either from the NIH Registry or from clinics or facilities where the cells were originally intended for reproductive purposes and are no longer needed (National Institutes of Health 2009). Consequently, even if one hesitates to confer "moral status" at the blastocyst stage, positions like the NIH's resonate with the "symbolic significance" of the human embryo, given its potential for eventually attaining inarguable moral status along with the fact that all of human life begins with a fertilized ovum that proceeds through the expression of its embryonic genes (Tauer et al. 2014).

In 2006, Shinya Yamanaka and his colleagues discovered how to develop human-induced pluripotent stem cells (iPSCs). iPSCs are derived from adult skin cells which Yamanaka and his colleagues figured out how to genetically manipulate and alter to behave like human embryonic stem cells (Takahashi and Yamanaka 2006). Because they do not require the creation of a human being, i.e., embryo, and exert no compelling claim to moral status, many (especially conservative) ethicists hailed the technological discovery and pronounced the problem of moral status resolved (Hyun 2014).

Unfortunately, that scientific enthusiasm and moral relief turned out to be premature. Some scientists pointed out that small, still poorly understood differences—which can nevertheless have colossal clinical repercussions—are known to occur in gene-expressed patterns of iPSCs. They may not behave enough like human embryonic stem cells and may ultimately yield clinically disappointing results. Other, albeit preliminary, studies have suggested that iPSCs retain some "epigenetic memories" of their original somatic cells that could interfere with their differentiating into preferred cell types (Kim et al. 2010). Still other researchers point to the amount of manipulation required in creating iPSCs and worry that those manipulations can cause undesirable and harmful mutations in resulting cells (Hyun 2014).

Only time will tell whether iPSCs represent therapeutically realistic alternatives to human embryonic stem cells, and we will probably witness trials comparing results from the two for quite some time. It nevertheless seems a virtual certainty that any ethical resolution of the more fundamental ideological debate regarding when to ascribe moral status is unlikely anytime soon. Obviously, the question isn't an empirical or factual one that can be scientifically settled but rather a valuative one to which different people will bring different points of view. Because those points of view are themselves nested in and informed by larger and even deeper networks of fundamental moral beliefs, feelings, and intuitions that resist scientific proof, the debate may go on endlessly (Lakoff 2008).

Interestingly, the use of mesenchymal stem cells for orthopedic conditions may avert these ethical worries as they are adult (often autologously donated) cells, do not require the creation of an embryo, and, hence, have no claim on moral status. However, because their ability to differentiate is not as elaborate as pluripotent, embryonic cells, only time will tell whether they can deliver satisfying, clinical outcomes. Some investigators, for example, have derived mesenchymal cells or "chondrogenically committed cells" from human embryonic cells for cartilage tissue regeneration (Hwang et al. 2008; Lian et al. 2007). If that technique consistently delivers functional outcomes that prove vastly superior to using only adult stem cells, it will reintroduce the violation of moral status controversy in orthopedic stem cell research. Alternatively, if adult stem cells can stand on their own without the supplementation of ethically controversial materials, the moral status question surrounding stem cells will seemingly be overcome.

6.3　Research Risks and Methodology

6.3.1　Informed Consent

If significant ideological differences revolve around whether neurologic research using human embryonic stem cells is or is not ethical, shouldn't investigators consider those differences in formulating their informed consent processes? Some patients who consider enrolling in trials might be concerned and perhaps decline treatment upon learning that they will receive biological products resulting from the intentional destruction of human embryos, while some clinicians might not wish to participate in research dependent on those materials. Consequently, ethicists urge that participants be informed of the origins of the transplanted materials (Lo and Parham 2010).

Furthermore, a number of stem cell investigations of neurological conditions have been launched including for stoke, Parkinson's, amyotrophic lateral sclerosis, pediatric neurodegenerative disease, and brain trauma (McMahon 2014). However, because a good deal of this research remains in early Phase I trials, research participants need to understand the nature and magnitude of risk that is present, especially given the difficulty of translating animal research to humans (Solbakk and Zoloth 2011). The literature on research risks typically observes that once stem cells (regardless of their origins) are implanted, they cannot be explanted; they might migrate to undesired and undesirable anatomical sites; the risk of oncogenesis cannot be discounted owing to the use of immunosuppressive medications and the possibility of cellular mutation; and complications such as seizures, pain, and worsened disability have been known to occur (Lo and Parham 2010). Reviewing a stem cell study of seven patients with spinal cord injury in China, neurologists from the USA noted that inappropriate sites were injected; that one patient suffered life-threatening complications including posttransplant meningitis, pneumonia, and gastrointestinal bleeding; and that no significant functional improvements could be detected in any of the patients (Dobkin et al. 2006). Another case study involving a child who had gone to a Moscow stem cell clinic reported that the child developed numerous tumors in his brain and spinal cord after receiving the transplants (Amariglio et al. 2009). Obviously, these kinds of risks are especially real and "more than minimal," prompting ethicists to call for multiple layers of research protections that will be discussed below.

6.3.2　Safety Risk in Orthopedic Research Using Mesenchymal Stem Cells

Because many stem cell trials in orthopedics not only use adult stem cells but autologous (self-donated) ones, much of the worrisome risk profile described above is increasingly thought to be eliminated. Indeed, various researchers reporting their experience with culture-expanded, adult stem cells have described an exceedingly positive safety profile.

Perhaps most compelling is the 2013 report by Hernigou et al. (2013) who followed 1873 patients treated from 1990 to 2006 with autologous bone marrow cell concentrate. The mean follow-up time was 12.5 years, and the study used magnetic resonance imaging and radiography to determine the appearance of cancer at the treatment site or elsewhere. The investigators found no tumor formation at the site of implantation. Although 53 cancers were diagnosed in other sites, it is unlikely that those tumors were treatment related, especially as that number was considerably below what the expected number would be for patients of the age and sex distribution of the research participants.

Likewise Centeno et al. (2010) found no neoplastic complications among 227 patients who were treated for various orthopedic conditions with autologous, mesenchymal cells. This is not to say, however, that no adverse events occur in such procedures, as Centeno et al. (2010) reported seven instances of probable procedure-related complications and three cases of possible stem cell complications. Yet, all turned out to be self-limiting or were treated with simple therapeutic measures. Perhaps this is not surprising as these cells are self-donated and are less likely to result in tumor formation. We might additionally mention that some types of mesenchymal cells appear to exert a tropic effect on other cells in the environment that is believed to promote tissue repair (Bahney and Miclau 2012). Also, autologous cells are manipulated considerably less than iPSCs such that the likelihood of contamination via procedure is probably dramatically reduced.

Nevertheless and as attested to by other articles in this volume, many details using these stem cells are unknown and need to be perfected, including choosing the best source for deriving them (which might vary depending on the pathology being treated), standardizing the optimum number of cells to inject and timing the injections, standardizing protocols for extracting and expanding stem cells from their original cellular environments, and standardizing the injection protocols themselves (Mautner and Blazuk 2015; Sepulveda et al. 2015).

6.3.3 The Reasonable Person Standard

A popular standard that ethicists have urged for determining the scope and content of informed consent communications is the "reasonable person" standard, meaning that disclosure of information should include what a "reasonable and prudent" person would wish to know, especially as that information might affect his or her decision to undergo some kind of treatment (Beauchamp and Childress 2013). However, because research trials with humans typically do not hold out therapeutic promise or value in their early stages but rather begin with determining safety without promising benefit, the informed consent process in research tends to be more scrupulous than it does in clinical care (Magnus 2010). As in any research trial, because human participants are being *used as a means* to gain generalizable knowledge rather than

Table 6.1 Informed consent elements in stem cell research (National Institutes of Health 2009; Hyun 2014; Macklin 1999)

Investigators should:
• Insure the participant's decision capacity and voluntariness regarding trial participation
• Describe the source or origin of the cells that will be used
• Confess uncertainty about the "fate" of the cells posttransplant, i.e., regarding the possibility of tissue rejection, undesirable cellular proliferation and migration, and tumor formation
• Describe other risks associated with the intervention such as those that are typically disclosed in neurosurgery
• Explain that the undertaking is experimental, i.e., intended to gain knowledge rather than intended to produce a therapeutic benefit
• Go over and explain the consent form's wording and vocabulary
• Characterize the intervention itself, explain its rationale, and admit that its long-term health effects are unknown
• Describe how "quality of outcome" is understood and will be measured
• Describe the study's adverse event reporting process
• Discuss the participant's likely need of long-term, invasive follow-up and that participants have the right to withdraw participation
• Note the uncertainty of long-term efficacy, even if a favorable outcome initially occurs
• Admit that the majority of stem cell participants have thus far not received a therapeutic benefit from their trial participation
• Note if the intervention is in early Phase I trials, it has never been used among persons, or it has been used in only a few
• Inform participants of the availability of participant advocates, clinical trial coordinators, and other personnel involved in the study
• Assess what the participants have understood, perhaps via post-consent quizzing or extended conversation

being treated as ends in themselves, the professional's duty to protect their autonomy is pronounced, especially given the inherent uncertainty of the research undertaking. Ethicists have pointed out that if the likely harms from participating in a stem cell trial are similar to the outcomes of the disease process itself, then the information gained from trial participation would have to be immensely useful to justify enrollment; otherwise, the trial should not proceed (Magnus 2010).

Of course, in clinical research in general, reasonable steps need to be taken to ameliorate the possibility of "therapeutic misconception," where participants might misconstrue a clinical trial as holding out a therapeutic benefit when it does not. Table 6.1 lists various items that are recommended in informed consent discussions with research participants in stem cell trials, and as will be discussed below, the specifics of an informed consent discussion will be heavily dependent on the contextual specifics and aims of the research trial itself (National Institutes of Health 2009; Hyun 2014; Kimmelman 2010).

6.3.4 Whom to Recruit and When to Intervene?

Because the possibility of serious harm befalling research participants in stem cell trials is well recognized and because the probability and magnitude of that harm might not be well known—often owing to a wide "translational gap" between non-human and human stem cell studies (Kimmelman 2010; Hess 2012)—clinical researchers may find themselves with a number of vexing decisions regarding their selection of subjects in Phase I trials. For example, some ethicists have urged that early Phase 1 stem cell trials only enroll participants for whom no current treatments are available, primarily because those trials pose the possibility of serious risk materialization that cannot be justified if some therapeutic remedy is available (Magnus 2010). The problem with that suggestion, however, is that in the process of exhausting the spectrum of standard clinical treatments, a patient's disease and its underlying pathophysiology can become more entrenched and resistant to intervention. Pascale Hess (2012) has reported on just such a scenario involving stem cell interventions among children with Batten diseases (infantile or late-infantile neuronal ceroid lipofuscinoses). Although the study's best case scenario would witness a child's disease process being stabilized, a final report on the study noted that no degree of efficacy could be measured, largely because the disease had decimated the participant's brain cells leaving only a limited number of cells left to protect (StemCells 2013).

Another problem associated with exhausting available clinical remedies is that the disease process might ultimately compromise the individual's capacity to give informed consent. While some study protocols might accept the consent of a legally authorized person to serve as a proxy for the individual, other studies might be required by their IRBs to only proceed with the participant's consent (HHS.gov Office for Human Research Protections 2009).

Alternatively, individuals who are in the earlier stages of a neurological disease might not be ideal candidates if their symptoms are not severe. Research ethicists commonly hold that the risk/benefit ratio of a risky trial requires enrolling sicker patients who are thought to have less to lose in the event of serious complications (Hess 2012). Yet, if a particular disease is aggressive and ultimately fatal (such as the Batten diseases), then an earlier rather than later intervention would seem reasonable—an observation that only confirms the recommendation of adjusting enrollment, interventional timing, and informed consent per contextual parameter of a study's focus and objectives.

It is worth pointing out that participant recruitment in orthopedic trials using stem cells can be controversial as well. Decisions over using autologous versus allogeneic cells can be difficult, given the (1) age and biological condition of patients, (2) their presenting conditions, and (3) the likely (or unpredictable) clinical course of their injuries or diseases. Furthermore, orthopedic pathologies can be enormously heterogeneous—ranging from modest tissue injury to extremely serious disease like osteogenesis imperfecta—not to mention the biogenetic uniqueness of each patient and how that uniqueness affects long-term outcomes (Mautner and Blazuk 2015; Sepulveda et al. 2015).

In timing a stem cell research intervention, all of these considerations would need to be taken into account, which only illustrates the importance of continuing research informing weighty ethical decisions at least involving a patient's disease severity, its projected rate of progression, and the participant's quality of life and (4) life expectancy (Hess 2012).

6.3.5 Efficacy Testing in Phase 1 Trials?

Given these decidedly risk-laden considerations which recall the fundamental and nonnegotiable duty of protecting research participants from harm, some ethicists have recommended that Phase I trials not only test for safety but for efficacy as well (Hess 2012; Gilbert et al. 2012). This would mean that these Phase I trials would instantiate *clinical* endpoints along with safety endpoints as part of the protocol's outcome measures. Of course, efficacy must be present in studies that present "more than minimal risk" to children (Code of Federal Regulations 2014).

A recommendation of combining efficacy testing with safety testing would require that animal studies show compelling and meaningful clinical endpoints that better "close" the translational gap between nonhuman and first-in-human studies (Kimmelman 2010). Indeed, the *Guidelines for the Clinical Translation of Stem Cells* drafted by the International Society for Stem Cell Research emphasizes that efficacy and safety after delivery of the cells need to be previously demonstrated in appropriate animal models (National Institutes of Health 2009). Furthermore, the validity of any stem cell trial would be in doubt if it fails to provide an adequate scientific rationale, has insufficient preclinical evidence of efficacy and safety, fails to describe and justify the characteristics of the cells that will be delivered, and fails to describe the mode of cell delivery and clinical follow-up (Hyun 2014).

Still for those insisting on closing the translational gap between animals and humans, it remains the case that the tissue morphology is different in humans than in animal models, disease and healing mechanisms are different, and the neural circuitries are different. Furthermore, laboratory-induced injury in animal models tends to be much more anatomically uniform and precise than the myriad of orthopedic and neurological injuries (and their underlying pathophysiologies) which humans present, while healing and recovery rates are different in humans and animals. For example, rats have been known to show a much greater degree of spontaneous recovery from neurological injury, especially spinal cord injury, than humans, whose degree of axonal regeneration required for functional improvement is considerably greater. Last, objectively comparing, assessing, and translating *functional* recovery from animal models to humans—considering that animal species evolved their functionalities in an environment of survival challenges much different from humans—might require Solomonic wisdom (Magnus 2010).

6.3.5.1 The "Tragedy of Translation"

These factors have prompted some scholars to characterize "first-in-human" use in embryonic stem cell research as a "tragedy of translation" (Solbakk and Zoloth 2011). Not only do we have a "leap into the dark" with first-in-human trials due to the translation gap, but investigators may import their subjective and inevitably self-interested judgments in assessing the degree of harm probability and magnitude present in a trial. For example, a diagnosis of amyotrophic lateral sclerosis may seem a death sentence and prompt patients and their research investigators to recklessly pursue dangerous or unprecedented trials, such as the above instances of "stem cell tourism" illustrate. Yet, that decision could turn out to aggravate these patients' conditions with other maladies and make them worse off than had they done nothing.

In light of these conundrums and the fact that functional recovery in humans enrolling in stem cell trials would probably not be evident for months or even years, some ethicists have recommended a duty of "fidelity" from the research community (Solbakk and Zoloth 2011). Such a duty would entail an abiding relationship, akin to the kind witnessed in prolonged courses of clinical treatment, that would be extended to research participants. Adding a duty of fidelity to the extant duty of nonmaleficence would alter the traditional understanding of research participants being "used as a means" to confirm a hypothesis or gaining generalizable knowledge to treating participants as ends in themselves who require a degree of understanding, care, and empathy that extends beyond the traditional investigator-participant relationship.

6.3.6 Sham Surgery?

A very interesting problem in designing stem cell trials in both neurologic and orthopedic research concerns the search for a realistic placebo, given the dramatic nature of these interventions. Some researchers oppose a sham surgery placebo (where burr holes or incisions are made into the body) as unnecessary and a needless invitation to increased risk, especially as functional recovery from stem cell implantation requires a much longer time than placebo-related benefits typically last (Macklin 1999). Other scholars argue that a placebo control nevertheless heightens the rigor of the study, especially if participants are receiving or undergoing other interventions such as exercise (Lo and Parham 2010). They worry that the absence of placebo controls would invite an unacceptably high risk of false-positive findings not to mention, of course, that double blinding would be impossible (Kim et al. 2005).

The issue is of particular interest in orthopedics. In a review of sham surgery in orthopedics, Mehta et al. (2007) favored a sham intervention if skepticism existed over the therapeutic value of a treatment (especially versus a placebo), the benefits of the intervention might be due to the "experience of surgery" or to postoperative care, no superior therapy was available, and the sham surgery risks were reduced as much as possible. These observations recall Sihvonen et al.'s (2013) study of 146 patients, 35–65 years of age, who had knee symptoms consistent with a degenerative medial meniscus tear and no knee osteoarthritis. The patients were randomly assigned to arthroscopic partial meniscectomy or sham surgery. The investigators reported that the outcomes after arthroscopic partial meniscectomy were no better than those after a sham surgical procedure. Of course, research participants enrolled in a trial that includes a sham surgery arm must be informed of such, so it is somewhat reassuring that at least one study of patients with Parkinson's disease reported that the majority were willing to participate in such a trial (Frank et al. 2008).

In any event, implementing a sham surgery control will require as much risk minimization as possible, which would primarily focus on reducing infection and anesthesia risks. Prudence might additionally dictate that Phase I and II studies use the best medical therapy available in control groups, while Phase III trials incorporate placebo surgery. Hurst et al. (2015) has raised interesting questions as to whether a (placebo) intracranial injection might itself affect the natural course of certain neurological disorders like Parkinson's. Nevertheless, she believes that the degenerative but prolonged course of certain diseases like Parkinson's seems to argue for a late stage, Phase III sham surgery so as to decisively counter concerns over controlling for a placebo effect and for effects from other treatments that patients will likely be having.

6.4 Hype and Monitoring Stem Cell Trials

6.4.1 Hype

Investigators doing stem cell research face critical ethical responsibilities deriving from the way such trials are highly innovative, little research experience among humans exists, and patients are sometimes enrolled with serious, untreatable disease and are desperate to try anything. Yet and despite the "frontier" quality of this research (Magnus 2010), commentators have complained that Phase I trials in gene transfer research—another frontier technology—have sometimes used misleading language that conveyed a hope of therapeutic benefit to prospective participants, e.g., "In this study, a team of physicians, and scientists will treat your [disease] by delivery of a pair genes to your [organ]" (Kimmelman and Levenstadt 2005) or "The investigators hope that gene therapy will be an effective treatment for your disease" (King et al. 2005). Of course, in neither instance was the clinical treatment of the research participant's disease the fundamental object of research concern. The phenomenon of "stem cell tourism" is particularly on point in that at least one study described how many of these clinics exaggerate the benefits of their therapies and dismiss associated risks (Lau et al. 2008). Perhaps not surprisingly, none of these clinics volunteered to an adequate peer review of their products' scientific rationale, safety, or efficacy (Hyun 2014).

Although clinical investigators may have learned to confine their reports to precisely what they have discovered and can scientifically demonstrate, innovative research can witness additional voices that don't always exhibit such restraint. Venture capitalists, for-profit technology development centers, disability advocates, and especially journalists might misunderstand, exaggerate, or distort investigators' findings that then shape public opinion. Given the likelihood that some desperate orthopedic and neurological patients may fall prey to reports that exaggerate research findings, investigators should take pains to explain their findings precisely and anticipate and correct misunderstandings. To the extent that scientific hype goes unaddressed, one can only expect a diminished trust in the scientific enterprise and an erosion in its claim to integrity.

6.5 Monitoring

An appropriate review and monitoring of neurological stem cell trials is ethically required to protect participants, assure the integrity of data, and protect institutions involved in the research. Investigators performing stem cell research, whether it is publicly or privately funded, are encouraged and sometimes required by their institution's institutional review board to have their proposals reviewed and approved by the Stem Cell Research Oversight (SCRO) committee (National Institutes of Health 2009). Such committees are usually comprised of an interdisciplinary team of professionals including scientists, ethicists, legal experts, and community members. Not only do such committees pay considerable attention to the quality of the informed consent process, but they also examine and assess safety issues such as cell processing and manufacture, standards for preclinical testing using animal models, fair and transparent enrollment procedures, the scientific rationale of the protocol, the translatability of in vitro and in vivo preclinical studies, and the risk of unexpected cell function, migration, proliferation, and tumorigenesis (Hyun 2014; Sugarman 2010). Obviously, committee members need to boast considerable expertise in stem cell trials and, as much as possible, be immune from conflicts of interest.

A significant monitoring problem concerns the coordination of monitoring sites, which can include an institution's IRB, a SCRO, and the FDA. Furthermore, some institutions might insist on additional and more refined reviews that, for example, will scrutinize a proposal's ethical implications and conflict of interest dimensions. Lo and Parham (2010) have suggested that institutions might consider a stem cell monitoring model akin to the Recombinant DNA Advisory Committee (RAC) that the NIH developed for gene transfer research. RAC review is required by the NIH for investigators using NIH funding or using vectors or transgenes developed with NIH funding. They are comprised of national experts in relevant scientific fields and occur in addition to IRB and FDA review. The emphasis of the RAC review is typically on the protection of research subjects including their selection, dose escalation, and selection of safety endpoints (Lo and Parham 2010).

Another monitoring model is the Centralized IRB Initiative (CIRB) developed by the National Cancer Institute (NCI). CIRBs typically review multisite oncology trials and provide facilitated reviews. IRBs can approach CIRBs for such a review and use it to inform how they will proceed, e.g., require a full IRB review or accept the CIRB review. The goal is to avoid duplicative reviews that can delay IRB approval. Furthermore, a centralized review entity, like the CIRB, can provide a highly reliable, consistent review; have greater transparency than local, individual IRBs; and especially deploy the benefits of an institutional memory of prior proposals, their strategies, challenges, resolutions, and follow-up considerations (Lo and Parham 2010).

6.6   Innovative Stem Cell Therapies Outside of Clinical Trials

Securing an ethically and institutionally adequate system of monitoring is heightened by FDA changes permitting a "compassionate use" and off-label prescribing of FDA-approved stem cell products (Hyun 2010). The FDA has enacted these allowances to improve access to investigational drugs for patients with serious life-threatening disease who lack therapeutic options. Physicians can request FDA permission to administer stem cell products as long as they are being tested elsewhere in a clinical trial and as long as such use will not interfere with research investigations. Importantly for monitoring purposes, however, physicians requesting expanded access must submit an application that describes the rationale for intended use, patient selection criteria, a description of the manufacturing facility, the method of administration to the patient, toxicology information, and assurance of IRB approval (Hyun 2010).

Notice that the intention in either compassionate use or off-label prescribing is a therapeutic one and is not considered research, even though the FDA requires IRB approval. Ethicists therefore worry as to whether local IRBs will have the expertise to perform such reviews, especially as the interventions involve the administration of innovative therapies. Obviously, the basic activity of an IRB is to evaluate and assure sound research, not patient care. Consequently, additional levels and varieties of review seem to be required in instances of stem cell interventions outside of the parameters of clinical trials, given the host of ethical vagaries of stem cell interventions present. Centeno and Bashir (2015) describe some of the ethical and regulatory challenges that almost inevitably beset frontier research in ways that blur the definitions of what is or isn't a "drug," medical practice versus research, and off-label or compassionate use prescribing.

6.7   Conflict of Interest

It would be extraordinarily naïve to think that investigators who are developing stem cell research technologies are blissfully unaware of the intellectual property rights their inventions and discoveries represent, along with the marketplace valuations their technologies or deliverables might attract. Because the last half-century has witnessed the emergence of the entrepreneurial scientist, massive (and, many think, oppressive) regulatory measures have been implemented by institutions conducting such research to insure that investigators' "conflicts of interest" are managed in ethically acceptable ways.

"Conflict of interest" is a term of art that characterizes a situation wherein a professional's secondary interests may compromise his or her primary interests (Thompson 1993). While these secondary interests can include friendship considerations, career advancement, or funding competition, they usually involve the potential promise of significant financial gain that can mar or obscure an individual's primary obligations (Schofferman and Banja 2008). In matters of research, those primary obligations at least include protecting the welfare of research participants, the integrity of research data, and the integrity and reputation of the investigator's institution (AAMC 2008).

Imagine the kinds of conflicts a clinician-researcher who is evolving a novel drug or device that has enormous market potential might encounter. The urge to present his or her data, discoveries, or inventions in the best light possible might be immense, such that the investigator might be tempted to enroll only those participants who will likely exhibit the best outcomes, fail to inform them about the risks inherent in trial participation so as to ensure their enrolling, fail to disclose his or her potential for economic gain, "torture" (or, worse, falsify or fabricate) data such that they impress with remarkably compelling findings, or simply overstate what the data actually show, reminiscent of the above discussion on hype.

While a treatment of these issues exceeds the scope of this article, we can note that institutions typically implement a number of management strategies when such conflicts are perceived. Usually, these strategies are calibrated according to the intersection of the investigator's potential material gain, the investigator's proximity to and involvement in the research activity itself, and the stage or maturity of the project's development (AAMC 2008). Thus, a clinician-scientist who owns a substantial equity in a stem cell start-up company might become tremendously wealthy from a successful product, but if he or she is entirely removed from the research endeavor itself, there is no conflict to be managed. Conversely, the investigator who resides in the research trenches and is enrolling subjects, doing informed consents, collecting and analyzing data, etc., but who has no financial interest whatsoever in the outcome would likewise have no conflict. Also, an investigator who has much to gain from a research project in which he or she has a substantial financial interest but is engaged in very basic (or "immature") research that must proceed well into the future before anything of marketplace value materializes would be deemed to have only a modest conflict if any. Consequently, only when the potential of personal or material gain significantly intersects with the researcher's performance of his or her research responsibilities will a formal or institutional concern about a conflict of interest arise (AAMC 2008).

The most common regulatory practices an institution imposes on conflicted scientists include requiring the latter to declare their conflict of interest to participants and on publications. Sometimes, conflict of interest committees might limit (or prohibit) the investigator's role in enrolling or consenting patients, gathering or interpreting data, or writing up research findings. Also, it is not uncommon for academic institutions to appoint independent reviewers to oversee a research project whose investigators might witness conflicts of interest, while in the most pronounced cases, an institution might insist that an investigator place his or her equity in escrow, take a leave of absence to do the research, divest his or her corporate interest, or even sever involvement with that conflict originating corporation (AAMC 2008). Although researchers are known to bristle at such regulatory measures—since they typically deny the possibility of succumbing to any influences that would compromise their scientific objectivity or their duty to protect research participants—there is an abundance of literature attesting to the power of such variables to compromise one's objectivity (Brody 2007). The best response from investigators is to manage their conflicts in a regulatory compliant fashion and, perhaps and paradoxically, take a certain amount of pride in those conflicts as indicative of a career that represents both scientific as well as material success.

6.8 Conclusion: Chimeras, Social Justice, and Living Forever

While the above provides an overview of the more salient dimensions of ethical considerations in clinical stem cell trials, many other problems persist whose discussion exceeds the scope of this article. For example, while the above mentions the pressing problem of closing the translational gap such that we should only undertake first-in-human trials upon securing "adequate" scientific and methodological confidence from animal research, it nevertheless omits the "animal rights" argument as to whether animals should be used *at all in any* kind of research. Furthermore, research that combines stem cell with genetic transfer research resulting in "chimeras" poses unprecedented ethical problems wherein animals may acquire human traits or capacities that raise the possibility of them acquiring moral status (Robertson 2010).

There is also a clutch of justice problems attaching to stem cell research, which are shared with other types of innovative research that promise enormous benefit. For example, especially when stem cell interventions evolve into standard therapies, will a disturbingly large number of persons be denied access to these technologies based on their cost? Will certain groups of persons having access to these technologies exploit them and create a wider gap between the functional haves and have nots? Will that seeming injustice also translate into a marketplace developing around "choice" embryonic cells especially affecting women who can donate their eggs? Notwithstanding the abovementioned concerns over the moral repugnance directed at creating human embryos solely for research purposes, possible harms to women who may be tempted by the income potential of egg donation include risks associated with ovarian hyperstimulation, the long-term risks of cancer caused by repeated ovulation attempts, and the risks of surgical retrieval (George 2007).

Of course, it is hardly beyond the pale that as stem cell therapies evolve for the treatment of lost, diminished, or declining function, they will also be used for the enhancement of "normal" functioning. These technologies will be functionally regenerative or enhancing in the sense of enabling optimal, possibly staggering, functional performance for an indefinite, perhaps limitless, period of time. Stem cells may ultimately serve as a "parts replacement" if not "refinement" technology where, as mentioned in the introduction of this essay, virtually any organ system can be restored to a reasonable if not astonishing functional level. And when the day comes that stem cell interventions can be combined with gene transfer or gene modification therapies, not only living without end becomes theoretically possible, but so too does enjoying functional levels that one's original DNA could not provide.

These will be among the ethical challenges confronting future generations. Presently, stem cell research is in its infancy, and, if we have learned anything from gene transfer research, these kinds of innovative technologies require more labor, thoughtful ethical analyses, patience, and financial investment than was originally supposed. It also seems likely that "frontier" research in the twenty-first century will be depressingly slow, highly multidisciplinary, and technically difficult to do well. Yet, it seems preposterous to think that once certain kinds of stem cell interventions yield significant and meaningful functional results, humans won't exert a relentless and unstoppable demand for them, regardless of their cost and ethical complexity. At that point, we can only hope that clinicians and potential stem cell consumers will be able to exercise a sufficient degree of rationality and ethics in the use of stem cell therapies, lest they invite even greater problems and burdens than anyone could ever imagine.

Acknowledgement Preparation of this Article is Supported by the National Center for Advancing Translational Sciences of the National Institutes of Health (Award No. UL1TR000454).

References

AAMC (2008) Protecting patients, preserving integrity, advancing health: accelerating the implementation of COI policies in human subjects research. http://www.research.uky.edu/ospa/info/docs/AAMCFeb2008COIReport.pdf. Accessed 13 Oct 2016

Amariglio N, Amariglio N, Hirshberg A, Scheithauer BW, Cohen Y, Loewenthal R et al (2009) Donor-driven brain tumor following neural stem cell transplantation in an ataxia telangiectasis patient. PLoS Med 6(2):e1000029

Bahney CS, Miclau T (2012) Therapeutic potential of stem cells in orthopedics. Indian J Orthop 46(1):4–9

Beauchamp T, Childress JF (2013) Principles of biomedical ethics, 7th edn. Oxford University Press, New York, NY, pp 125–127

Bliss T, Guzman R, Daadi M, Steinberg GK (2007) Cell transplantation therapy for stroke. Stroke 38(Pt 2):817–826

Brody H (2007) Hooked: ethics, the medical profession, and the pharmaceutical industry. Rowman & Littlefield Publishers, Lanham, MD

Centeno C, Bashir J (2015) Safety and regulatory issues regarding stem cell therapies: one clinic's perspective. PM R 7(4):S4–S7

Centeno CJ, Schultz JR, Cheever M, Robinson B, Freeman M, Marasco W (2010) Safety and complications reporting on the re-implantation of culture-expanded mesenchymal stem cells using autologous platelet lysate technique. Cur Stem Cell Res Ther 5:81–93

Code of Federal Regulations, 46.404 and 46.405 (2014). Available at http://www.hhs.gov/ohrp/humansubjects/guidance/45cfr46.html. Accessed Oct 6, 2014

Dobkin BH, Curt A, Guest J (2006) Cellular transplants in China: observational study from the largest human experiment in chronic spinal cord injury. Neurorehabil Neural Repair 20(1):5–13

Frank SA, Wilson R, Holloway R, Zimmerman C, Peterson DR, Kieburtz K, Kim SY (2008) Ethics of sham surgery: perspectives of patients. Mov Disord 23:63–68

George K (2007) What about the women? Ethical and policy aspects of egg supply for cloning research. Reprod Biomed Online 15(2):127–133

Gilbert F, Harris AR, Kapsa RMI (2012) Efficacy testing as a primary purpose in Phase 1 clinical trials: is it applicable to first in humans bionics and optogenetics trials? AJOB Neurosci 3(2):20–22

Hernigou P, Homma Y, Flouzat-Lachaniette C, Poignard A, Chevallier N, Rouard H (2013) Cancer risk is not increased in patients treatment for orthopaedic diseases with autologous bone marrow cell concentrate. J Bone Joint Surg Am 95:2215–2221

Hess P (2012) Intracranial stem cell-based transplantation: reconsidering the purpose and ethical justification of Phase 1 clinical trials in light of irreversible interventions in the brain. AJOB Neurosci 3(2):3–13

HHS.gov Office for Human Research Protections (2009) Recommendations from the subcommittee for the inclusion of individuals with impaired decision making in research. http://www.hhs.gov/ohrp/sachrp-committee/recommendations/2009-july-15-letter-attachment/index.html. Accessed 13 Oct 2016

Hurst S, Mauron A, Momjian S, Burkhard PR (2015) Ethical criteria for human trials of stem cell-derived dopaminergic neurons in Parkinson's disease. AJOB Neurosci 6(1):52–60

Hwang NS, Varghese S, Elisseeff J (2008) Derivation of chondrogenically committed cells from human embryonic cells for cartilage tissue regeneration. PLoS One 3:e2498

Hyun I (2010) Allowing innovative stem cell-based therapies outside of clinical trials: ethical and policy challenges. J Law Med Ethics 38(2):277–285

Hyun I (2014) Embryonic stem cell research. In: Jennings B (ed) Bioethics, 4th edn. Gale Cengage Learning, Farmington Hills, MI, pp 935–942

Kim K, Doi A, Wen B (2010) Epigenetic memory of induced pluripotent stem cells. Nature 467(7313):285–290

Kim SY, Frank S, Holloway R, Zimmerman C, Wilson R, Kieburtz K (2005) Science and ethics of sham surgery: a survey of Parkinson disease clinical researchers. Arch Neurol 62(9):1357–1360

Kimmelman J (2010) Gene transfer and the ethics of first-in-human research. Cambridge University Press, New York, NY

Kimmelman J, Levenstadt A (2005) Elements of style: consent form language and the therapeutic misconception in Phase gene transfer trials. Hum Gene Ther 16(4):502–508

King NM, Henderson GE, Churchill LR, Davis AM, Hull SC, Nelson DK et al (2005) Consent forms and the therapeutic misconception: the example of gene transfer research. IRB 27:1–8

Lakoff G (2008) The political mind: why you can't understand 21st-century American politics with an 18th-century brain. Viking, New York, NY

Lau D, Oqboqu U, Taylor B, Stafinski T, Menon D, Caulfield T (2008) Stem cell clinics online: the direct-to-consumer portrayal of stem cell medicine. Cell Stem Cell 3(6):591–594

Lian Q, Lye E, Suan YK, Khia Way Tan E, Salto-Tellez M, Liu TM et al (2007) Derivation of clinical compliant MSCs from CD105+, CD24- differentiated human ESCs. Stem Cells 25:425–436

Lo B, Parham L (2010) Resolving ethical issues in stem cell clinical trials: the example of Parkinson disease. J Law Med Ethics 38(2):257–266

Macklin R (1999) The ethical problems with sham surgery in clinical research. N Engl J Med 341(13):992–996

Magnus D (2010) Translating stem cell research: challenges at the research frontier. J Law Med Ethics 38(2):267–276

Mautner K, Blazuk J (2015) Where do injectable stem cell transplants apply in treatment of muscle, tendon and ligament injuries? PM R 7(4):S33–S40

McMahon DS (2014) Regenerative medicine. In: Jennings B (ed) Bioethics, 4th edn. Cengage Learning Center, Farmington Hills, MI, pp 2693–2698

Mehta S, Myers TG, Lonner JH, Huffman R, Sennett BJ (2007) The ethics of sham surgery in clinical orthopedic research. J Bone Joint Surg Am 89(7):1650–1653

National Institutes of Health (2009) National Institutes of Health guidelines on human stem cell research. http://stemcells.nih.gov/policy/pages/2009guidelines.aspx. Accessed 13 Oct 2016

Pharmaceutical Research and Manufacturers of America (2013) Medicines in development. Biologics 2013 report. http://www.phrma.org/sites/default/files/pdf/biologics2013.pdf. Accessed 13 Oct 2016

Robertson JA (2010) Embryo stem cell research: ten years of controversy. J Law Med Ethics 38(2):191–203

Sargent B (2013) Fifteen cell therapies/stem cell therapies in Phase III clinical trials. Cell Culture Dish http://www.cirm.ca.gov/sites/default/files/files/agenda/Fifteen_Cell_Therapies_PhaseIII_Clinical_Trials.pdf. Accessed 13 Oct 2016

Schmitt A, van Griensven M, Imhoff AB, Buchmann S (2012) Application of stem cells in orthopedics. Stem Cells Int 2012:394962

Schofferman J, Banja J (2008) Conflicts of interest in pain medicine: practice patterns and relationships with industry. Pain 139:494–497

Sepulveda F, Baerge L, Micheo W (2015) The role of physiatry in regenerative medicine: how will physiatry change the world of regenerative medicine? How will physiatrists fit in? PM R: 7(4 Suppl):S76–S80

Sihvonen R, Paavola M, Malmivaava A, Itala A, Joukainen A, Nurmi H et al (2013) Arthroscopic partial meniscectomy versus sham surgery for a degenerative meniscal tear. N Engl J Med 369:2515–2524

Solbakk J, Zoloth L (2011) The tragedy of translation: the case of "first use" in human embryonic stem cell research. Cell Stem Cell 8(5):479–481

StemCells (2013) StemCells, Inc. announces results of long-term follow-up study in Batten Disease. http://investor.stemcellsinc.com/phoenix.zhtml?c=86230&p=irol-newsArticle&id=1866312. Accessed 13 Oct 2016

Sugarman J (2010) Reflections on governance models for the clinical translation of stem cells. J Law Med Ethics 38(2):251–256

Takahashi K, Yamanaka S (2006) Induction of pluripotent stem cells from mouse embryonic and adult fibroblast cultures by defined factors. Cell 126(4):663–676

Tauer C, Master Z, Campo-Engelstein L (2014) Embryo research. In: Jennings B (ed) Bioethics, 4th edn. Cengage Learning, Farmington Hills, MI, pp 923–935

Thompson D (1993) Understanding financial conflicts of interest. N Engl J Med 329:573–576

Chapter 7
New Regulatory Pathways for Stem Cell-Based Therapies: Comparison and Critique of Potential Models

Barbara von Tigerstrom

7.1 Introduction

The regulation of stem cell-based therapies is challenging in many respects. The relevant science is complex and rapidly evolving, and traditional regulatory frameworks need to be adapted to address their unique safety, efficacy and quality issues. At the same time, public interest in obtaining faster access to these innovative therapies has led some to question the appropriateness and even the legal authority of US Food and Drug Administration (FDA) regulation of stem cell-based therapies, while others urge strict regulation and stronger enforcement. Given these tensions and the complexity of these therapies, it is difficult to find a regulatory balance that will adequately promote safety, efficacy and quality while at the same time allowing responsible innovation and access to therapies. Although FDA's authority to regulate in this area was reaffirmed by a court decision, debate continues as to whether and how it should exercise this authority.

Within the context of this broader debate, this article examines recent attempts in other jurisdictions to craft specific provisions allowing additional flexibility in regulating cell and tissue therapies. Europe now has a specialized regulation for "advanced"

An earlier version of this chapter was previously published by von Tigerstrom B (2015) "Revising the Regulation of Stem Cell-Based Therapies: Critical Assessment of Potential Models". *Food and Drug Law Journal* 70(2):315–337. Material from that article is reprinted with permission from the Food and Drug Law Institute. This work was funded by the Stem Cell Network (Public Policy Core Project Grant "From Banking to International Governance: Fostering Innovation in Stem Cell Research") and by the Government of Canada through Genome Canada and the Ontario Genomics Institute (OGI-064). The author gratefully acknowledges research assistance by University of Saskatchewan law students Jordan Richards (J.D. 2014), Emily Harris (J.D. 2015), Kate Rattray (J.D. 2015) and Kimberley Osemlak (J.D. 2017) and helpful suggestions from anonymous reviewers.

B. von Tigerstrom (✉)
College of Law, University of Saskatchewan, Saskatoon, SK S7N 5A6, Canada
e-mail: barbara.vontigerstrom@usask.ca

© Springer International Publishing AG 2017
P.V. Pham, A. Rosemann (eds.), *Safety, Ethics and Regulations*, Stem Cells in Clinical Applications, DOI 10.1007/978-3-319-59165-0_7

therapies, including cell therapies, and in particular a specific exemption, known as the "hospital exemption", which was intended to leave room for small-scale innovations to be regulated at a local level. At first glance, at least, this approach seems to be an attractive model for countries like the United States and Canada. Another model can be found in Australia, which exempted certain cell and tissue therapies from the regulatory scheme for biological products that was adopted in 2011. The discussion that follows will explore these examples and the lessons that can be drawn from them for the North American context.

7.2 Regulation of Stem Cell-Based Therapies in North America

7.2.1 *Overview of Regulatory Framework*

Under US law, stem cell-based therapies would be considered "human cells, tissues, or cellular or tissue-based products" (HCT/Ps), which are defined as "articles containing or consisting of human cells or tissues that are intended for implantation, transplantation, infusion, or transfer into a human recipient" (21 CFR § 1271.3(d) (2015)). These are divided into two main categories, known as "Section 351" and "Section 361" HCT/Ps. (Public Health Service Act, ch. 373, §§ 351, 361, 58 Stat. 682, 702–03 (1944), codified as amended at 42 USC §§ 262(a), 264 (2012)). Section 361 HCT/Ps are subject only to the requirements in the HCT/Ps regulations, which include establishment registration and product listing, donor eligibility and current Good Tissue Practice (cGTP), while Section 351 HCT/Ps must also comply with these requirements but are subject to additional regulation as drugs or biological products.

Generally, any article or substance that is intended for a diagnostic or therapeutic purpose or intended to affect bodily structure or function is a "drug" as defined in the Federal Food, Drug, and Cosmetic Act (FDCA) (21 USC § 321(g)(1) (2012)). A cell therapy product would meet the definition of a drug and is also considered a biological product, which is defined in the Public Health Service Act (PHSA) to include "a virus, therapeutic serum, toxin, antitoxin, vaccine, blood, blood component or derivative, allergenic product, protein ... or any analogous product... applicable to the prevention, treatment, or cure of a disease or condition of human beings" (42 USC § 262(i)(1) (2012)). Both drugs and biological products are subject to strict controls in the form of premarket authorization and quality standards. In the case of a biological product, a biologics license must be approved before the product can be marketed, and this requires evidence (obtained from preclinical testing and clinical trials) that the product is "safe, pure and potent" (Public Health Service Act, ch. 373, § 351, 58 Stat. 682, 702 (1944), codified as amended at 42 USC § 262(a) (2012)). In addition, manufacturers of drugs and biological products are required to comply with current Good Manufacturing Practice (cGMP) standards (21 USC § 351(a)(B)).

In order to qualify as a Section 361 HCT/P, a product must:

- Be only minimally manipulated
- Be intended for homologous use (i.e., intended to perform the same function in the recipient as in the donor)
- Not be combined with another drug or device
- Not have a systemic effect or depend on the metabolic activity of living cells for its primary function, unless it is for autologous use or allogeneic use in a first- or second-degree relative or for reproductive use (21 CFR § 1271.10(a) (2015))

All of these criteria must be fulfilled for a product to be classified as a Section 361 HCT/P, so whenever the cells are more than minimally manipulated, for example, the product will be subject to the more extensive regulatory requirements applicable to drugs and biological products. Minimal manipulation is defined as "processing that does not alter the relevant biological characteristics of cells or tissues" (21 CFR § 1271.3(f)(2)). A recent draft guidance document indicates that FDA interprets this to mean that expansion of cells in culture or processing of adipose tissue to isolate cellular or nonstructural components would be more than minimal manipulation (United States Food and Drug Administration 2014a, b). Another recent draft guidance document aims to clarify the meaning of "homologous use" (United States Food and Drug Administration 2015). The regulations do not generally distinguish between autologous use (in the same individual from which the cells or tissues were removed) and allogeneic use (in a different human individual), but autologous use during the same surgical procedure is exempt from the regulations (21 CFR § 1271.15(b)). A 2014 draft guidance document states that three criteria must be met for this exemption to apply: the HCT/P must be implanted into the same person from whom they were removed, the implantation must take place within the same surgical procedure (usually understood as a single operation) and the HCT/Ps remain in their original form, without processing other than rinsing, cleansing and sizing (United States Food and Drug Administration 2014c).

Only Section 351 HCT/Ps, which are more than minimally manipulated and/or intended for non-homologous use, are subject to the full range of FDA requirements including authorization of investigational use in clinical trials, premarket approval based on clinical trial data and compliance with cGMP. Even within this classification, various forms of flexibility are available, as they are with any other drug or biologic. For example, it is possible to apply for "expanded access" to investigational drugs for individual patients or groups of patients with serious or life-threatening diseases or conditions (United States Food and Drug Administration 2016c). There are also several mechanisms to expedite approval of drugs and biological products for serious conditions (United States Food and Drug Administration 2014d).

7.2.2   Concerns and Proposals for Change

Critics of the current US regime have argued that FDA cannot or should not continue to exercise authority over all stem cell-based therapies, especially autologous therapies. Some argue that the relevant legislation should be interpreted in a way that would significantly narrow FDA's authority in this area. The current position depends on the interpretation of the term "drug" as including cell and tissue therapies and the scope of FDA authority under the federal power over interstate commerce and the PHSA. These questions were considered in litigation between FDA and a clinic formerly known as Regenerative Sciences (now the Centeno-Schultz Clinic) regarding a treatment called Regenexx-C, which involved the injection of cultured autologous mesenchymal stem cells (MSCs) for various orthopaedic purposes (von Tigerstrom 2011; Chirba and Garfield 2011; Drabiak-Syed 2013; Pivarnik 2014). FDA classified this treatment as a drug and biological product. It sought to enjoin the clinic from selling a product that it considered to be misbranded and adulterated because of the clinic's failure to comply with labelling and quality requirements. The clinic challenged FDA authority to regulate this product, arguing that it was not a drug or biological product, but constituted the practice of medicine, over which FDA has no jurisdiction. However, its arguments were rejected both by the District Court (United States District Court for the District of Columbia 2012) and by the D.C. Circuit Court of Appeals (United States Court of Appeals for the District of Columbia Circuit 2014).

In a decision rendered in February 2014, the Court of Appeals held that the mixture of cells and other substances used in the Regenexx-C therapy fell within the definition of a drug on its plain meaning and that this conclusion was not affected by the fact that there was also a procedure used to administer the mixture or that FDA regulation does affect medical practice through its control of the availability of drugs. Furthermore, the court ruled that jurisprudence on the Commerce Clause and the relevant statutory provisions (the PHSA and FDCA provisions described above) supported FDA regulation of the mixture even though it was only used within the State of Colorado. Finally, the court rejected arguments that the Regenexx-C therapy should be exempt from labelling and manufacturing requirements. The record, including the appellants' own concessions, showed that expanding the cells in culture did affect their characteristics and, therefore, constituted more than minimal manipulation; accordingly, classification under 21 CFR 1271(10)(a) was not available. The application of the Part 1271 Regulations to autologous stem cell therapies was upheld, since, contrary to the appellants' claim, there was evidence that these do carry a risk of transmission of communicable diseases.

The appeal decision in this leading case seems to settle the major questions that have been raised about the scope of FDA authority in this context. However, some also argue that FDA *should not* exercise authority over these products, even if it can lawfully do so, or that significant changes to the regulations applied by FDA are needed (Chirba and Garfield 2011; Pivarnik 2014; Bipartisan Policy Center 2015). Recommendations include general suggestions that the regulatory regime should be more flexible or that stem cell-based products—at least autologous products—should not be subject to the usual requirements for drugs or biologics, including premarket approval based on clinical trials (Chirba and Garfield 2011; Epstein 2013; Freeman and Fuerst 2012). Some argue that autologous therapies should be removed from the scope of the PHSA altogether or that they should not be regulated by FDA at all (Chirba and Garfield 2011; Epstein 2013). Another suggestion is that all autologous cell therapies should be regulated as Section 361 HCT/Ps, which would mean that they would still be subject to the regulations that are directed at minimizing communicable disease risk but would not require premarket approval (Pivarnik 2014).

One proposal that seems to be gaining some traction in the USA is the creation of a new regulatory pathway for regenerative medicine therapies that would allow for conditional approval based on preliminary evidence. Caplan and West (2014) propose a new "progressive approval" regulatory pathway for promising new regenerative medicine therapies, which would allow approval subject to post-market trials for efficacy. A similar proposal was put forward in a recent report of the Bipartisan Policy Center (2015) and taken up in a bill introduced in the US Senate in the spring of 2016 (United States Senate 2016). This new regulatory pathway would allow cell therapies to receive conditional approval for a fixed period (5 years in the Senate bill), based on preliminary evidence of safety and "a reasonable expectation of effectiveness", without doing phase III trials. During the period of conditional approval, the sponsor would be required to submit regular reports and an application for marketing approval (Bipartisan Policy Center 2015; United States Senate 2016). Those providing the therapies to patients under a conditional approval would be able to be reimbursed (Bipartisan Policy Center 2015), unlike expanded access or clinical trials, in which patients are provided with experimental treatments free of charge or on a cost-recovery basis (United States Food and Drug Administration 2016a).

In support of proposals for reform, critics argue that clinical trials (often assumed to be large-scale double-blind trials, similar to what would be required for pharmaceutical products, although most trials of cell-based therapies are quite small) are too slow and cumbersome, are inappropriate for testing these products and would be prohibitively expensive for the small operations that are developing many autologous therapies (Epstein 2013; Chirba and Garfield 2011; Caplan and West 2014). Instead, Epstein suggests that innovation should proceed through "large amounts of trial and error" and allowing "entrepreneurs (including physicians and surgeons) to take their best shot at a particular problem" without centralized oversight or control (Epstein 2013). The Bipartisan Policy Center report (2015) argues that allowing physicians to provide and be reimbursed for therapies supported by preliminary evidence will lower "financial barriers to entry" to the field, thereby "increasing the pace of innovation". Regulatory requirements imposed by FDA are seen to be impeding innovation by physicians and access to new therapies by patients (Chirba and Garfield 2011; Bipartisan Policy Center 2015). The requirement to comply with cGMP is also said to be unduly onerous in this context (Freeman and Fuerst 2012; Centeno et al. 2011).

It is difficult to gather empirical evidence that would support or refute these claims that FDA regulation is impeding innovation or access in the area of stem cell-based therapies. It has been suggested that certain types of stem cell-based therapies, such as those for neurological diseases like Parkinson's disease, may not be well suited to the traditional clinical trials model (Hyun 2010; Lindvall 2012), such that their development could be impeded if the regulatory requirements are not sufficiently flexible. The experience of Geron's abandoned clinical trial of a human embryonic stem cell-based product for spinal cord injury could also be taken to support arguments that the delay and expense associated with clinical trials pose a significant hurdle to development in this context (Baker 2011; Hayden 2014).

Another argument is that, given the nature of these products and the risks associated with them, local regulation of medical professionals and facilities is sufficient and more appropriate than federal regulation of the products themselves. Pivarnik (2014) suggests that a central rationale for FDA regulation, the information asymmetry between the provider and recipient of a treatment, is less of a concern in the context of autologous stem cell therapies, because they do not involve mass manufacture but a direct relationship between doctor and patient. Similarly, Freeman and Fuerst (2012) distinguish between an "individual consented risk", which typically exists in the context of medical procedures and requires informed consent by a patient following disclosure by the treating physician, and "mass production risk", where FDA regulation is relied upon by the physician and patient to ensure that minimal standards have been met. The risks associated with procedures used to harvest or implant cell and tissue therapies and contamination or quality issues in the facilities where they are processed are said to be similar to other medical procedures, which fall outside FDA authority, and more appropriately dealt with through local regulation (Chirba and Garfield 2011; Epstein 2013; Centeno et al. 2011). Regulation of medical professionals and facilities at the state level, along with potential tort liability in the event of injury to patients, could address many of the same risks to which FDA regulation is directed (Epstein 2013; Freeman and Fuerst 2012).

There has been much less controversy surrounding the regulation of stem cell-based therapies in Canada, where the regulatory structure is broadly similar to that of the USA, in that it includes a category of "cells, tissues, and organs" analogous to HCT/Ps and regulates products that are more than minimally manipulated or for non-homologous use as biologics (von Tigerstrom et al. 2012; von Tigerstrom and Schroh 2007). However, some have suggested that existing regulations may be too restrictive and that greater flexibility is required, in particular to balance the risks associated with novel therapies with the risks already facing seriously ill patients who have few other options (von Tigerstrom et al. 2012).

Although there are a number of outspoken critics of the current regulatory approach, there are also many voices supporting this approach and even calling for stricter enforcement (International Society for Stem Cell Research 2013; International Society for Stem Cell Research 2016a, Sipp and Turner 2012; Bianco et al. 2013; Lysaght et al. 2013), arguing that rigorous regulation is necessary and in fact serves the interests of both patients and those developing new therapies. They point to the harms that are caused by allowing unproven therapies to be marketed to patients without adequate oversight, including risks to patients' safety and the potential for financial exploitation when patients pay large sums of money for therapies of dubious value (Enserink 2006; McLean et al. 2014; Bianco and Sipp 2014; Bianco et al. 2013; International Society for Stem Cell Research 2013). Recent examples of serious harms suffered by patients following stem cell-based therapies have added weight to these concerns (Sweet 2016). Studies have shown that many clinics are making unsubstantiated claims about their products' safety and efficacy (Lau et al. 2008; Regenberg et al. 2009; McLean et al. 2014; Bianco et al. 2013; Turner and Knoepfler 2016) and that public perceptions and expectations about stem cell therapies are often out of step with current scientific evidence (Bubela et al. 2012). Although those who advocate for rigorous oversight have been criticized as lacking compassion for patients (Cattaneo and Corbellini 2014), the counterargument is that compassion is only relevant where there is a safe and effective therapy being offered (Bianco et al. 2013). Criticism of FDA regulation as a barrier to access tends to assume that effective therapies are being kept from patients, but there is little basis for this assumption (Nature 2016). Very limited preliminary evidence is sometimes uncritically offered in support of arguments for faster access; for example, the Bipartisan Policy Center report advocating for a new conditional approval pathway includes small, uncontrolled studies among those cited as demonstrating the promise of cell therapies (Bipartisan Policy Center 2015).

In addition, many are concerned that allowing ad hoc experimental treatment without adequate oversight and outside of well-designed clinical trials will undermine the advancement of stem cell science by compromising the legitimacy of field and wasting opportunities to generate high-quality evidence in an area where so much remains unknown (International Society for Stem Cell Research 2013; Bianco et al. 2013; Sipp and Turner 2012; International Society for Stem Cell Research 2016b). Allowing therapies to be marketed without adequate evidence of efficacy undermines the incentive to invest in therapies that are actually effective (Nature 2016). The enormous potential of stem cell-based therapies can only be realized if progress in the field is based on "solid scientific evidence" (Weissman 2012). In this view, dispensing with the need for FDA oversight and approval, far from enabling progress, would be "a giant step backward" (Weissman 2012).

7.3   Models from Other Jurisdictions

In calling for reform to the US approach, some have pointed to special provisions for cell therapies in other jurisdictions as possible models (Bipartisan Policy Center 2015). For example, one of the proposals for a new conditional approval pathway cited recent changes to the regulatory framework in Japan as an example which allowed that country to "take the lead" in this field (Bipartisan Policy Center 2015). Similar to the new model proposed in the USA (Bipartisan Policy Center 2015; United States Senate 2016), the Japanese law allows for conditional approval of regenerative medicine products based on preliminary evidence of safety and a reasonable likelihood of clinical benefit, and products are eligible for reimbursement during the period of conditional approval (Sipp 2015).

The new regulatory pathway in Japan is relatively new, with only two conditional approvals as of mid-2016 (Konishi et al. 2016). More time will be needed to evaluate its impact, but concerns have already been raised that suggest the USA and others would do well to exercise caution in imitating it as a model (Nature 2016). The safety of therapies receiving conditional approval may be in question, given that preliminary evidence cannot be conclusive of safety and will not take account of long-term safety concerns (Sipp 2015). Even if the products are safe, evaluating their efficacy during the conditional approval period will be challenging and prone to bias (Sipp 2015). The cost of therapies can be reimbursed by the public healthcare system, which may involve patients paying up to 30% of the cost (Konishi et al. 2016; Nature 2015). It is not clear whether the cost—over US$100,000 for one of the first therapies approved—is justified given the limited evidence of efficacy (McCabe and Sipp 2016). The new scheme has been said to effectively shift the risk and cost of developing new therapies to patients and taxpayers in Japan and to create incentives that may actually damage the industry in the longer term (Nature 2015; Nature 2016; McCabe and Sipp 2016).

Other jurisdictions, notably Australia and Europe, have special exemptions for certain types of cell and tissue therapies that have been in place for a number of years and have been the subject of recent consultations. Lessons from these experiences could inform debate elsewhere about potential regulatory reforms.

7.3.1 Australia's Biologicals Framework and the Exemption for Autologous Cells and Tissues

In 2011, Australia adopted a new regulatory framework for biologicals, which applies to products made from or containing human tissues and cells (Therapeutic Goods Administration 2011a). Previously some of these products were excluded from regulation, while others were regulated as medicines or medical devices; it was believed that those regulations were not "a good fit" due to the "unique properties and risks" of biological products (Therapeutic Goods Administration 2011a). Some cells and tissues, such as blood and blood components or haematopoietic progenitor cells (used for haematopoietic reconstitution), are still regulated separately (Therapeutic Goods Administration 2011a; Therapeutic Goods Administration 2011d), but most cell and tissue products used for therapeutic purposes are now regulated as biologicals.

Biologicals are divided into four classes, each of which is subject to different requirements. The classification is intended to capture varying levels of risk and is based on the extent of manipulation involved and the closeness of the intended use of cells or tissues to their original biological function (Therapeutic Goods Administration 2011a). Classes 1 and 2 are the lowest-risk categories. Class 1 biologicals may be designated as such by regulation and simply require submission of a statement of safety and compliance with applicable standards (*Therapeutic Goods Act 1989* (Cth) s 32DA). There are no biologicals designated as Class 1; Trickett and Wall (2011) explain that it is "an empty category ... previously populated by some fresh products which now are defined as excluded products". Class 2 biologicals are minimally manipulated cells and tissues for homologous use; Class 3 are more than minimally manipulated (for homologous or non-homologous use) and Class 4 are manipulated in a way that alters "an inherent biochemical, physiological or immunological property" (again for either homologous or non-homologous use) (Therapeutic Goods Administration 2011a). Classes 2, 3 and 4 must be evaluated before registration based on a dossier submitted by the manufacturer (*Therapeutic Goods Act 1989* (Cth) s 32DD), which for all three classes will include compliance with GMP (Therapeutic Goods Administration 2013a). Classes 3 and 4 are also evaluated for safety, efficacy and quality (Therapeutic Goods Administration 2011a); therefore, their product dossiers are much more extensive; these two classes essentially form a "spectrum ... where the position (between Class 3 and Class 4) is somewhat vague but ... considered on a case by case basis by the TGA", with the dossier requirements increasing according to the perceived level of risk (Trickett and Wall 2011). Donor selection and testing standards also apply to all of these classes of human cell and tissue products (Therapeutic Goods Administration 2013b).

Even with this flexible, risk-based framework, it was determined that certain products should be excluded from the scope of the new provisions. Among the excluded goods are "fresh viable human organs or parts of human organs" and "fresh viable human haematopoietic progenitor cells" to be used for "direct donor-to-host transplantation", as well as "reproductive tissue for use in assisted reproductive therapy" (Therapeutic Goods Administration 2011c). In addition, there is an exclusion for human cells and tissues that are:

i. Collected from a patient who is under the clinical care and treatment of a medical practitioner registered under a law of a state or an internal territory.

ii. Manufactured by that medical practitioner or by a person or persons under the professional supervision of that medical practitioner, for therapeutic application and in a single course of treatment of that patient by the same medical practitioner or by a person or persons under the professional supervision of the same medical practitioner (Therapeutic Goods Administration 2011c).

According to the Therapeutic Goods Administration (TGA), this exclusion "reflects the Australian Health Ministers' Conference (AHMC) agreement that single surgical procedures and medical practice should not be regulated by the TGA" (Therapeutic Goods Administration 2011b). It applies to autologous use in a single course of treatment for a single clinical indication, which may, however, entail more than one dose or administration (Therapeutic Goods Administration 2011b). Professional supervision by a medical practitioner "requires that the medical practitioner with primary responsibility for the clinical care of the patient is party to all manufacturing steps that are performed in a formal governance arrangement with the person or persons undertaking the manufacturing. This would include input into the protocols and quality systems used in the manufacturing process" (Therapeutic Goods Administration 2011b).

The exemption means that these products are not subject to any TGA oversight whatsoever, leaving protection of patients wholly dependent on the regulation and potential liability of medical professionals and facilities. The TGA's guideline document stresses that the exclusion from TGA regulation "has no effect on the professional obligations of medical practitioners to maintain satisfactory standards of practice that are appropriate to their profession", including standards of practice and informed consent requirements (Therapeutic Goods Administration 2011b). Advertising is also regulated by the health practitioners' regulatory body and under national consumer protection legislation (Therapeutic Goods Administration 2011b).

Although this exemption has received far less attention than the European hospital exemption clause discussed below, it has given rise to significant concerns. Unlike the other exclusions (HPC, organs for transplant and reproductive material), which relate to established practices "which are already overseen by existing peer-review and accreditation processes" (Munsie and Pera 2014), the exclusion of autologous cells and tissues captures a broad range of uses, some of which are untested or unproven and carry potentially significant risks. The original intent of the exclusion was to preserve the boundary between TGA regulation and the medical profession's regulation of practice and procedures and apparently to "exempt straightforward procedures" such as grafts during surgery (Munsie and Pera 2014). However, the exclusion order as presently drafted is very broad—applying to any autologous cells and tissues, regardless of the type or degree of manipulation and the intended use—and this has given rise to consequences that presumably were unintended.

Since 2011, when the new biologicals framework and the exclusion for autologous cell and tissue products were established, there has been an "exponential" growth in private clinics in Australia marketing "stem cell" therapies for a wide range of indications, including some serious conditions (Munsie and Pera 2014; Munsie and Hyun 2014; Sipp and Turner 2012; McLean et al. 2014; McLean et al. 2015). The therapies are offered at a significant cost—up to $10,000 or more—but are not necessarily supported by medical evidence and are not subject to even minimal quality standards (Munsie and Pera 2014; McLean et al. 2014; McLean et al. 2015). As many observers have noted, there are risks associated with any unproven therapies, including financial exploitation and foregone opportunities for alternative forms of treatment or legitimate clinical trials (Munsie and Pera 2014; McLean et al. 2015). Furthermore, the exclusion in Australia could allow even therapies involving a high degree of manipulation such as reprogrammed cells (Munsie and Pera 2014) and/or non-homologous use, which would carry a significant direct risk of harm for patients, to be provided with minimal oversight. In theory, at least, professional regulation and potential liability can help to protect patients' interests in such situations (McLean et al. 2014; Lysaght et al. 2013; Zarzeczny et al. 2014), but the weaknesses of these, including lack of consistent enforcement and the post hoc nature of these mechanisms, are well recognized (Munsie and Pera 2014; von Tigerstrom 2011; Zarzeczny et al. 2014; Lysaght et al. 2013; Chan 2013; McLean et al. 2015). Certainly, moves to further develop self-regulation such as a proposed code of conduct are welcome (Tuch and Wall 2014), but they are not likely to be sufficient to address the concerns that have been identified.

Concerned members of the scientific community in Australia and elsewhere have called for stronger enforcement of professional standards and for the exclusion to be removed or more carefully tailored (Munsie and Pera 2014; Munsie and Hyun 2014; McLean et al. 2014; McLean et al. 2015). A recent coroner's report on the death of a patient following autologous stem cell therapy at an Australian clinic added weight to these concerns, recommending consideration of how to manage experimental procedures and issues of informed consent and potential conflicts of interest (New South Wales Coroners Court 2016). The TGA has begun to respond, holding two rounds of consultations on the autologous cell and tissue therapy exemption. The first discussion paper, released for consultation in 2015, acknowledged concerns regarding safety, lack of evidence of efficacy, cost, lack of adverse event reporting and inappropriate advertising (Therapeutic Goods Administration 2015). It proposed five options for regulation, including the status quo or a narrower exemption, regulating autologous cell and tissue therapies as various classes of biologicals or regulating them under the Act but exempt from registration and manufacturing requirements (Therapeutic Goods Administration 2015).

After receiving over 80 submissions expressing a range of views, the TGA released a second consultation document in August 2016 (Therapeutic Goods Administration 2016). The second paper presented revised options and discussed some broader questions such as the definition of "minimal manipulation" and implications for the biologicals framework as a whole (Therapeutic Goods Administration 2016). It noted the lack of reliable evidence about the nature of the therapies being provided, their efficacy and safety and who was providing them and acknowledged that this presented a challenge in terms of identifying a problem; this did not necessarily indicate that there was no problem and the status quo should be maintained, however, given international reports of potentially significant risks. Elaborating on the risks identified in the 2015 document, it noted that some therapies being provided involved a significant degree of manipulation and/or higher risk routes of administration and that therapies were being offered and advertised with little or no evidence supporting efficacy (Therapeutic Goods Administration 2016). The revised options address not only the scope of the exclusion and form of regulation but also whether advertising would be permitted (Therapeutic Goods Administration 2016). Interested parties were invited to comment on these during the consultation period, which closed in October 2016.

7.3.2 Europe's ATMP Regulation and the Hospital Exemption

Europe arguably has been a leader in developing specialized regulations and processes to deal with new categories of innovative cell- and tissue-based therapies, and its framework for these therapies contains some unique features. Not surprisingly, then, some North American commentators have suggested that we look to Europe for models and new strategies (Chirba and Garfield 2011; Centeno 2014).

Regulation (EC) 1394/2007 on Advanced Therapy Medicinal Products (ATMP Regulation) was adopted in 2007 and came into force at the end of 2008. It applies to three types of advanced therapy medicinal products (ATMP): gene therapy, somatic cell therapy and tissue-engineered products (Regulation (EC) 1394/2007 art. 2(1)(a)). These classifications capture cell and tissue therapies that are substantially manipulated (in a way that alters their biological characteristics) or for non-homologous use (the use other than the function the cells or tissues performed in the donor) (Regulation (EC) 1394/2007 art. 2(1)(c); Directive 2001/83/EC, 2001 O.J. (L 311), Annex 1, Part IV, para. 2). Products that fall within the definition of ATMP are required to have centralized marketing authorization from the European Medicines Agency (EMA) (Regulation (EC) 726/2004, 2004 O.J. (L136) 1, Annex), although authorization of clinical trials remains at the national level, as is the case for all medicinal products (Regulation (EU) No 536/2014, 2014 O.J. (L 158)). For the purposes of the EMA authorization, ATMP are treated as medicinal products (or combined products if they contain elements of both medicinal products and medical devices), but a specialized committee, the Committee on Advanced Therapies, provides a scientific assessment to the Committee for Medicinal Products for Human Use, which the latter committee is required to take into account in making its decision (Regulation (EC) 1394/2007 art. 8). The donation, procurement and testing of cells and tissues used in ATMP must comply with the EC Tissues and Cells Directive (Directive 2004/23/EC, 2004 O.J. (L 102) 48), but later stages such as processing of cells and tissues must comply with the requirements for medicinal products, including specific Good Manufacturing Practice (GMP) requirements for ATMP (Regulation (EC) 1394/2007 art. 5).

One of the challenges with the regulation of ATMP is that they are typically developed and produced on a small scale, often in academic settings or small enterprises where regulatory compliance may be disproportionately burdensome (Blasimme and Rial-Sebbag 2013; von Tigerstrom 2008; Pearce et al. 2014). The preamble to the ATMP Regulation states that it is intended to regulate products that are "prepared industrially or manufactured by a method involving an industrial process" (Regulation (EC) 1394/2007 para. 6). The regulation contains a "hospital exemption" that is designed to leave room for innovative use of ATMP in treating individual patients, without the need for centralized authorization. According to Article 28(2) of the regulation, the hospital exemption (HE) applies to an ATMP that is:

> Prepared on a non-routine basis according to specific quality standards, and used within the same Member State in a hospital under the exclusive professional responsibility of a medical practitioner, in order to comply with an individual medical prescription for a custom-made product for an individual patient.

The article also states that an ATMP falling within this exemption "shall be authorised by the competent authority of the Member State", and member states must ensure that "national traceability and pharmacovigilance requirements" as well as specified quality standards are equivalent to those applicable to centrally authorized ATMP. Essentially, this provision leaves regulation of products within the HE to member states, subject to minimum requirements for traceability, pharmacovigilance and quality.

Although the impetus behind it has broad appeal, the HE has emerged as one of the most contentious and difficult aspects of the ATMP Regulation (Forgó and Hildebrandt 2013). There had been opposition to the HE clause during the negotiation and adoption of the regulation (Forgó and Hildebrandt 2013; Pirnay et al. 2013), and the controversy has continued and even grown with the implementation and the use of the exemption. In a recent public consultation on the ATMP Regulation, the HE was "the topic in the consultation that triggered most responses and … where more conflicting views were manifested" (EC Health and Consumers Directorate-General 2013).

Implementation of the exemption in EU member states is still ongoing and has been quite variable. Some member states are still developing their national laws (Alliance for Advanced Therapies 2013), have only implemented the regulation in part (Forgó and Hildebrandt 2013) or have not implemented the provisions regarding HE licensing (Pearce et al. 2014; Rial-Sebbag and Blasimme 2014; Cuende et al. 2014). Some have adopted new or amended laws to give effect to the HE. In France, for example, a new law provides for the authorization of establishments or organizations to prepare ATMP under the HE; it requires applicants to show that the treatment falls within the HE, to provide information on the number of patients who might use it and to comply with practices applicable to medicines (Lucas-Samuel 2013). The UK adopted regulations in 2010 that apply to "exempt" (i.e. HE) ATMPs (United Kingdom 2010). The regulations provide for licensing of exempt ATMPs and impose requirements regarding traceability, manufacturer licensing, reporting and compliance with GMP and other safety and quality standards—a framework that is in many respects comparable to the regulation of such products by the EMA or FDA, apart from the requirement for centralized premarket authorization. The regulations also impose some restrictions that go beyond the requirements of the ATMP Regulation: they specify labelling requirements and prohibit advertising or soliciting orders for exempt ATMPs. The UK regulatory agency has released guidance documents that provide clarification on several issues, such as the relationship between the exempt ATMP category and what are referred to in the UK as "specials" (unlicensed products that are permitted to be used to meet the needs of individual patients) (Medicines and Healthcare products Regulatory Agency n.d.a) and the interpretation of "nonroutine" in the definition of the HE/exempt ATMP category (Medicines and Healthcare products Regulatory Agency n.d.b).

The variation in national implementation is one of several issues that have caused concern in relation to the HE. In a recent study of academic ATMP facilities in Europe, the "most contentious issue … was the lack of harmonization of implementation of the Hospital Exemption Clause" (Pearce et al. 2014). One point on which there does seem to be general consensus regarding the HE is that greater clarity and consistency in its application would be desirable (EC Health and Consumers Directorate-General 2013; European Medicines Agency 2016a). Apart from this, the specific concerns regarding the HE and its implementation, which are discussed below, can be grouped around three interrelated themes: questions regarding the standards that should apply to products falling within the HE, definition of the scope of the HE and the intended and unintended effects of the HE on the development and the use of ATMP in Europe.

Strict compliance with manufacturing standards such as GMP is challenging for ATMPs generally, given the nature of the products and the fact that they are often developed on a small scale by academic or other small organizations, not mass-produced by large pharmaceutical companies. These challenges are arguably even greater for ATMP falling within the HE, since they will be individualized therapies, produced on a nonroutine basis for specific patients who may have rare diseases or other specialized needs. This raises the question of how strictly the quality standards that normally apply to medicinal products should be applied to ATMP and in particular to those falling within the HE. For ATMP generally, arguments can be made in favour of a risk-based approach, which "could allow for the manufacture of a certain product even when GMP compliance is not entirely maintained as long as the safety of the product is not affected" (Forgó and Hildebrandt 2013). It has been suggested that GMP requirements applicable to ATMP were "designed for and in collaboration with pharmaceutical companies, which typically produce large batches of drugs" (Pirnay et al. 2013). However, others argue that all relevant safety and quality standards, including GMP, should apply to HE therapies (Alliance for Advanced Therapies 2013; Pfizer 2013; European Association of Hospital Pharmacists 2013). The ATMP Regulation does stipulate that quality standards (along with traceability and pharmacovigilance) under the HE should be equivalent to those for medicinal products; presumably this would include GMP as adapted for ATMP. In most member states, manufacturers of exempt ATMP must comply with GMP, though it is applied with some flexibility (Medicines and Healthcare products Regulatory Agency n.d.b, Flory and Reinhardt 2013; Cuende et al. 2014). It seems to be uncertain whether full GMP compliance should be required for HE therapies, and responses in the 2012–2013 public consultation indicated a need for greater clarity on this point (Italian National Transplant Center 2013; International Society for Cellular Therapy 2013).

For products within the scope of the HE, evidence of safety or efficacy is not required by the regulation. This accommodates the reality that in many cases, it will not be possible to have done clinical trials due to the small numbers of patients involved (Lucas-Samuel 2013). However, some individual member states have imposed safety and efficacy requirements on HE products in their national laws (Cuende et al. 2014), which could limit the exemption's use. Where efficacy and safety requirements are not applied to products within the scope of the HE, however, these products can be given to patients without having "*demonstrated* quality, efficacy, and safety" (Van Wilder 2012), which are required (albeit evaluated with some flexibility) for centrally authorized products.

A diversity of approaches and views can also be found regarding the scope of the HE. Presently, national laws and policies take different positions on several key aspects of the HE's scope and definition, leading to considerable variation across Europe (Alliance for Advanced Therapies 2013; Pearce et al. 2014; Blasimme and Rial-Sebbag 2013). In particular, the interpretation of the requirements that HE ATMP be produced on a "nonroutine basis" and "custom-made" for individual patients is unsettled. A recent "white paper" on the HE suggested that some EU member states apply liberal or "stretched" interpretations of the HE criteria (Alliance for Advanced Therapies 2013). A 2013 study found that most member states "have applied annual limits to the numbers of a specific product type that can be manufactured under an HEC licence, presumably in response to the stated requirement for 'non-routine' production in the Regulation, whereas others apply no limits" (Pearce et al. 2014). Limiting the number of products allowed under the HE has been criticized as having the potential to lead to negative consequences (Pearce et al. 2014). However, some clear and consistent way of defining "nonroutine" production is needed (Cell Therapy Catapult Ltd. 2013). The UK guidance on this point states that whether production is nonroutine will be considered separately for each individual product prepared by a given operator and will take into account the scale and frequency of production (including both the numbers of products and the time period over which they are produced) and the wider context (Medicines and Healthcare products Regulatory Agency n.d.b). This avoids the problems associated with setting fixed numbers, though the increased flexibility comes at a cost of greater uncertainty.

One specific question is whether autologous ATMP—those in which the patient's own cells and tissues used—should generally fall within the scope of the HE. It can be argued that since autologous products are by definition custom-made for individual patients, they meet the requirements for the HE (European Association of Tissue Banks 2013; Van Wilder 2012). Some submissions to the ATMP public consultation suggested that all autologous products should be exempt under the HE or a general exemption (European Association of Tissue Banks 2013; European Confederation of Pharmaceutical Entrepreneurs 2013). Much depends, however, on the interpretation of "nonroutine" production, since it is possible for autologous products to be made repeatedly according to standard protocols, for substantial numbers of patients, even though each one is custom-made in the sense that it uses the individual patient's own cells or tissues. As a result, it can be argued that "autologous products that are prepared on a regular basis fall under the ATMP Regulation and cannot be exempted only because the product itself is produced from unique material from one person" (Alliance for Advanced Therapies 2013).

The final issue regarding the HE's scope is the question of how the HE category relates to investigational ATMP and the use of ATMP in research, as well as compassionate use. The HE does not create an exemption from the laws governing clinical trials; therefore, ATMP produced under the HE should not be used in clinical trials. However, it has been reported that some national regulatory authorities "are encouraging the use of [the HE] to produce ATMP for first-in-man cases, allowing the data arising from these to be used as part of the investigational medicinal product dossier for subsequent clinical trial applications … Indeed, some [regulatory authorities] are referring to these first-in-man, compassionate-use cases as a new 'phase 0' type of clinical study" (Pearce et al. 2014). On one hand, it makes good sense to allow experiences with innovative or experimental ATMP under the HE to inform more systematic clinical investigation at a later stage. It has recently been suggested, for example, that clinical efficacy and safety data from hospital exemption ATMP should be systematically collated (European Medicines Agency 2016a). On the other hand, this creates a risk that the HE could be used to conduct small-scale studies without the oversight normally required for clinical trials or to "bypass the generation of valid clinical data" (European Organisation for Rare Diseases 2013; Erben et al. 2014).

There is also the question of how the HE relates to compassionate use (expanded access). It has been noted that the HE has been little used in the UK, where the "specials" scheme has often been used instead for access to unlicensed ATMPs (Cuende et al. 2014). The lack of clarity about the relationship between the HE and compassionate use has been referred to as "a serious and worrisome ambiguity" (Rial-Sebbag and Blasimme 2014).

The debates about the HE's standards and scope relate to larger questions about the objectives and effects of the exemption on the development of ATMP. The HE is intended to allow for small-scale innovation in response to immediate needs of individual patients, without undermining the general requirement for centralized marketing authorizations for ATMP (Lucas-Samuel 2013; EC 2014). Depending on how HE products are defined and regulated, however, it is possible that the exemption could allow those developing ATMP to circumvent the usual requirement for marketing authorizations. This was a widespread concern among industry respondents in the public consultation (EC Health and Consumers Directorate-General 2013) and was expressed again by participants at a recent multi-stakeholder meeting (European Medicines Agency 2016a). If unlicensed products under the HE are competing with licensed products, this is perceived to be unfair, particularly when the standards applied to HE products are less stringent (Alliance for Advanced Therapies 2013; EC Health and Consumers Directorate-General 2013); it is also possible that exempt treatments may have lower prices (Van Wilder 2012), which would make it even harder for licensed products to compete. This will reduce the incentive for industry to develop ATMP and apply for marketing authorizations (EC Health and Consumers Directorate-General 2012; European Medicines Agency 2016a, Blasimme and Rial-Sebbag 2013).

There are also concerns about impacts on patients, including detrimental effects on patients' access to therapies and on the safety and quality of products they receive. If the use of the HE has the effect of limiting the market for ATMP, this might "make it unaffordable to develop a centrally approved product", meaning that "certain advanced therapies will only remain available for a limited number of patients in a Member State" (Alliance for Advanced Therapies 2013). This will impede access for patients elsewhere, since ATMP produced under the HE may only be used in the same member state. Inconsistencies in regulating ATMP may "lead to patients traveling to other jurisdictions for novel therapies, both within and beyond the EU" (Pearce et al. 2014). Where member states implement the "nonroutine" requirement by limiting the number of products under the HE, patients may be forced to seek treatment in another country "simply because an arbitrary maximum number of patients have been treated in a single centre in 1 year" (Pearce et al. 2014). Furthermore, it has been suggested that "it is very hard to obtain the experience and training necessary to guarantee the best quality of work when production is only sporadic (e.g. less than 10 applications per year)" (Pirnay et al. 2013). Patients may also be put at risk due to the fact that HE ATMP are not required to meet the safety and efficacy standards that apply to centrally authorized products (Alliance for Advanced Therapies 2013; Van Wilder 2012).

Many of these risks could be mitigated by defining the HE's scope restrictively and applying rigorous standards and oversight to ATMP produced under the exemption—hence the calls for the "nonroutine" requirement to be clearly, consistently and narrowly defined and for the same standards, including GMP, to apply to HE products. However, an unduly restrictive approach could undermine the very purpose of the HE if it limits the use too much or imposes standards that are too onerous for many hospitals to meet. Different stakeholders are likely to have different views of the ideal balance. There are some specific restrictions that would likely have broad support, such as not allowing the use of the HE where an approved ATMP is available for the same indication (EC Health and Consumers Directorate-General 2013; Cell Therapy Catapult Ltd. 2013; Alliance for Advanced Therapies 2013; International Society for Cellular Therapy 2013; European Medicines Agency 2016a). However, the diversity of approaches and views on many aspects of the HE reflects a lack of consensus on what the purpose of the HE is or should be and how it should relate to other parts of the regulatory framework.

7.4 Lessons for North America

In articulating the lessons that can be drawn from these two experiences, a preliminary point that needs to be addressed is the differences in context that would affect the way that either of these options would function in the North American environment. The legal structure of each jurisdiction is distinct in important ways. First, and most obviously, the European HE would not translate well into either the United States or Canada given the lack of any equivalent at the state or provincial level of the competent authorities that regulate medicines in EU member states. In Europe, "a main motive of classification comes from the principle of sharing competence between the EU and its member states. The EU can only regulate where it has competency, in this case, in cross-border movement of products ... and where action at the EU level has added value in accordance with the subsidiarity principle applying in the field of public health" (Mahalatchimy et al. 2012). Rather than exempting products from regulation altogether, the HE provision leaves them to be regulated at the national level, a responsibility that is undertaken by the regulatory agencies in each state.

As discussed above, the variation that results from this has been problematic; in North America there is also the more fundamental difficulty that the states and provinces to which responsibility might be devolved in a federal system have laws and agencies to regulate medical practitioners and facilities but generally not to regulate products themselves. Even if the creation of such a regulatory framework at the state or provincial level was possible in the constitutional structure of each country, it would be unwise given the duplication of resources and fragmentation of expertise that such an attempt would entail. If an exemption with a scope similar to the HE were to exist in this context, it would need to function quite differently.

A possible response to this is that products within the scope of the exemption should not be regulated *as products* at all, but rather their use could be regulated only indirectly through authority over practitioners and facilities, which does exist at the state and provincial level. The result would be the withdrawal of federal regulatory authority over a larger class of products, as has been suggested by some critics of the current US framework and has resulted from the Australian exemption. The Australian experience suggests the need for caution regarding this approach, given the lack of effective and consistent enforcement that many believe is putting patients at risk. Furthermore, there is reason to believe that these challenges would only be exacerbated in the North American context. In Australia, professional regulation is somewhat more centralized, due to a national scheme that has been in place since 2010 for regulation of the medical and other health professions (McLean et al. 2014; Australian Government Department of Health 2016). This allows for a greater degree of consistency across the country, despite the fact that legal authority to regulate the professions rests at the state and territory level. In the United States and Canada, each state medical board or provincial regulatory body would be both setting and enforcing its own standards. The controversy surrounding the Texas Medical Board's Standards for Use of Investigational Agents (State of Texas 2015; Levine 2012; Drabiak-Syed 2013; Cyranoski 2011) illustrates how variations in local standards could be problematic.

One thing that seems clear is that even if an exemption from federal drug regulation were to be contemplated, the scope of the Australian autologous use exemption is too broad to be acceptable. Exempting all autologous cell and tissue products from safety, efficacy and quality oversight, regardless of their intended function or degree of manipulation, is a troubling anomaly in the context of a risk-based regulatory scheme. Even if we accept some role for clinical trial and error in the development of innovative therapies (Epstein 2013), a blanket exemption for autologous use goes too far, in terms of undermining systematic clinical research and putting patients at risk of physical harm and financial exploitation. If the exemption was initially intended to capture procedures that fall within the US "same surgical procedure" exemption (United States Food and Drug Administration 2014c), it has become clear by now that as drafted, it is unduly broad, particularly in that it allows even products that are highly manipulated or for non-homologous use without sufficient safeguards. The limits placed on this exemption in existing North American regulations are preferable and should be maintained. Recent moves by the Australian TGA to revisit the exemption and the observations made during the consultations strengthen the argument that this is not a model to be emulated. More generally, the Australian experience stands as a cautionary tale illustrating the potential risks and unintended consequences of deregulation.

The lessons from the European experience with the HE are more complex. At first glance, this experience illustrates the risks inherent in introducing a new category or exemption that creates uncertainty and variability. The amount of controversy and discussion surrounding the interpretation of the HE and its application is remarkable. When considering any change to a regulatory framework, it is always worth bearing in mind the added transaction costs of introducing additional complexity to an already complex system. This is not to say that such changes are never warranted, only that their consequences must be carefully weighed to ensure that a particular change is indeed needed and its benefits are likely to outweigh its costs.

More fundamentally, the debates regarding the terms and scope of the HE reflect two important dynamics that are at play in this context. The first is the challenge of ensuring "access" to promising therapies. Critics of the existing regulatory framework who propose to remove or exclude requirements for regulatory approval often seem to assume that access to therapies is necessarily best promoted by allowing them to be provided to individual patients with a minimum of regulatory oversight. This rather narrow and simplistic view ignores the fact that access has multiple dimensions: short-term and medium- or long-term and individual and national/regional or population wide. Each component of the regulatory framework can be seen as representing a trade-off between these dimensions. Allowing individual access to one patient in the short term can indirectly compromise access for others in the future (Chan 2013). This can be seen in recent debates regarding expanded access in the United States, where manufacturers may be reluctant to provide access to an investigational product on an individual basis if that could compromise ongoing clinical trials (Sanghavi et al. 2014; Caplan and Moch 2014), which, if successful, would ensure broad access in the future. One of the concerns about the new Japanese regulatory pathway is that it could weaken incentives to invest in the development of regenerative medicine therapies (Nature 2016). In the European context, one of the most important concerns regarding the HE is that allowing new therapies to be provided individually at the local level will provide access for those individuals while undermining future access for other patients. This can occur if widespread use of the HE removes incentives to apply for centralized marketing authorization and if localized use does not generate the systematic data needed to assess safety and efficacy. Those who assume that deregulation will improve access tend to ignore this very important point.

Modern regulatory frameworks do recognize that there are certain instances in which immediate or short-term local access can or should be prioritized over population-wide access in the longer term or where other compelling arguments can be made to vary some aspect of the usual requirements. The second, related dynamic in the European experience is the way the HE and the impetus behind it relate to recognized challenges in the regulatory scheme and other attempts to address them. The rationales behind the HE could be described as some combination of allowing individual access in compelling cases of urgent unmet medical needs; providing a faster and simpler alternative pathway for promising therapies, as compared to the usual process for marketing authorizations; allowing initial, small-scale testing before moving forward with larger clinical trials; and accommodating therapies whose anticipated market is small enough that investing in their development is not financially viable. Some of the debates surrounding the HE's use—for example, the questions of whether data collected in that context should be used in a later application for marketing authorization or whether the HE should be available as a permanent alternative to marketing authorization or simply a means of allowing preliminary use at an early stage—reflect a lack of clarity or consensus regarding the exemption's purpose.

More importantly, each of those potential purposes is reflected, albeit imperfectly, in mechanisms that already exist both in the European regulatory framework and in other jurisdictions, including the United States. Compassionate or expanded (special) access regimes and various forms of accelerated and priority approval speak directly to the issues of unmet medical needs and faster approval or individual access for promising therapies. Orphan drug policies aim to address the challenges of developing products for small markets. Recent efforts to accelerate the development of treatments and vaccines for the Ebola virus illustrate the range of existing mechanisms that can be used to facilitate faster development of high-priority products—including scientific and protocol advice, orphan drug designation, expedited review, emergency use authorizations and expanded access (United States Food and Drug Administration 2016b, European Medicines Agency 2016b)—as well as the potential for enhanced collaboration among regulatory authorities (International Coalition of Medicines Regulatory Authorities 2014). A number of jurisdictions are also developing progressive or adaptive licensing regimes, which could allow initial marketing approval based on preliminary data, followed by ongoing study and broader use (European Medicines Agency 2016c, Yeates et al. 2007; Eichler et al. 2012). Meanwhile, regulatory agencies are continuing their efforts to clarify and tailor the type and amount of evidence that is required for clinical trials or marketing of novel products like cell therapies (Committee for Advanced Therapies 2010; Kooijman et al. 2013).

Given the overlap between the HE and these various initiatives, it is not surprising that questions have arisen about how the HE should relate to mechanisms like compassionate use and that in some cases, the HE has been little used because those other mechanisms are filling the needs for which it was designed. For jurisdictions outside Europe, the central question becomes whether there is a distinct role for a specialized exemption like the HE that would provide sufficient added value to justify adding another layer of complexity to the regulatory regime. For the time being at least, the arguments for creating a new exemption or discrete regulatory pathway for stem cell-based therapies are not compelling. Informed by the Australian and European experiences—along with early indications of concern about the new Japanese conditional approval pathway—the disadvantages of formally modifying the current regime seem to outweigh the advantages. This is particularly so if we consider the forms of flexibility that already exist and could be used in this context. It is notable that even in the context of the Ebola outbreak, in which the international community faced a serious and urgent public health threat with no known effective treatment, most of the special mechanisms called for are ones that already exist, with a few modest innovations that can now be added to the list of possibilities for the use in appropriate circumstances.

A critical distinction between the proposed REGROW Act in the United States and Japan's conditional approval scheme, on one hand, and existing mechanisms such as expanded access, on the other, is that providers are able to charge or be reimbursed for therapies before they are proven to be effective. Although there are significant concerns about the availability of adequate funding for the development of innovative cell and tissue therapies (Dodson and Levine 2015), allowing payment for therapies supported by limited evidence is not the solution to those difficulties. As some have argued in the Japanese context, this simply shifts the burden from sponsors to patients, insurers, and the healthcare system and distorts incentives within the regenerative medicine sector (Nature 2015; Nature 2016; Sipp 2015; McCabe and Sipp 2016).

Few would argue that the existing mechanisms are perfect, and some—particularly expanded access and the orphan drug regime—are the subject of considerable debate and ongoing reform efforts. The way these operate, generally and in the regenerative medicine context, and whether they are meeting perceived needs adequately are matters that certainly deserve attention. A full discussion of these issues is beyond the scope of this article; here, the key point, based on an examination of the arguments and leading alternatives, is that attention should be focused on identifying and addressing room for improvement within the existing framework, rather than formulating new stem cell-specific exceptions or special regimes. There are distinctive challenges associated with developing stem cell-based therapies, but it is far from clear that this means existing mechanisms cannot be used or adapted to address concerns about access or development costs. Already some scholars have begun to examine the specific issues that might arise in using mechanisms like expanded access for stem cell-based therapies (Hyun 2010; Knoepfler 2015). We can build on this work to analyze and improve existing forms of regulatory flexibility, with potential benefits for stem cell research and beyond.

7.5   Conclusion

No regulatory framework is perfect, and in this context—as in so many others—it is tempting to assume the grass must be greener elsewhere. This examination of recent examples of regulatory flexibility for stem cell-based therapies in Australia and Europe casts doubt on this assumption; although the rationales behind each of them have some appeal, experience has shown that they have their own difficulties. Upon closer inspection it is difficult to say that either of these two examples is superior to the status quo in North America.

The Australian exemption for autologous cell and tissue therapies has created a regime that is very similar to what some critics of the US regulations have been calling for. By ignoring the level of risk associated with higher degrees of manipulation and non-homologous use, however, the exemption has been framed in a way that is too broad to be acceptable. Some might argue that there is insufficient evidence of resulting harm, but as recently acknowledged by the TGA, a responsible government need not wait for those serious harms to occur before considering corrective action. The response of regulatory bodies that might exert control over practitioners and clinics has not given much confidence that these can be relied on to prevent and redress harms. There does not seem, either, to be any evidence of a benefit to health or science that can be attributed to the exemption.

In Europe, the difficulties with interpreting and applying the HE seem almost to have overshadowed the potential benefits of the exemption. Even setting aside the complexities involved in translating the HE into the North American context, it is unclear that an exemption for small-scale hospital-based innovative therapies would be a useful addition to the various forms of regulatory flexibility that already exist. The purposes that animate the HE are reflected in other mechanisms that exist both in Europe and other jurisdictions including the United States and Canada. If these are perceived as not adequately meeting their objectives, careful examination of any shortcomings, along with ongoing efforts to apply flexible, risk-based assessment to the unique properties of cell-based therapies should be the focus of attention.

References

Alliance for Advanced Therapies (2013) Focus hospital exemption on developing innovative and safe treatments for patients. Regen Med 8:121–123

Australian Government Department of Health (2016) National registration and accreditation scheme (NRAS). http://www.health.gov.au/internet/main/publishing.nsf/Content/work-nras. Accessed 15 Oct 2016

Baker M (2011) Stem-cell pioneer bows out. Nature 479:459

Bianco P, Sipp D (2014) Sell help not hope. Nature 510:336–337

Bianco P, Barker R, Brustle O et al (2013) Regulation of stem cell therapies under attack in Europe: for whom the bell tolls. EMBO J 32:1489–1495

Bipartisan Policy Center (2015) Advancing regenerative cellular therapy: medical innovation for healthier Americans. http://cdn.bipartisanpolicy.org/wp-content/uploads/2015/12/BPC-Advancing-Regenerative-Cellular-Therapies.pdf. Accessed 15 Oct 2016

Blasimme A, Rial-Sebbag E (2013) Regulation of cell-based therapies in Europe: current challenges and emerging issues. Stem Cells Dev 22:14–19

Bubela T, Li MD, Hafez M et al (2012) Is belief larger than fact: expectations, optimism and reality for translational stem cell research. BMC Med 10:133. doi:10.1186/1741-7015-10-133

Caplan A, Moch K (2014) Rescue me: the challenge of compassionate use in the social media era. Health Aff Blog. http://healthaffairs.org/blog/2014/08/27/rescue-me-the-challenge-of--compassionate-use-in-the-social-media-era/. Accessed 15 Oct 2016

Caplan AI, West MD (2014) Progressive approval: a proposal for a new regulatory pathway for regenerative medicine. Stem Cells Transl Med 3:560–563

Cattaneo E, Corbellini G (2014) Stem cells: taking a stand against pseudoscience. Nature 510:333–335

Cell Therapy Catapult Ltd (2013) EC review of the ATMP regulation—cell therapy catapult responses. http://ec.europa.eu/health/files/advtherapies/2013_05_pc_atmp/13_pc_atmp_2013.pdf. Accessed 15 Oct 2016

Centeno CJ (2014) Clinical challenges and opportunities of mesenchymal stem cells in musculo-skeletal medicine. PM R 6:74–77

Centeno CJ, Fuerst M, Faulkner SJ et al (2011) Is cosmetic platelet-rich plasma a drug to be regulated by the Food and Drug Administration. J Cosmet Dermatol 10:171–173

Chan TE (2013) Legal and regulatory responses to innovative treatment. Med L Rev 21:92–130

Chirba MA, Garfield SM (2011) FDA oversight of autologous stem cell therapies: legitimate regulation of drugs and devices or groundless interference with the practice of medicine? J Health Biomed Law 7:233–272

Committee for Advanced Therapies (2010) Challenges with advanced therapy medicinal products and how to meet them. Nat Rev Drug Discov 9:195–201

Cuende N, Boniface C, Bravery C et al (2014) The puzzling situation of hospital exemption for advanced therapy medicinal products in Europe and stakeholders' concerns. Cytotherapy 16:1597–1600

Cyranoski D (2011) Texas prepares to fight for stem cells. Nature 477:377–378

Dodson BP, Levine AD (2015) Challenges in the translation and commercialization of cell therapies. BMC Biotechnol 15:70. doi:10.1186/s12896-015-0190-4.

Drabiak-Syed K (2013) Challenging the FDA's authority to regulate autologous adult stem cells for therapeutic use: celltex therapeutics' partnership with RNL bio, substantial medical risks, and the implications of United States v. Regenerative sciences. Health Matrix 23:493–535

EC (2014) Report from the Commission to the European Parliament and the Council in accordance with Article 25 of Regulation (EC) No 1394/2007 of the European Parliament and of the Council on advanced therapy of medicinal products and amending Directive 2001/83/EC and Regulation (EC) No 726/2004

EC Health and Consumers Directorate-General (2012) Public consultation paper on the regulation on advanced therapy medicinal products. http://ec.europa.eu/health/files/advtherapies/2012_12_12__public_consultation.pdf. Accessed 15 Oct 2016

EC Health and Consumers Directorate-General (2013) Regulation (EC) No. 1394/2007 on advanced therapy medicinal products: summary of the responses to the public consultation. http://ec.europa.eu/health/files/advtherapies/2013_05_pc_atmp/2013_04_03_pc_summary.pdf. Accessed 15 Oct 2016

Eichler H-G, Oye K, Baird LG et al (2012) Adaptive licensing: taking the next step in the evolution of drug approval. Clin Pharmacol Ther 91:426–437

Enserink M (2006) Selling the stem cell dream. Science 313:160–163

Epstein FA (2013) The FDA's misguided regulation of stem cell procedures: how administrative overreach blocks medical innovation. http://www.manhattan-institute.org/pdf/lpr_17.pdf. Accessed 15 Oct 2016

Erben RG, Silva-Lima B, Reischl I et al (2014) White paper on how to go forward with cell-based advanced therapies in Europe. Tissue Eng 20:2549–2554

European Association of Hospital Pharmacists (2013) Consultation response: public consultation paper on the regulation on advanced therapy medicinal products. http://ec.europa.eu/health/files/advtherapies/2013_05_pc_atmp/20_pc_atmp_2013.pdf. Accessed 15 Oct 2016

European Association of Tissue Banks (2013) Public consultation paper on the regulation on advanced therapy medicinal products. http://ec.europa.eu/health/files/advtherapies/2013_05_pc_atmp/21_pc_atmp_2013.pdf. Accessed 15 Oct 2016

European Confederation of Pharmaceutical Entrepreneurs (2013) Public consultation paper on the regulation on advanced therapy medicinal products: comments of the european confederation of pharmaceutical entrepreneurs. http://ec.europa.eu/health/files/advtherapies/2013_05_pc_atmp/23_pc_atmp_2013.pdf. Accessed 15 Oct 2016

European Medicines Agency (2016a) Advanced therapy medicines: exploring solutions to foster development and expand patient access in Europe. http://www.ema.europa.eu/docs/en_GB/document_library/Report/2016/06/WC500208080.pdf. Accessed 15 Oct 2016

European Medicines Agency (2016b) Ebola. http://www.ema.europa.eu/ema/index.jsp?curl=pages/regulation/general/general_content_000624.jsp. Accessed 15 Oct 2016

European Medicines Agency (2016c) Final report on the adaptive pathways pilot. http://www.ema.europa.eu/docs/en_GB/document_library/Report/2016/08/WC500211526.pdf. Accessed 15 Oct 2016

European Organisation for Rare Diseases (2013) Public consultation paper on the regulation on advanced therapy medicinal products http://ec.europa.eu/health/files/advtherapies/2013_05_pc_atmp/27_pc_atmp_2013.pdf. Accessed 15 Oct 2016

Flory E, Reinhardt J (2013) European regulatory tools for advanced therapy medicinal products. Transfus Med Hemoth 40:409–412

Forgó N, Hildebrandt M (2013) Joint conference on the impact of EU legislation on therapeutic advance. Cytotherapy 15:1444–1448

Freeman M, Fuerst M (2012) Does the FDA have regulatory authority over adult autologous stem cell therapies? 21 CFR 1271 and the emperor's new clothes. J Transl Med. doi:10.1186/1479-5876-10-60

Hayden EC (2014) Funding windfall rescues abandoned stem-cell trial. Nature 510:18

Hyun I (2010) Allowing innovative stem cell-based therapies outside of clinical trials: ethical and policy challenges. J Law Med Ethics 38:277–285

International Coalition of Medicines Regulatory Authorities (2014) .Medicine regulators to work together internationally to find innovative solutions to facilitate evaluation of and access to potential new medicines to counter Ebola outbreaks http://www.fda.gov/downloads/EmergencyPreparedness/Counterterrorism/MedicalCountermeasures/UCM412791.pdf. Accessed 15 Oct 2016

International Society for Cellular Therapy (2013) Public consultation paper on the regulations of advanced therapy medicinal products: comments from the ISCT EU Legal & Regulatory Affairs (LRA) Committee. http://ec.europa.eu/health/files/advtherapies/2013_05_pc_atmp/40_pc_atmp_2013.pdf. Accessed 15 Oct 2016

International Society for Stem Cell Research (2013) ISSCR statement on delivery of unproven autologous cell-based interventions to patients. http://www.isscr.org/docs/default-source/isscr-statements/isscr-acbistatement-091113-fl.pdf. Accessed 15 Oct 2016

International Society for Stem Cell Research (2016a) Guidelines for stem cell research and clinical translation.http://www.isscr.org/docs/default-source/guidelines/isscr-guidelines-for-stem-cell--research-and-clinical-translation.pdf?sfvrsn=2. Accessed 15 Oct 2016

International Society for Stem Cell Research (2016b) ISSCR opposes the REGROW Act. http://www.isscr.org/home/about-us/news-press-releases/2016/2016/09/15/isscr-opposes-the-regrow-act. Accessed 15 Oct 2016

Italian National Transplant Center (2013) Public consultation on regulation (EC) 1394/2007. http://ec.europa.eu/health/files/advtherapies/2013_05_pc_atmp/41_pc_atmp_2013.pdf. Accessed 15 Oct 2016

Knoepfler PS (2015) From bench to FDA to bedside: US regulatory trends for new stem cell therapies. Adv Drug Deliver Rev 82–83:192–196

Konishi A, Sakushima K, Isobe S et al (2016) First approval of regenerative medicinal products under the PMD Act in Japan. Cell Stem Cell 18:434–435

Kooijman M, van Meer PJK, Gispen-de Wied CC et al (2013) The risk-based approach to ATMP development—generally accepted by regulators but infrequently used by companies. Regul Toxicol Pharmacol 67:221–225

Lau D, Ogbogu U, Taylor B et al (2008) Stem cell clinics online: the direct-to-consumer portrayal of stem cell medicine. Cell Stem Cell 3:591–594

Levine A (2012) Improving oversight of innovative medical interventions in Texas, USA. Regen Med 7:451–453

Lindvall O (2012) Why is it taking so long to develop clinically competitive stem cell therapies for CNS disorders? Cell Stem Cell 10:660–662

Lucas-Samuel S (2013) Thérapie innovante: du cadre réglementaire européen au cadre réglementaire national. Transfus Clin Biol 20:221–224

Lysaght T, Kerridge I, Sipp D (2013) Oversight for clinical uses of autologous adult stem cells: lessons from international regulations. Cell Stem Cell 13:647–651

Mahalatchimy A, Rial-Sebbag E, Tournay V et al (2012) The legal landscape for advanced therapies: material and institutional implementation of European Union rules in France and the United Kingdom. J Law Soc 39:131–149

McCabe C, Sipp D (2016) Undertested and overpriced: Japan issues first conditional approval of stem cell product. Cell Stem Cell 18:436–437

McLean AK, Stewart C, Kerridge I (2014) The emergence and popularisation of autologous somatic cellular therapies in Australia: therapeutic innovation or regulatory failure? J Law Med 22:65–89

McLean AK, Stewart C, Kerridge I (2015) Untested, unproven, and unethical: the promotion and provision of autologous stem cell therapies in Australia. Stem Cell Res Ther 6:33–40

Medicines and Healthcare products Regulatory Agency. (n.d.a) Guidance on the UK's arrangements under the hospital exemption scheme. https://www.gov.uk/government/uploads/system/uploads/attachment_data/file/397738/Guidance_on_the_UK_s_arrangements_under_the_hospital_exemption_scheme.pdf. Accessed 15 Oct 2016

Medicines and Healthcare products Regulatory Agency. (n.d.b) Guidance on "non routine". https://www.gov.uk/government/uploads/system/uploads/attachment_data/file/397739/Non-routine_guidance_on_ATMPs.pdf. Accessed 15 Oct 2016

Munsie M, Hyun I (2014) A question of ethics: selling autologous stem cell therapies flaunts professional standards. Stem Cell Res 13:647–653

Munsie M, Pera M (2014) Regulatory loophole enables unproven autologous cell therapies to thrive in Australia. Stem Cells Dev 23(Suppl 1):34–38

Nature (2015) Stem the tide (Editorial). Nature 528:163–164

Nature (2016) False assumptions (Editorial). Nature 535:7–8

New South Wales Coroners Court (2016) Inquest into the death of Sheila Drysdale. http://www.coroners.justice.nsw.gov.au/Documents/Findings%20Drysdale.pdf. Accessed 15 Oct 2016

Pearce KF, Dickinson AM, Lowdell MW et al (2014) Regulation of advanced therapy medicinal products in Europe and the role of academia. Cytotherapy 16:289–297

Pfizer (2013) Regulations on advanced therapy medicinal products: Pfizer feedback to the European Commission on the public consultation paper. http://ec.europa.eu/health/files/adv-therapies/2013_05_pc_atmp/48_pc_atmp_2013.pdf. Accessed 15 Oct 2016

Pirnay J-P, Vanderkelen A, De Vos D et al (2013) Business oriented EU human cell and tissue product legislation will adversely impact member states' health care systems. Cell Tissue Bank 14:525–560

Pivarnik G (2014) Cells as drugs?: Regulating the future of medicine. Am J Law Med 40:298–321

Regenberg AC, Hutchinson LA, Schanker B et al (2009) Medicine on the fringe: stem cell-based interventions in advance of evidence. Stem Cells 27:2312–2319

Rial-Sebbag E, Blasimme A (2014) The European Court Of Human Rights' ruling on unproven stem cell therapies: a missed opportunity? Stem Cells Dev 23(Supp 1):39–43

Sanghavi D, George M, Bencic S (2014) Health policy issue brief: the four A's of expanding patient access to life-saving treatments and the regulatory implications. Brookings Institution, Washington DC. https://www.brookings.edu/wp-content/uploads/2016/06/The-Four-As-of-Expanded-Access-Sanghavi-2014.pdf. Accessed 15 Oct 2016

Sipp D (2015) Conditional approval: Japan lowers the bar for regenerative medicine products. Cell Stem Cell 16:353–356

Sipp D, Turner L (2012) U.S. regulation of stem cells as medical products. Science 338:1296–1297

State of Texas (2015) 22 Texas Administrative Code §§ 198.1–198.3

Sweet DJ (2016) If you see something, say something. Cell Stem Cell 19:139–140

Therapeutic Goods Administration (2011a) Australian regulatory guidelines for biologicals: Part 1—introduction to the Australian regulatory guidelines for biologicals. http://www.tga.gov.au/sites/default/files/biologicals-argb-p1.pdf. Accessed 15 Oct 2016

Therapeutic Goods Administration (2011b) Therapeutic goods (excluded goods) order no. 1 of 2011. http://www.tga.gov.au/industry/legislation-excluded-goods-order-1101.htm. Accessed 15 Oct 2016

Therapeutic Goods Administration (2011c) Therapeutic goods (things that are not biologicals) determination no. 1 of 2011. http://www.comlaw.gov.au/Details/F2011L00894. Accessed 15 Oct 2016

Therapeutic Goods Administration (2011d) Excluded goods order no. 1 of 2011: guideline for items 4(o), 4(p), 4(q) and 4(r). http://www.tga.gov.au/excluded-goods-order-no-1-2011-guideline-items-4o-4p-4q-and-4r. Accessed 15 Oct 2016

Therapeutic Goods Administration (2013a) Australian code of good manufacturing practice for human blood and blood components, human tissues and human cellular therapy products.

http://www.tga.gov.au/revised-code-gmp-human-blood-and-tissues (follow the link with the source title at the bottom-center part of the main page). Accessed 15 Oct 2016

Therapeutic Goods Administration (2013b) Therapeutic goods order no. 88—standards for donor selection, testing, and minimising infectious disease transmission via therapeutic goods that are human blood and blood components, human tissues and human cellular therapy products. http://www.comlaw.gov.au/Details/F2013L00854. Accessed 15 Oct 2016

Therapeutic Goods Administration (2015) Regulation of autologous stem cell therapies: discussion paper for consultation. https://www.tga.gov.au/sites/default/files/consult-autologous-stem-cell-150106.pdf. Accessed 15 Oct 2016

Therapeutic Goods Administration (2016) Consultation: regulation of autologous cell and tissue products and proposed consequential changes to the classification of biologicals. https://www.tga.gov.au/sites/default/files/consultation-regulation-autologous-cell-and-tissue-products.pdf. Accessed 15 Oct 2016

Trickett AE, Wall DM (2011) Regulation of cellular therapy in Australia. Pathology 43:627–634

Tuch BE, Wall DM (2014) Self-regulation of autologous cell therapies. Med J Australia 200:196

Turner L, Knoepfler P (2016) Selling stem cells in the USA: assessing the direct-to-consumer industry. Cell Stem Cell 19:154–157

United Kingdom (2010) The medicines for human use (advanced therapy medicinal products and miscellaneous amendments) Regulations 2010, 2010 No. 1882

United States Court of Appeals for the District of Columbia Circuit (2014) United States v. Regenerative Sciences, LLC, 741 F.3d 1314, 1317

United States District Court for the District of Columbia (2012) United States v. Regenerative Sciences, LLC, 878 F. Supp. 2d 248, 251

United States Food and Drug Administration (2014a) Minimal manipulation of human cells, tissues, and cellular and tissue-based products—draft guidance for industry and food and drug administration staff. http://www.fda.gov/downloads/BiologicsBloodVaccines/GuidanceComplianceRegulatoryInformation/Guidances/CellularandGeneTherapy/UCM427746.pdf. Accessed 15 Oct 2016

United States Food and Drug Administration (2014b) Human cells, tissues, and cellular and tissue-based products (HCT/Ps) from adipose tissue: regulatory considerations—draft guidance for industry. http://www.fda.gov/downloads/BiologicsBloodVaccines/GuidanceComplianceRegulatoryInformation/Guidances/Tissue/UCM427811.pdf. Accessed 15 Oct 2016

United States Food and Drug Administration (2014c) Same surgical procedure exception under 21 CFR 1271.15(b): questions and answers regarding the scope of the exception—draft guidance for industry. http://www.fda.gov/downloads/BiologicsBloodVaccines/GuidanceComplianceRegulatoryInformation/Guidances/Tissue/UCM419926.pdf. Accessed 15 Oct 2016

United States Food and Drug Administration (2014d) Expedited programs for serious conditions—drugs and biologics—guidance for industry. http://www.fda.gov/downloads/Drugs/GuidanceComplianceRegulatoryInformation/Guidances/UCM358301.pdf. Accessed 15 Oct 2016

United States Food and Drug Administration (2015) Homologous use of human cells, tissues, and cellular and tissue-based products—draft guidance for industry and FDA staff. http://www.fda.gov/downloads/BiologicsBloodVaccines/GuidanceComplianceRegulatoryInformation/Guidances/Tissue/UCM469751.pdf. Accessed 15 Oct 2016

United States Food and Drug Administration (2016a) Charging for investigational drugs under an IND—questions and answers. http://www.fda.gov/ucm/groups/fdagov-public/@fdagov-drugs-gen/documents/document/ucm351264.pdf. Accessed 15 Oct 2016

United States Food and Drug Administration (2016b) Ebola response updates from FDA. http://www.fda.gov/EmergencyPreparedness/Counterterrorism/MedicalCountermeasures/ucm410308.htm. Accessed 15 Oct 2016

United States Food and Drug Administration (2016c) Expanded access to investigational drugs for treatment use—questions and answers—guidance for industry. http://www.fda.gov/downloads/Drugs/GuidanceComplianceRegulatoryInformation/Guidances/UCM351261.pdf. Accessed 15 Oct 2016

United States Senate (2016) Bill S. 2689: A Bill to amend the Federal Food, Drug, and Cosmetic Act with respect to cellular therapies

Van Wilder P (2012) Advanced therapy medicinal products and exemptions to the regulation 1394/2007: how confident can we be? An exploratory analysis. Front Pharmacol 3:12

von Tigerstrom B (2008) The challenges of regulating stem cell-based products. Trends Biotechnol 26:653–658

von Tigerstrom B (2011) The Food and Drug Administration, regenerative sciences, and the regulation of autologous stem cell therapies. Food Drug Law J 66:479–506

von Tigerstrom B, Schroh E (2007) Regulation of stem cell-based products. Health Law J 15:175–220

von Tigerstrom B, Nguyen T, Knoppers B (2012) Regulation of stem cell-based therapies in Canada: current issues and concerns. Stem Cell Rev Rep 8:623–628

Weissman I (2012) Stem cell therapies could change medicine... If they get the chance. Cell Stem Cell 10:663–665

Yeates N, Lee DK, Maher M (2007) Health Canada's progressive licensing framework. Can Med Assoc J 176:1845–1847

Zarzeczny A, Caulfield T, Ogbogu U et al (2014) Professional regulation: a potentially valuable tool in responding to "Stem Cell Tourism". Stem Cell Rep 3:379–384

Chapter 8
Travelling Cells: Harmonized European Regulation and the BAMI Stem Cell Trial

Christine Hauskeller and Nicole Baur

8.1 Introduction

RCTs and Regulation in Europe

Randomized controlled trials (RCTs) are widely used as the gold standard of clinical innovation. Although the limits of the deductive method have often been criticized over the past decades (Cartwright 2010), and other protocols have been developed to test new potential therapies, such as hospital exemptions in the case of rare diseases (Salter et al. 2014), the RCT is still the dominant method of scientific validation in the clinic.

Europe, as a harmonized space and unified economic zone, aims for an even provision of basic services such as health and access to medicines. Likewise, all member states should be given the chance to benefit from developments in the growing biomedical sector, both in scientific and in industrial terms. Every year, ethics committees across Europe receive 4000–6000 applications for RCTs (EC 2009). Variations in the assessment criteria and practices and consequently the committees' responses to applications have raised concerns about the future of clinical trials (Cressy 2010; Hartmann and Hartmann-Vareilles 2006; Hunter 2011). Over the past 25 years, the EU has made much effort to harmonize practices in biological and medical research across its member states. Such harmonization is expected to foster equality in patient care and treatment standards but also to facilitate the even introduction of new therapeutic options and drugs as well as closer cooperation between researchers and clinicians across Europe (EC 2016). The steps taken include setting up regulatory frameworks and institutions that oversee their application and implementation.

C. Hauskeller (✉) • N. Baur
Department of Sociology, Philosophy and Anthropology, Egenis, Centre for the Study of the Life Sciences, University of Exeter, Byrne House, Exeter EX4 4PJ, UK
e-mail: c.hauskeller@exeter.ac.uk

© Springer International Publishing AG 2017
P.V. Pham, A. Rosemann (eds.), *Safety, Ethics and Regulations*, Stem Cells in Clinical Applications, DOI 10.1007/978-3-319-59165-0_8

First attempts to harmonize clinical trials across the European Union date back to the EU Clinical Trials Directive 2001/20/European Commission (Amexis and Schmitt 2011; EC 2001). Harmonization across EU member states is particularly important for multinational trials. In order to overcome the diversity of clinical ethics application outcomes, national competent authorities were set up in all European countries to streamline approval practices (Hartmann 2012). This equalized the decision-making processes at a national level, but significant differences remained between European countries. A further step towards standardization and simplification of clinical approval processes was intended by introducing the so-called Voluntary Harmonisation Procedure (VHP). This procedure in its current form aims to overcome the heterogeneous responses to a trial protocol by the different national competent authorities. It involved two of the many national competent authorities submitting a jointly agreed protocol that would then be scrutinized and eventually approved by the Paul-Ehrlich-Institut (PEI), Langen, Germany, which initiated and now conducts the VHP on behalf of the European Medicines Agency (EMA). The so-called national competent authorities, national clinical trial approval institutions established in EU countries in accordance with the Clinical Trials Directive, then only take note and approve the VHP agreed trial protocol without requesting further changes.

Furthermore, clinical trials involving stem cells within the EU have to adhere to the regulation of Advanced Therapy Investigational Medicinal Products (ATIMP). This is an amendment to the European Tissue and Cells Directive (EC 2004), which classifies all novel uses of adult stem cells, including autologous bone marrow-derived stem cells, as Advanced Therapy Medicinal Products (ATMPs). This regulation passed the EU regulator in 2007 but only came into force in 2010. All RCTs conducted in Europe are subject to oversight by the European Medicines Agency (EMA), founded in 1995, whose role is the scientific evaluation of medicines. EMA's regulatory power has gradually increased since its inception. While EMA does not eclipse the power of national regulatory bodies, it is 'increasingly responsible for regulation of the risks and benefits of newly invented pharmaceutical products in Europe' (Davis and Abraham 2011). This includes overseeing the implementation of these Directives and the scientific approval of the protocols for European clinical trials.

Our sociological study on the implementation of a clinical trial in practice identified many specific local differences between European countries that affect daily operations, but most importantly it showed how harmonized regulations affect the logistics, technical details of practice and patient recruitment in a trial. Below we illustrate examples from these findings, especially:

- A difficult process of developing the trial protocol for approval
- Problems with both accessing nearby GMP-compliant laboratory processing facilities and transporting cells to them and back to the hospital
- Issues with the new Voluntary Harmonisation Procedure for national competent authority approval

Europe's attempt to harmonize medical research and therapeutic innovation across the many different cultures and health-care systems in its constituent nations is an enormous task. It can only succeed if the side effects and pitfalls are overcome. At present, harmonization struggles to reach its aims of opening up a unified area for effective research and speedy approval of new medicine and therapies.

8.2   Research Project and Methodology

The lead author has been studying the effects in one particular phase III stem cell RCT from development and funding application to its future completion. At the time of writing in 2016, the trial is still in the patient recruitment phase.

The trial we report from is entitled 'The Effect of Intracoronary Reinfusion of Bone Marrow-Derived Mononuclear Cells (BM-MNC) on All-Cause Mortality in Acute Myocardial Infarction' abbreviated as BAMI. It is a 3000-patient multinational clinical trial conducted across nine countries in Europe, funded under the EU Framework Programme 7 Health scheme. BAMI has scientific, regulatory and clinically driven objectives focusing on the standardization of the treatment and the testing of the treatment's benefits for all-cause mortality. The aims include (1) the creation of a pre-clinical dossier for a standardization method for the preparation of bone marrow mononuclear cells for intracoronary injection, (2) the standardization of an intracoronary infusion method that allows for the safe and efficient delivery of cell therapy to the coronary artery and (3) the designing of a randomized controlled clinical trial that addresses whether the standardized mononuclear cell product confers an all-cause mortality benefit in patients with acute myocardial infarction who undergo primary angioplasty (25% reduction on top of standard therapy).

'Towards Harmonised Ethical Standards' (WP7, 2011–2017) is a subproject in BAMI, conducting a sociological and medical ethics study alongside the clinical trial. As researchers in this subproject, we study the effects of unified protocols on trials conducted across the participating clinics, how the different NCCs[1] proceeded in gaining ethical approval for BAMI, and how the trial has been implemented in the participating countries. The research gathers data on the difficulties and hurdles the BAMI teams encountered in bringing the trial from funding approval to clinical implementation under the new harmonized ethical and regulatory framework in the EU. The focus on trial implementation has highlighted that recruiting the required patients for such a phase III stem cell trial in acute myocardial infarction in Europe is difficult under any circumstances. The standard treatment for this condition is so effective that the severe postsurgical patient conditions that are commonly selected for regenerative medicine trials become quite rare. Thus, the number of eligible patients can only be recruited from a large catchment area covering many hospitals in many countries. The efficacy of the BAMI intervention itself cannot be reported on at the time of writing, but trialling it has been evaluated as worthwhile by the

---

[1]NCC stands for National Coordinating Centre staff. In each country, there is one hospital that leads the trial in that country. The NCC recruits, contracts and assesses the work at other satellite hospitals that also recruit patients to BAMI.

Task Force of the European Society for Cardiology (Bartunek et al. 2006), and an update of their review of this line of stem cell treatments for heart repair is in press (Mathur et al. 2017).

Areas of investigation included the effects of regulatory or other practical problems affecting on-site on patient recruitment as well as any positive or negative effects of cross-national regulation. The aim of the subproject is to harness the experiences of the clinical teams with trial implementation while complying with the regulations in order to inform about the effects of the current regulatory set-up on clinical trial practice in cell therapy across the EU.

Our findings illustrate that multinational stem cell trials are very difficult to conduct in practice, even if a shared regulatory framework is in place. ATIMP compliance in particular severely affected BAMI's logistics and finances, resulting in the actual withdrawal of three original partners from patient recruitment and the joining of a new one. In this chapter, we outline the regulatory environment in which BAMI operates before drawing on our empirical data to describe the effects these regulations had on the day-to-day running of the BAMI RCT.

Methodology and Data Collection

Changes in regulation between the planning of BAMI and its start mean that the project's timeline must be part of our reporting. BAMI grew out of a long-standing prior cooperation between cardiologists in different EU countries. The principal investigators (PIs) each had conducted smaller trials on autologous BM stem cell injection to repair heart tissue following AMI. The lead author has been party to this collaborative effort. It took a while before BAMI received EU funding in 2011 and started in the same year. The subproject 'Towards Harmonised Standards' was introduced to all participating BAMI PIs at the 'kick-off meeting' in London on 1 December 2011. Using a brief questionnaire, which was also mailed to non-attending members, we collected information on previously encountered problems with approval procedures in the small local trials and any problems that BAMI might face. In total 14 BAMI PIs completed and returned the questionnaire representing eight countries—Denmark, Spain, the UK, Germany, Italy, Norway, Belgium and the Czech Republic. Responses to the questions were collated, compared and analyzed.

It took over 2 years before patient recruitment in BAMI began, and consequently we started in April 2014 to conduct semi-structured interviews—face-to-face in Germany and Spain and via Skype and telephone with staff from BAMI partners in other countries—on the process of implementation. We conducted 28 interviews, 25 with clinical staff (e.g. PIs, NCCs, cardiologists, study nurses) and 3 with project managers responsible for trial-wide infrastructure (insurance, centralized echocardiography, patient randomisation). Interviews lasted for 30–60 minutes, followed by a predefined topic guide. The interviews were recorded on tape and transcribed. Some key agents contributed follow-up interviews. The topics included issues that

Table 8.1 Number of BAMI staff participating in questionnaires, interviews and surveys

Country	Pre-study questionnaire	Interviews	Mini survey
Belgium	1	2	4
Czech Republic	0	1	1
Denmark	1	0	1
France	0	1	1
Germany	4	6	2
Italy	1	1	2
Poland	0	0	1
Spain	2	7	3
UK	3	5	7

had emerged from participant observation of regulatory approval processes and pilot interviews. Data collection and analysis proceeded in parallel as initial findings required further data collection and the going back to early interviewees. Interview materials were transcribed and the data analyzed using conventional methods supported by NVivo (Version 10) where keyword searches helped to identify themes and categories. We decided against an entirely electronic-based analysis not the least because some interviewees were not native speakers of English and therefore will have used language in nonstandardized ways. We conducted some interviews in other languages that we are able to speak (viz. German and Spanish).

After the interview and data analysis, we used another BAMI consortium meeting in February 2015 for a finding-check mini survey. This was distributed to all participants and again mailed to BAMI members who had been unable to attend the meeting, including many study nurses. In total, 22 participants from nine different countries responded to the finding-check mini survey. Its aim was to validate our interpretation of the findings and to receive comments to them by the clinical teams involved. The finding-check exercise confirmed that we had correctly identified all the problems that individual team members had encountered and highlighted the dramatic effect of the regulatory changes described above on the implementation of BAMI (Table 8.1).

One key aim of our study on BAMI was to trace the process of implementation and to identify and report on the problems that BAMI staff and patients encountered. More specifically, we studied how the harmonization of clinical protocols and ethics approvals actually proceeded and how the participants perceived the process, locally and overall between and across the participating institutions.

On the basis of our observations, mini survey and interview data over the trial period until April 2015, we report below about the most prominent issues that delayed patient recruitment. These challenges include issues of a logistical or technical nature and of ethical approval. We shall begin with a brief overview of the regulatory situation BAMI encountered in 2011. Based on our empirical findings, we then present how BAMI staff responded to the challenges arising from the ATIMP classification of the trial. The focus lies on the effects of the recently altered regulatory requirements and also teething problems with VHP. BAMI was the first phase III trial applying the VHP. The late start of patient recruitment has become increasingly problematic for the completion of the BAMI trial in its original design, and it already expanded the timeframe for completion beyond the envisaged 5 years (Mathur et al. 2017).

8.3 The BAMI RCT and Its Situation in This Regulatory Environment

BAMI operates in the standardized regulatory environment of Europe for multinational stem cell RCTs. It was prepared before and during 2009 and started late in 2011, a time of change in the implementation of the European Clinical Trials Directive and the European Tissue and Cells Directive (EUTCD). The implementation rules of this latter directive had also been amended to ensure that all national competent authorities follow the same rules when classifying stem cell therapies and research, after great variations in interpretation had been found across Europe (Veerus et al. 2014). This variation had also included RCTs in Europe that used autologous stem cells which were the precursors of BAMI (Weber et al. 2011; Wilson-Kovacs et al. 2010). Generally, since 2007 all new RCTs using stem cells in innovative ways, autologous or not, have had to comply with regulations for ATIMPs, according to the changed implementation rules for the Tissues and Cells Directive. Specific decisions were left to the newly set up Committee for Advanced Therapies (CAT). At its meeting on 14–15 October 2010, this committee formally classified the intervention used in BAMI as tissue-engineered product, thus requiring ATMP processing. CAT explicitly stated that 'autologous bone marrow-derived progenitor cells intended for treatment of patients with failed left ventricular recovery despite successful reperfusion therapy post-acute myocardial infraction, chronic ischaemic heart disease, peripheral vascular disease and Buerger's syndrome' are to be treated as 'tissue-engineered products' (CAT 2010). The effect of this formal classification has since had significant impact on the field of autologous stem cell uses in research and clinical trials, and would also affect potential future therapy roll-out, as we will illustrate. The classification of autologous bone marrow-derived stem cells in heart repair now requires GMP licensed laboratories to handle the cells and detailed monitoring of research partners to ensure compliance with the trial protocol. This has caused major problems in BAMI, as the requirement to use such laboratories had not seemed necessary in all partner countries at the time of designing the trial and it was thus not budgeted for.

Concerning the EU Clinical Trials Directive from 2001, ethical approval for multinational clinical trials still kept facing challenges owing to different interpretations of these frameworks in different countries. As a consequence, the VHP was set up by the Clinical Trials Facilitation Group (CTFG) (Krafft et al. 2012). The Tissue and Cells Directive and its clarifications regulate the handling of tissues and cells as well as laboratory practices. The Clinical Trial Directive and the VHP, however, aim to streamline ethical approval of trial protocols. VHP is a protocol developed and implemented in 2010 through the PEI. According to this procedure, the country leading the clinical trial submits the clinical and ethics application to the PEI. The application is then approved by the PEI in cooperation with one or two other national competent authorities. After approval by the PEI, the other national competent authorities in the multinational trial—in an 11-partner trial this may be 8 or 9 competent authorities across Europe—are expected to approve the application without demanding any further changes to the protocol. This simple VHP was altered in 2013. The PEI now expects to invite two or more competent authorities to assess and decide on the application from the lead institution in its approval process. The aim is to avoid later stage delays arising from additional requests for changes to the protocol. This modified procedure has been in place since 2013. From the observation of the BAMI application process, we have strong reasons to believe that it is a response to the problems BAMI encountered. Unfortunately, despite several requests, PEI representatives were unavailable for interview or comment on this issue. Yet, BAMI, as the first phase III multinational trial that applied through the initial design of the VHP, had not been served well by it—although, except for Poland and Finland, the BAMI partner countries are signatories to the VHP.

VHP was introduced because attempts to harmonize practice through Directives and scientific standardization encountered difficulties in gaining the required national competent authority approvals. Different authorities requested different protocol changes, and in consequence approval processes were protracted regularly. EU member states have very different cultural and ethical traditions, and their health-care systems operate differently. Consequently, policies on the funding and clinical application of stem cell research, genetics, reproductive medicine and the introduction of new treatment protocols and drugs have been difficult to agree upon if many different competent authorities process a protocol simultaneously. VHP can be seen as an expression of trust that a select set of national competent authorities will develop guidance acceptable to the other member states. BAMI is the first multinational phase III clinical trial approved via VHP and affected by the described regulatory changes.

8.4 Findings and Empirical Analyses

8.4.1 Expectations Based on Previous Experience with RCTs

The kick-off survey of BAMI consortium members showed that some had previously encountered a variety of problems relating to regulatory and ethical issues with phase I local trials designed very similarly to BAMI. In spite of this, every respondent expected that their national competent authorities and local ethics committees would respond favourably to the BAMI RCT. A unified approach was envisaged, although some differences in the area of regulation were anticipated (Fig. 8.1).

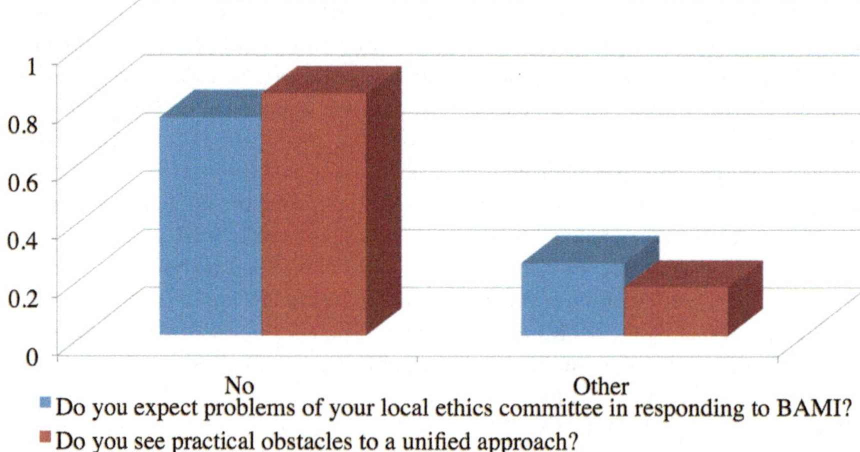

Fig. 8.1 Expected problems with clinical trial and ethics approval in the BAMI trial

This confidence was partly based on previous experience, when the PIs had each been able to resolve any issues with the smaller trials locally. Because such issues appear during the preparation of any trial, they affect the trial protocol, but are not as such part of it nor are they commonly reported in the findings from the trials. They are externalities. Probably for this reason, the local problems individual PIs had faced prior to BAMI and solved were also not discussed among the PIs before this meeting. In this initial meeting and in several subsequent meetings, the gulf between the scientific protocol and its merits to deliver evidence for the efficacy of autologous BMSC in AMI, on the one hand, and the clinical protocols and detailed implementation plans required for RCT approval and VHP, on the other hand, had to be discussed and squared. In preparations for BAMI, the scientific tasks had been to design the best trial and obtain funding for it based on the insights to be gained. The next step was turning this scientific protocol into a clinical protocol that would be approved and could be implemented across 11 different EU countries with different health-care systems and cultural conventions affecting clinical work. Following the mini survey and exposure and discussion of the local challenges, it became obvious that in preparing the VIIP submission of the BAMI clinical protocol, all the lessons from the smaller early trials with approval institutions had to be considered and addressed as far as possible, in order to prevent encountering old criticisms from national competent authorities and clinical research ethics committees. It also became obvious that this inclusive approach towards local understandings of best clinical practice in the clinical protocol for all countries might raise the workload of NCCs and clinical teams and slow down patient recruitment. Also, two partner countries, Poland and Finland, have not signed up to VHP, so even if clinical approval via VHP did simplify the process for most partners, two would require separate national competent authority and local clinical ethics committee approval.

8.4.2   Regulatory Issues Affecting the Implementation of BAMI

In this chapter, we can report in detail only on a select number of issues and provide the empirical data to support our points. We focus on the effects of the ATMP amendment to the Human Tissue and Cells Directive which has changed conditions for the stem cell procedure used in BAMI since October 2010 and the VHP.

8.4.2.1   Implementation of EUTCD, ATMP and ATIMP

A major problem for BAMI centres became gaining access to the newly required ATMP-certified laboratories for the processing of the BMSC. In the preparation phase, seven centres in six countries (Germany, Spain, Denmark, UK [two centres], France, Belgium) indicated that they could act as BAMI cell processing centres. The 2010 ATMP amendment to the Human Tissue and Cells Directive required that the stem cell processing centres used in BAMI hold an IMP[2] manufacturing licence and were GMP[3] compliant. Only the laboratories related to the German, Spanish and Danish centres held all the required licences and documentation. This caused substantial logistical and financial problems, as cells cannot be transported by plane (owing to the inevitable X-ray) and time-compliant transport by courier is expensive. The options were either to gain special accreditation for the local laboratory or to transport the cells to an established fully certified stem cell laboratory elsewhere.

Figure 8.2 illustrates how the lack of a national cell processing lab was perceived by BAMI staff.

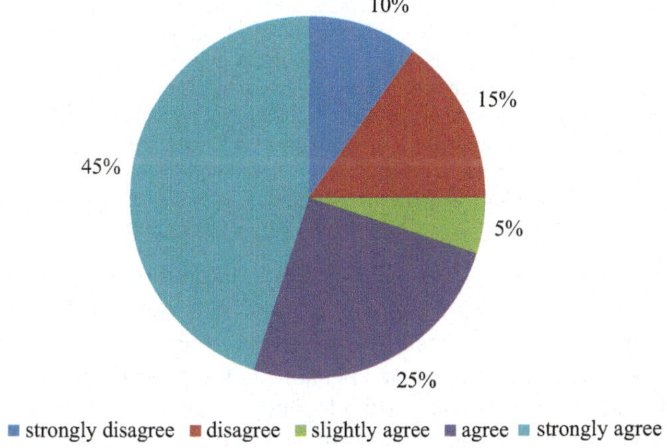

Fig. 8.2 Has the lack of access to your own cell processing lab caused problems?

---

[2] Investigational medicinal product.

[3] Good manufacturing practice.

As can be seen from the diagram, 70% of BAMI partners strongly agree/agree that lack of access to their own cell processing lab caused problems. These problems, which affected the lead centre in the UK and partners in the Czech Republic, Belgium and France, are financial and logistical. All these participants had to ship the cells to the nearest laboratory, situated in Frankfurt. This was not allowed to interfere with other aspects of the protocol, especially the reinjection of the cells 24 hours after retrieval, and it turned out to be a main cost factor. The ATMP amendment and ATIMP classification of BAMI had a serious impact on timing and budget and, as the following quotation illustrates, came as a surprise to some PIs:

> What slightly took us back is that some of the cell processing centres were not certified to do the processing on the trial and that was part of the VHP procedure. […] Because it's VHP then it's all under EMA regulations, therefore we have to send the cells to Frankfurt, so massive issue in terms of prices and all that. […] VHP created a problem for the UK, because we would have been able to do it otherwise, without the EMA re-classification. (NCC, UK, 18/8/14)

The fact that some BAMI PIs expressed surprise at this shows that despite being well versed in conducting stem cell research, they were not up-to-date on the regulatory reclassification of what they deemed a perfectly established use of BMSC. As a consequence of this regulation, many participating countries had to find alternative ways of compliantly processing their cells. The above-mentioned centres, for example, made plans to transport the cells to the most accessible ATIMP-certified laboratory in a partner country, which is in Frankfurt. This return transportation of the cells for processing had to occur within the allocated cell processing times, i.e. 24 hours from bone marrow aspiration to the laboratory and again 24 hours between the release from the laboratory and reinjection. Given the distance and logistics that now needed to be covered, this turned into a problem:

> We still haven't got our own lab, so the cell processing centre is still over there in Germany in Frankfurt. […] So, usually, what happens, they go by plane, they can fly there, so they go by plane, with the bone marrow samples. […] On the way back, stem cells should not go through X-ray. So, they are driven back. So, you've got to get the times right from Frankfurt to here. We had, I remember, once an incident that they had on the motorway in Germany, a very bad accident, so they were stuck in traffic for quite some time […] And we had the patient here who was already waiting. I think we've been lucky because it managed to reach us within time, within the time window that we have. […] It will be much easier once we have the licence for the cell processing centre here. (Study nurse, UK)

In 2016, London is still sending the BM cells to Frankfurt, as UK-based ATIMP-certified laboratories would be much more expensive to use. The French and Czech teams were hampered additionally by national implementation requirements demanding that only a specially trained person can transport human stem cells. For the French team, this became one of the logistical and economic obstacles which ended their participation as a recruiting partner. The Czech team resorted to an ingenious solution, namely, ensuring that some drivers of the respective courier service were trained accordingly and thus no additional person was required to transport the cells to Frankfurt and back, as the quotation below illustrates:

> I had to find a courier who would deliver the cells each time to Frankfurt and back to Brno. … [S]omeone who would do it and would be trained by our tissue people who have the

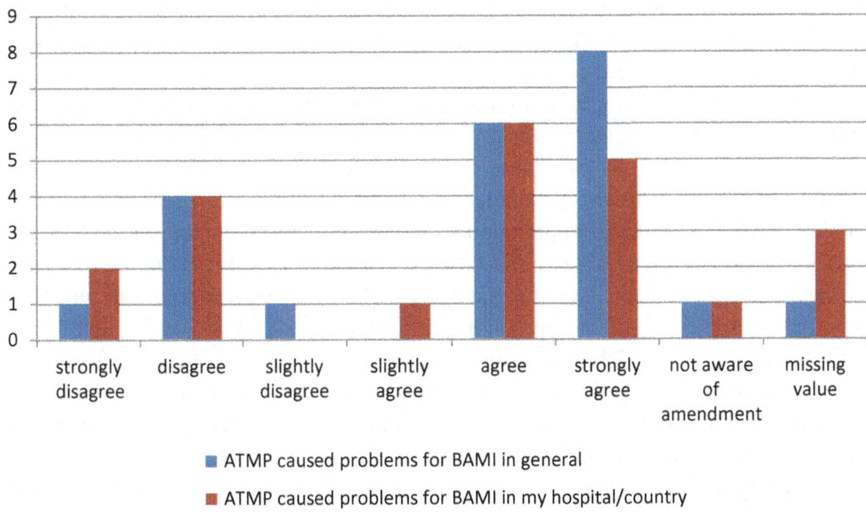

Fig. 8.3 ATMP caused problems for BAMI

permission by ZUKL, so they can train other people and, you know, sign contract and stuff. So, I found a good group of taxi drivers, and they have a contract with the hospital and they underwent the training programme that took about two hours. They know what's in the package, they have to go directly to Frankfurt, no stops, right temperature in the car. Everything is done and organised by ZUKL, so they follow the law. So, they deliver the cells to Frankfurt, wait for like two or three hours, get a phone call saying 'you can pick them up' and they pick them up and go back to Brno. They deliver here, it's safe. (NCC, Czech Republic)

Figure 8.3 shows the difference in perception of ATMP causing problems in BAMI in general and in the respondents' own hospital or national competent authority approval process.

The diagram illustrates that the majority of respondents considered the ATMP amendment responsible for at least some of the major problems occurring in BAMI, although some did not encounter any problems themselves, because their national competent authorities had classified autologous BMSC as 'tissue-engineered products' (ATMP) when initially implementing the EUTCD, before the amendment enforced this classification throughout Europe. The discrepancy between the number of BAMI staff claiming that the amendment has affected BAMI in general ($n = 14$) and those stating that their own country or hospital has been affected by it ($n = 11$) is explained by the differences in the initial implementation of the EUTCD between European countries (Wilson-Kovacs et al. 2010).

8.4.2.2 Voluntary Harmonisation Procedure (VHP)

VHP was introduced as a pilot project in 2009. It has been developed by the Clinical Trial Facilitating Group (CTFG) which was formed in 2004 by the Heads of Medicines Agencies (HMA) with the aim of obtaining a more uniform interpretation of EU regulatory requirements in national competent authority decisions on clinical trials. The specific aims of VHP are to facilitate the process of assessment of clinical studies that involve several countries.

When we gauged some first opinions about VHP, many members said that they consider it an important step, as it can act as a guarantee for authorities and local ethics committees, thereby facilitating and speeding up the approval process. Deeper questioning, however, revealed that there were and are substantial problems with the VHP in BAMI. The initial application to VHP had to be prepared taking care to address previously encountered national authority and local ethics committee sensitivities. This affected aspects of the trial design, but the adaptation was quite smooth. Other issues reported include the unfamiliarity with VHP (BAMI is the first study of this size to run under VHP) and difficulties in complying with VHP requirements which resulted in delays of centre initiations and subsequent delays in patient recruitment. The political structure of a country has also been suggested as an obstacle. Some countries are divided into federal states, with more than one competent authority and clinical ethics committee. The UK reported that VHP raised a lot of country-specific issues which led to an overall delay in obtaining VHP, as the UK could not go forward before these issues had been resolved. However, BAMI has applied and gained VHP relatively quickly, within a year of starting, and has been conducted under this protocol since.

Figure 8.4a, b highlights a noteworthy discrepancy in the perception of VHP speeding up the process of gaining clinical approval in BAMI in general and ethical approval in the participants' hospitals. While almost half of the respondents to our finding-check mini survey in 2014 thought that VHP speeded up the approval process in BAMI in general, only about 35% stated that this was the case concerning their own hospital or country. Even more strikingly, while none of the respondents disagreed with the notion that VHP speeded up the approval process in BAMI in general, 10% thought that it delayed it in their own hospital or country.

One of the positive things reported about VHP—and opposed to the local ethics committees—concerns the timeliness of dealing with requests. The idea of VHP is that applicants obtain VHP approval and that, within 10 days, this is followed swiftly by approval from the local competent authority. This, however, did not work in BAMI. Although the UK received VHP approval and approval from their national competent authorities within 5 days, other countries had queries coming back from the local competent authority, although the PEI had considered those issues when approving the protocol in the course of VHP.

8.4.2.3 Amendments to the BAMI Protocol

Several national competent authorities demanded amendments to the PEI-approved protocol, which in turn had to be applied in each participating country. For example, the Paul-Ehrlich-Institut, then in its other role as one of the national competent authorities in Germany, requested that patients be recalled to the trial centre 30 days after their discharge from the hospital. This raised various issues—even for the

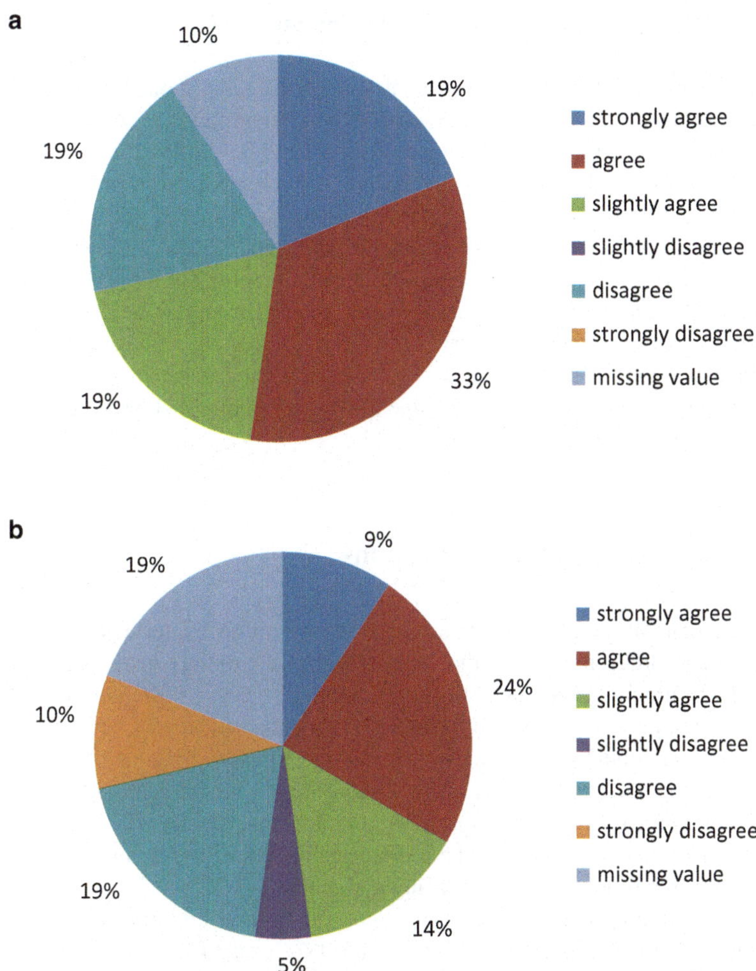

Fig. 8.4 (**a**) VHP speeded up the approval process in BAMI in general. (**b**) VHP speeded up the approval process in my country/hospital

German clinics, where many patients perceive this recall as a nuisance, interfering with their lives during the rehabilitation phase after the clinic stay. Similar problems with the 30-day recall have not been reported from the UK or Spain. Differences in times of clinic and rehabilitation centre stays are likely related to this difference.

Because of the resulting initial time delays, several amendments to the approved trial protocol for BAMI have been submitted or considered in the course of the 4 years since the trial started. In some countries, Italy for example, the competent authorities request that patient recruitment is halted while such reapproval is sought. The protracted procedures of reapproval through VHP can mean long recruitment stops in these cases. This counteracts the intent to maintain the trial's research validity and imposes additional costs due to the harmonized regulatory requirements.

8.4.2.4   Other External Differences Affecting the Trial

Our study also identified other issues that arise from harmonizing increasingly many trial-related practices without—obviously—planning also to harmonize other cultural practices in the participating countries that are not directly related to the clinical trial but which affect it. They include differences in conventions in insurance practices, expected content of patient information sheets, and even national drug approval between partner countries. We shall report about those elsewhere in more detail, but we mention them here to indicate the full picture of potential stumbling blocks encountered disproportionally in an academic trial such as BAMI. As an academic phase III trial with very limited funding, BAMI received little support from industry sponsors—sponsors who have rich experience in managing such complex multinational trials for therapeutic interventions that promise not only clinical but also economic potential.

8.5   Concluding Recommendations

Our participant observation, mini surveys and qualitative interviews with many BAMI staff in different roles in the trial present a rich picture of the struggles for which any future multinational European academic trials must prepare. If the route of RCTs remains in place as the ultimate test for the efficacy of new treatments, which may be in doubt, the move towards multiple sites for trials involves complex and intersecting societal factors. One must consider both external and lifestyle factors affecting efficacy in order to retain validity. Increasingly specific diagnoses and inclusion criteria, respectively, make it increasingly harder to recruit for a phase III trial in a suitable timeframe within a single country.

The EU's commitment to standardize, ease and speed up clinical research and drug innovation over the past 25 years has led to a set of regulations with good intentions and speedy amendments where initial procedures did not fulfil their goals. However, establishing one set of regulations and institutions that works for both academic and industrial research requires the simultaneous creation of sufficient laboratory capacity for multiple research trajectories. Moreover, competent authorities and other approval institutions have to develop trust for the centralized process so that the new European regulatory platform can work. At present, the structure is not settled and a trial that is conceived in one regulatory setting and then falls under another experiences challenges during implementation and conduct. If academic clinical trials are valued as in the public interest, then support is needed to bridge periods of regulatory overhaul that will predictably affect specific research pathways and academic trials especially hard. Financial and administrative support by the EU institutions should be considered to avoid undermining EU-funded medical research in order to achieve better harmonized medical research practice in the long term. BAMI would certainly have run more smoothly if it had not been unfortunately timed or if such support had been offered. Cooperation to remedy the fallout of regulatory transition between the institutions overseeing and managing the regulation and the Directorate for Research, who gave millions in funding for the BAMI trial, might be worth considering for similar instances in the future. Facing both expected and unexpected challenges, the BAMI clinical teams have shown ingenuity, commitment and team spirit to keep going and expressed clear support in principle for the regulatory ambitions to introduce centralized approval procedures. It is obvious to them that such harmonization can overcome some of the obstacles from national and cultural variations in practices and customs that have previously made the conduct of multinational trials very difficult.

Acknowledgements This research would not have been possible without the continuous support and team spirit within the BAMI team as a whole and especially the help of both the BAMI coordinator, Anthony Mathur, and Sheik Dowlut at the BAMI Trial Office. The authors thank Jean Harrington, in 2011 Research Assistant to Christine Hauskeller, for contributing to the first phase of data collection in BAMI, which included help with the kick-off meeting questionnaire, analysis of its findings, and reporting them to the funder. For our research, we received funding from the European Union's Seventh Framework Programme for research, technological development and demonstration under grant agreement no. 278967.

References

Amexis G, Schmitt E (2011) A sponsor's experience with the Voluntary Harmonization Procedure for clinical trial applications in the European Union. Nat Rev Drug Discov 10:393

Bartunek J, Dimmeler S, Drexler H et al (2006) The consensus of the task force of the European Society of Cardiology concerning the clinical investigation of the use of autologous adult stem cells for repair of the heart. Eur Heart J 27(11):1338–1340

Cartwright N (2010) What are randomised controlled trials good for? Philos Stud 147:59–70

CAT (2010) EMA/CAT657770/2010. Monthly report, http://www.ema.europa.eu/ema/index.jsp?curl=pages/news_and_events/document_listing/document_listing_000374.jsp&mid=WC0b01ac058071e6d8

Cressy D (2010) UK doctors demand research reform. Nat News. doi:10.1038/news.2010.243

Davis C, Abraham J (2011) A comparative analysis of risk management strategies in European Union and United States pharmaceutical regulation. Health Risk Soc 13:413–431. doi:10.1080/13698575.2011.596191

EC (2001) Directive 2001/20/EC of the European Parliament and of the Council of 4 April 2001. Official Journal of the European Union 1-5-2001, L 121/3M4

EC (2004) Directive 2004/23/EC of the European Parliament and of the Council of 31 March 2004. Official Journal of the European Union 7-4-2004, L 102/48

EC (2009) Directive 2009/120/EC of the Parliament and of the Council of 14 September 2009. Official Journal of the European Union 15-9-2009, L 242/3

EC (2016) http://ec.europa.eu/growth/single-market/european-standards/policy/benefits/index_en.htm. Accessed 30 May 2016

Hartmann M (2012) Impact assessment of the European Clinical Trials Directive: a longitudinal, prospective, observational study analyzing patterns and trends in clinical drug trial applications submitted since 2001 to regulatory agencies in six EU countries. Trials 13:53. doi:10.1186/1745-6215-13-53

Hartmann M, Hartmann-Vareilles F (2006) The clinical trials directive: how is it affecting Europe's noncommercial research? PLoS Clin Trials 1:e13. doi:10.1371/journal.pctr.0010013

Hunter P (2011) The health of European medical research. EMBO Rep 12:110–112

Krafft H, Bélorgey C, Szalay G (2012) Experience and further development with the Voluntary Harmonization Procedure for multinational clinical trials in the European Union. Nat Rev Drug Discov 11:419

Mathur A, Fernández-Avilés F, Dimmeler S et al (2017) The consensus of the task force of the European Society of Cardiology concerning the clinical investigation of the use of autologous adult stem cells for the treatment of acute myocardial infarction and heart failure—update 2016. Eur Heart J. doi:10.1093/eurheartj/ehw640

Salter B, Zhou Y, Datta S (2014) Making choices: health consumers, regulation and the global stem cell therapy market. BioDrugs 28:461–464. doi:10.1007/s40259-014-0105-2

Veerus P, Lexchin J, Hemminki E (2014) Legislative regulation and ethical governance of medical research in different European Union countries. J Med Ethics 40:409–413. doi:10.1136/medethics-2012-101282

Weber S, Wilson-Kovacs D, Hauskeller C (2011) The regulation of autologous stem cells in heart repair: comparing the UK and Germany. In: Lenk C, Hoppe N, Beier K, Wiesemann C (eds) Human tissue repair: a European perspective on the ethical legal challenges. Oxford University Press, Oxford, pp 159–164

Wilson-Kovacs DM, Weber S, Hauskeller C (2010) Stem cells clinical trials for cardiac repair: regulation as practical accomplishment. Sociol Health Illn 32:89–105

Chapter 9
Oversight and Evidence in Stem Cell Innovation: An Examination of International Guidelines

Tamra Lysaght

Abbreviations

ASC	Adult stem cell
EBM	Evidence-based medicine
ESC	Embryonic stem cell
FDA	Food and Drug Administration
HCT/Ps	Human cell- and tissue-based products
ICMS	International Cellular Medicine Society
IRB	Institutional Review Board
ISSCR	International Society for Stem Cell Research
PHS Act	Public Health Service Act
RCT	Randomised controlled trials
SCBIs	Stem cell-based interventions

9.1 Introduction

Stem cell-based interventions (SCBIs) offer enormous potential in the treatment of many significant diseases, illnesses and conditions. However, with the exception of a small number of malignant and non-malignant disorders of the blood and immune systems, the clinical use of stem cells thus far has mostly been limited to Phase I and Phase II clinical trials (Power and Rasko 2011). Despite the experimental status of many proposed approaches, some physicians in the USA and abroad have begun to

An earlier version of this chapter was previously published as: Lysaght T (2014) Oversight and evidence in stem cell innovation: an examination of international guidelines. Law Innov Technol 6(1):30–50. Material from that article is reprinted with permission from Taylor & Francis.

T. Lysaght (✉)
Centre for Biomedical Ethics, Yong Loo Lin School of Medicine,
National University of Singapore, Singapore, Singapore
e-mail: tlysaght@nus.edu.sg

© Springer International Publishing AG 2017
P.V. Pham, A. Rosemann (eds.), *Safety, Ethics and Regulations*, Stem Cells in Clinical Applications, DOI 10.1007/978-3-319-59165-0_9

offer interventions with autologous adult stem cells (ASCs) for a variety of diseases and conditions outside the context of formal clinical trials (Lysaght and Campbell 2013). For example, numerous high-profile professional athletes have reportedly been administered with autologous ASCs for sporting injuries (Caulfield and McGuire 2012). The Governor of Texas, Rick Perry, also underwent a similar procedure for back pain in 2011, and the Texas Medical Board has since established guidelines that allow physicians to administer stem cells in certain clinical settings (Cyranoski 2012).

The stem cells used in these procedures are apparently isolated from the bone marrow and/or adipose tissue of patients and expanded to unnaturally large numbers in culture before being administered back into the patient. These interventions are controversial because, even though early clinical studies have suggested that certain types of autologous ASC cultures may be relatively safe, evidence of efficacy outside of their current uses in certain blood and immune disorders has not yet been established (Trounson et al. 2011). At issue is whether interventions with autologous ASCs should be offered to patients as 'innovative therapies' without first establishing their safety and effectiveness in formal clinical trials.

From a legal perspective, regulatory agencies typically require evidence of safety and efficacy from formal clinical trials before licences are granted for the marketing and distribution of biological drugs for use in humans. Exemptions to this requirement are available for stem cell-based products that are for autologous use providing that they are only used in tissues that perform the same basic functions, are not mixed with other agents and are minimally manipulated. Licenced medical practitioners may thus innovate with stem cells and use other licenced cell-based products 'off-label', without initiating formal clinical trials to demonstrate their safety or effectiveness (Lindvall and Hyun 2009). Under the expanded access programme of the Food and Drug Administration (FDA), physicians in the USA may also apply for the so-called compassionate use of biological drug products that are being investigated in registered clinical trials. Providing there are no compatible alternatives, this 'humanitarian' provision allows physicians to offer investigational stem cell-based products to patients in the context of clinical care.

Even if some types of stem cells may legally be offered to certain patients as an innovative therapy without undergoing formal clinical trials, opinions differ over the need for scientific and ethical oversight. Two professional organisations, the International Society for Stem Cell Research (ISSCR) and the International Cellular Medicine Society (ICMS), have issued guidelines that reflect different perspectives on when it may be justifiable to introduce stem cell innovations into clinical settings without evidence from clinical trials that demonstrates their safety and effectiveness. In this paper, I examine these guidelines for differences in how they each frame stem cell innovation and define what constitutes as evidence and oversight for such innovations. Specifically, I analyse how they each define key terms and concepts and prescribe the level of oversight and standards of evidence that should be required before introducing a novel SCBI into clinical settings.

9.2 Clinical Guidelines of the ISSCR and ICMS

Before proceeding, it is useful to provide a brief background on the two organisations as both guidelines are aimed at particular audiences in ways that reflect their overall mission and objectives. The ISSCR represents an international membership of mostly scientists and clinical researchers working within the broader field of stem cell science and is widely regarded as the leading authority for stem cell research. According to its mission statement, the organisation was established to 'encourage the general field of research involving stem cells and to promote professional and public education in all areas of stem cell research and application' (ISSCR 2002). Since its inception, the organisation has also played an important advocacy role in promoting a less restrictive regulatory environment for human embryonic stem cell (ESC) research in the USA. Specifically, the ISSCR has actively supported the relaxation of federal funding restrictions that were placed on human ESC research by President George W. Bush in 2001 and lifted by President Barack Obama in 2008.

In 2008, the ISSCR convened a taskforce to develop guidelines in response to the 'growing number of centers throughout the world that are testing stem cell interventions and [...] claiming to offer stem cell treatments for a variety of conditions without clear evidence of safety or efficacy' (Daley et al. 2008). This response was made in the context of escalating concerns over patients travelling across international borders to access 'unproven' treatments in a phenomenon often referred to as 'stem cell tourism' (Regenberg et al. 2009, Lau et al. 2008, Kiatpongsan and Sipp 2008, Kiatpongsan and Sipp 2009). The taskforce was composed of scientists, clinicians, law academics, bioethicists and patient advocates who, in consultation with the FDA, published their recommendations in the *Guidelines for the Clinical Translation of Stem Cells* (Hyun 2008).

While the ISSCR guidelines are generally targeted towards clinician-scientists working in translational areas of stem cell research, the ICMS guidelines are targeted towards a much narrower audience of American medical practitioners who they claim are engaged specifically in 'adult stem cell medicine' (ICMS 2010): the membership is not, however, representative of physicians that specialise in the currently accepted clinical uses of stem cells (e.g. haematopoietic stem cell transplantation) nor is the organisation aligned with other relevant bodies such as the Foundation for the Accreditation of Cellular Therapy or the International Society for Cellular Therapy. Rather, the ICMS originally formed as the American Stem Cell Therapy Association (2009) in response to claims made by the FDA that autologous ASCs were subject to regulation as biological drugs. These claims presumably arose from legal actions taken against a Colorado-based company, whose medical director (and now CEO) is a founding member of ICMS and its predecessor. In these proceedings, which were recently addressed by the US District Court of Columbia,[1] the FDA claimed that the company's expanded culture of bone

---

[1] The Court has since found in favour of the FDA. See *United States v Regenerative Sciences, LLC et al.* (2012) DC Cir. 47.

marrow-derived mesenchymal stem cells was an adulterated and misbranded drug being entered into interstate commerce (Lysaght and Campbell 2011). The company denied these claims and instead argued (unsuccessfully) that the autologous use of their cultured stem cell product was a medical procedure that constituted as the practice of medicine and thus fell outside the regulator's jurisdiction.

As part of its mission, the ICMS has since established a registry and an accreditation programme to evaluate member clinics and operates on the premise that the autologous use of cultured ASCs is a medical procedure and not drug manufacturing (ICMS 2010). The ICMS (2012) has also released draft guidelines on 'best practice standards' for cell-based medicines, which focus exclusively on autologous ASCs. The clinical guidelines, which are presented in Section VII, were adopted from an earlier version (ICMS 2009) that was likely drafted by founding members during the legal proceedings against the FDA in the US District Court and is available on the ICMS website. In the following analysis, I compare this version with the ISSCR's (2008) guidelines, starting with their definitions of key terms and concepts.

9.2.1   Defining Key Terms and Concepts

The definition of key terms and concepts is important because it helps to establish when an innovative use of stem cells would fall within the scope of medical practice and when it would require more rigorous oversight and regulation as research. Both guidelines agree that 'the level of regulation and oversight should be proportionate to the degree of risk raised by a particular cell product and intended use' (ISSCR 2008, p. 7; ICMS 2012, p. 8). The ISSCR (2008, p. 6) guidelines also recognise that minimally manipulated products need not require the same level of oversight as 'cell products subjected to extensive manipulations ex vivo'. However, the ICMS (2012, p. 3) guidelines are framed around addressing a gap in 'existing guidelines' that focus on the 'relatively greater risks of allogeneic transplants' rather than the 'substantially smaller risk techniques' used for the 'minimal expansion' of autologous ASCs.

This framing has significance within the regulatory context of the USA, where both organisations are based. In the USA, ASCs are classified as human cell- and tissue-based products (HCT/Ps) that fall under the purview of Sections 351 or 361 of the *Public Health Service Act* (PHS Act) and Title 21 of the *Code of Federal Regulations for Food and Drugs*. Products that fall within Section 351 are regulated by the Center for Biologics Evaluation and Research Office of Cellular, Tissue, and Gene Therapies, a division of the FDA, which grants licences to manufacturers for the marketing and distribution of HCT/Ps within the USA. These licences are usually only granted after evidence of safety and efficacy has been demonstrated in clinical trials. Exemptions to this rule are permitted under Section 361 for products that are for homologous use, are not combined with other agents, are minimally manipulated and either have no systemic or metabolic effects or are either for reproductive use, allogeneic use in the first or second degree relative or *autologous use*[2] that is, cells must be implanted into the same individual from which they were recovered.[3] These products are regulated solely under the PHS Act which only requires registration with the CBER and compliance with Current Good Tissue Practices for the prevention of communicable diseases.

Key to gaining this exemption, however, is the definition of minimal manipulation. According to the regulations, minimal manipulation is defined as 'processing that does not alter the relevant biological characteristics of cells',[4] and the FDA considers expanded cultures of ASCs to be more than minimally manipulated.[5] The ISSCR (2008, p. 6) guidelines, which are aligned with these regulations, define minimal manipulation as 'cells maintained in culture under non-proliferating conditions for short periods of time, normally less than 48 h'. For these cells, the ISSCR (2008, p. 7) guidelines recommend adherence to any relevant laws that regulate the manufacture and use of HCT/Ps. They also recommend that 'scientists and regulators [...] work together to develop common reference standards for minimally acceptable changes during cell culture'.

The ICMS guidelines, on the other hand, limit the concept of cell manipulation to genetic modification and distance themselves from the regulator's definitions by focusing instead on the notion of 'minimal culture expansion'. In framing this concept, the ICMS (2012, p. 3) guidelines draw an analogy between the 'minimal culture expansion' of autologous ASCs and the techniques that have developed in culturing human blastocysts for in vitro fertilisation. Minimal culture expansion is defined as cells cultured for no longer than 60 days and not exceeding ten passages after colony formation (2012, p. 8). According to the guidelines, the autologous use of ASCs that meet this definition should constitute the practice of medicine and be overseen by medical professionals, rather than the delegated regulatory authorities. Indeed, there is no mention of laws or regulations in the guidelines whatsoever.

In the USA, medical practice is not regulated by a federal agency but is instead governed by a complex framework of state licencing boards, professional accreditation bodies, third-party payers and torts law (Taylor 2010). Medical and clinical research is subject to the US Health and Human Services Department's *Federal Policy for the Protection of Human Subjects* (the 'Common Rule'), which stipulates that research protocols must be reviewed and approved by an Institutional Review Board (IRB). However, this rule only applies to research that is conducted or supported by federal departments and agencies. Therefore, a stem cell innovation would fall solely within the practice of medicine providing that the procedure is not done in the context of research in a publicly funded institution and does not involve the use of an unlicenced biological drug product.

---

[2] CFR Section 21 §§ Part 1271.10, revised 2012.

[3] CFR Section 21 §§ Part 1271.3(a), revised 2012.

[4] CFR Section 21 §§ Part 1271.3(f)(2), revised 2012.

[5] 'We do not agree that the expansion of mesenchymal cells in culture or the use of growth factors to expand umbilical cord blood stem cells are minimal manipulation.' Final rule: 66 Fed. Reg. 5447 (Apr 4, 2001) (to be codified at 21 CFR pt. 207, 807, and 1271).

The concept of minimal manipulation in the ICMS guidelines has been framed in such a way that, if accepted, the autologous use of 'minimally expanded' ASC cultures would be excluded from regulation as a Section 361 product. The multiple passages and cell growth would thus provide physicians with the much greater latitude to administer autologous ASCs without needing to apply for costly licences or register formal clinical trials, whereas the non-proliferating conditions and short culturing period endorsed in the ISSCR guidelines would see many, if not most types, of stem cell innovations falling under the purview of the regulatory authorities as formal clinical research. This framing, along with recommendations that scientists work with regulators in generating clearer definitions of minimal manipulation, discursively aligns the scientific and clinical research community with the regulatory agencies and situates the translation of stem cell medicine within the context of research. These framing strategies are also reflected in the level of oversight and standards of evidence that are prescribed in each guideline for the introduction of stem cell innovations into clinical settings.

9.2.2   *Establishing Evidence and Oversight Provisions*

In the USA and elsewhere, medical interventions are generally introduced into clinical settings through practice standards, codes and guidelines. International guidelines and 'gold standards' in clinical care are developed according to principles of evidence-based medicine (EBM) (Timmermans and Berg 2003). EBM is defined as the systematic evaluation of safety and efficacy through 'the conscientious, explicit, and judicious use of current best evidence' (Sackett et al. 1996). This paradigm, which now dominates the practice of modern medicine, is based on a hierarchical framework that evaluates the strength of evidence (Guyatt et al. 1995) and prioritises the meta-analysis and systematic review of randomised controlled trials (RCTs) over other evidence such as case-control studies and case reports (Greenhalgh and Rogers 2007). According to this approach, scientific evidence should be gathered in formal research protocols, preferably in statistically significant double-blinded RCTs across multiple sites, before novel interventions are introduced into clinical settings.

The evidence base prescribed in the ISSCR guidelines is most consistent with these principles. While the guidelines do not explicitly prioritise RCTs, they assert that, with few exceptions, evidence to support the safety and efficacy of stem cells, from any source, outside of their currently accepted uses should be done within formal clinical trials and with appropriate regulatory oversight. It states that pre-clinical studies 'in an appropriate in vitro and/or animal model' must first demonstrate proof of principle and provide 'persuasive evidence' of the toxicity and efficacy of novel stem cell-based approaches (ISSCR 2008, p. 8). These studies should be carried out 'prior to the initiation of clinical trials' and be 'subject to rigorous and independent peer review and regulatory oversight' (ISSCR 2008, p. 8). Human clinical research should then proceed under 'internationally accepted principles that govern the ethical conduct of clinical research and the protection of human subjects' (ISSCR 2008, p. 11). All studies must also comply with local and national regulatory processes and be:

> subject to independent review, approval, and ongoing monitoring by human subjects research oversight bodies with supplemental appropriate expertise to evaluate the unique aspects of stem cell research and its application in a variety of clinical disciplines. (ISSCR 2008, p. 8)

This framework reflects the traditional paradigm of translational clinical research that dominates EBM. The *in principle* demonstration that an intervention is safe and effective in peer reviewed scientific studies is the beginning of a legitimation process that validates evidence to support the introduction of novel medical interventions through a pathway of Phase I–IV clinical trials. Efficacy is typically established during Phase III in RCTs. This process is not only epistemologically functional, but it has important legitimating effects in maintaining the professional credibility and accountability of clinicians working in the stem cell field (Wilson-Kovacs et al. 2010). Incorporating the relevant laws and regulations, and conducting clinical trials in accordance with principles of EBM, reinforces the authority of science and scientific expertise over the collection, validation and oversight of evidence. It also ensures that this evidence is established in the context of research and not practice.

This approach differs substantially from the evidence base prescribed in the ICMS guidelines. Instead of establishing evidence through the translational pathway, the ICMS guidelines asserts that the introduction of autologous ASCs be facilitated through an industry-managed registry and accreditation programme. According to the guidelines, the accreditation process would require members to submit information about their ASC lines (and the patients they are administered to) as evidence to the ICMS registry. The cells would then be graded according to a matrix that classifies them as either having been established or unestablished 'in prior human testing' (ICMS 2012, p. 9). Established cells are considered to be 'clinical grade', while unestablished ASCs are broken down further into five categories (ICMS 2012, pp. 9–10):

1. 'Pre-investigational' cell lines where no animal data are available. These lines should be tested in animals with IRB approval before being used in humans.
2. 'Early investigational' cell lines where there are several animal models to show efficacy and safety but no human data exists. These cells may be used in early human studies with 5–10 patients with IRB approval.
3. 'Late investigational' cell lines which are being tested in larger groups of 20–50 patients with IRB approval.
4. 'Early clinical' cell lines which are being administered clinically in 50–200 patients.
5. 'Late clinical' cell lines which are being administered clinically in 100–300 patients.

The matrix contradicts the underlying premise that physicians are not manufacturing biological drugs when 'cell lines' are created and maintained in culture for continued testing in more than one patient. Furthermore, the evidence prescribed in this matrix would likely be limited to case series, which are descriptive studies that do not rank highly in the hierarchy of evidence in EBM. While members are encouraged to conduct 'blinded clinical trials', they are not considered necessary as 'ICMS physicians will pool resources to allow these trials to be undertaken so that insurance reimbursement can be sought for therapies that pass this level of evidence' (ICMS 2012, p. 22). IRB approval is only required until an intervention has been tested in 50 patients and exemptions are available for 'late investigational cell lines' that are used 'within the same tissue category' (i.e. homogeneous) of an established cell line (ICMS 2012, p. 22). Some cell lines may be 'grand fathered' into the registry if safety data exists that meets the above requirements with the ICMS (2012, p. 11) acting as the 'final arbiter of the number cases followed in each stage and if those cases have shown significant complications'. The ICMS also 'takes no position on how research is funded' and approves the use of 'pay for trial' type research (2012, p. 22).

The last statement implies that members may charge patients to take part in the prescribed investigational studies as a treatment option, regardless of whether a cell line has been established as safe or effective. It is reflective of a consumer-driven approach in which evidence is collected, validated and disseminated by medical professionals practicing within the context of the private healthcare industry in the USA. The registry, which is administered and overseen by the ICMS leadership, focuses largely on safety data, leaving market forces to determine whether a procedure is effective or better than the available alternatives. Where properly controlled studies are encouraged, they are aimed at satisfying the evidentiary requirements of insurance providers. As third-party payers generally do not provide coverage for interventions that are considered as experimental, or outside the standard of care, establishing an ASC intervention as clinical practice, rather than research, not only has clear commercial imperatives, but it is an important framing strategy that inserts the authority of the medical profession over the legitimation of innovative medicines with autologous ASCs.

9.2.3 Framing Stem Cell Innovation as Research or Practice

The analysis has thus far examined how each guideline defines key terms and concepts, and prescribes the standards of oversight and evidence that should be required to introduce SCBI into clinical settings. However, both guidelines have additional provisions that allow members to administer stem cells outside these prescribed settings as innovative therapies. While these provisions imply different standards of evidence and oversight, the underlying concepts are consistent with an overall approach that frames stem cell innovation as either research or practice.

The ICMS guidelines do not explicitly address the concept of innovation per se, but it is implied in Section K, which allows for 'compassionate uses' of untested cell lines in terminally ill patients. Under this section, unestablished cell lines, including pre-investigational cells, may be administered 'without restriction' provided that the patient is at end stage with a likely fatal illness or incurable disease and 'no other types of care are available or other reasonable alternatives have failed, and the patient's condition is expected to worsen' (ICMS 2012, p. 30). This prognosis should be confirmed in written statement from a 'board certified physician in the same area of specialty' and the intervention must be approved by an IRB (ICMS 2012, p. 30). The guidance contradicts the classification matrix by stipulating that 'multiple animal models must be available showing efficacy' for pre-investigational cells (ICMS 2012, p. 30), which may be a drafting error as these cells should then qualify as early investigational lines. In any case, it is implied that the use of these measures would arise as a variation in the care of an individual patient rather than as a formal research strategy.

The ISSCR (2008, p. 5) guidelines take a more direct approach in Section 7 under the heading titled *Stem Cell-Based Medical Innovation*, which outlines the 'exceptional circumstances' in which members may administer stem cells 'outside the context of a formal clinical trial'. In this section, medical innovation is defined as an activity 'aimed primarily at providing new forms of clinical care that have a reasonable chance of success for patients with few or no acceptable medical alternatives' and is distinguished from clinical research as being focused on improving 'an individual patient's condition' rather than producing 'generalisable knowledge' (ISSCR 2008, p. 15). The guidance recognises that 'responsible clinician-scientists may have an interest in providing medically innovative care' with stem cells 'prior to proceeding to a formal clinical trial' (ISSCR 2008, p. 5). According to recommendations that follow:

> Clinician-scientists may provide unproven stem cell-based interventions to at most a very small number of patients outside the context of a formal clinical trial, provided that:

(a) there is a written plan for the procedure that includes [...] scientific rationale and justification explaining why the procedure has a reasonable chance of success, including any preclinical evidence of proof-of- principle for efficacy and safety [...]
(b) the written plan is approved through a peer review process by appropriate experts who have no vested interest in the proposed procedure. (ISSCR 2008, p. 15)

The guidelines also recommend that there be 'a commitment by the clinician-scientist to use their experience to contribute to generalisable knowledge' by 'ascertaining outcomes in a systematic and objective manner' and 'moving to formal clinical trials in a timely manner' (ISSCR 2008, p. 16). This provision implies that while innovation may occur outside the context of research in a limited set of certain circumstances, the 'clinician-scientist' should remain committed to the translational model of EBM. The provision thus reinforces the role of science and scientific expertise in the oversight of stem cell innovation even when it occurs in the context of clinical practice.

The provision also has an important function in setting out criteria that, if not adhered to, should 'call into question the legitimacy of the purported attempts at medical innovation' (ISSCR 2008, p. 16). Indeed, the guidelines explicitly draw a distinction between 'legitimate attempts at medical innovation' and the 'commercial purveyance of unproven stem cell interventions' where 'a large series of patients' are administered with stem cells 'outside a clinical trial' and charged for such services (ISSCR 2008, p. 5). This distinction is important because it reflects the stated purpose of the ISSCR guidelines as the organisation's response to clinical establishments offering SCBI outside the standard of care on a commercial basis. Members of the ISSCR taskforce reason that negative public reactions to the stem cell tourism industry could potentially undermine the credibility of stem cell science if these establishments were to be associated with the broader field of stem cell research and innovation (Hyun 2008). By discursively distancing the scientific community from the perceived threat of stem cell tourism, the guidelines thus perform a degree of 'boundary work' (Gieryn 1999).

The ISSCR distinction is also in sharp contrast with the framework adopted in the ICMS guidelines. This framework not only endorses the administration of autologous ASCs to many patients outside clinical trials but allows members to charge patients throughout the investigational stages of the innovation process. The ICMS guidelines thus implicitly endorse a commercial model of stem cell innovation that the ISSCR guidelines explicitly reject. While it is not a norm for researchers to charge patients to take part in clinical trials, medical practitioners working in the private healthcare sector may administer non-standard/experimental procedures on a fee for service basis, even though patients are unlikely to be reimbursed from their insurance provider. These features, along with the evidence prescribed in each guideline, reinforces the respective framing of innovation and the authority of scientific and clinical expertise in the emerging field of stem cell medicine.

9.3 Implications for Conceptual and Oversight Frameworks

In the above analysis, two policy documents from two organisations were examined to compare how they each frame the concept of innovation and apply standards of evidence and oversight for use of such innovations in clinical contexts. While the two organisations differ greatly in size, scope and global influence, this comparison nonetheless sheds light on how the framing strategies employed in these documents reflects their institutional interests. The framing of stem cell innovation as either research or practice normatively prescribes the contexts in which novel uses of autologous ASCs should be introduced into the clinical settings and who should oversee this process. This framing has important implications for who should have authority over the collection, validation and dissemination of evidence and what should count as evidence. Framing stem cell innovation as medical practice locates authority to oversee the introduction of stem cells into clinical contexts with the medical profession and privileges case studies/series in establishing evidence, while innovations that are framed as translational research are placed within the domain of the scientific community and prioritise the scientific method in generating evidence. In this section, I will review some of the conceptual and practical implications of these findings for establishing oversight mechanisms and an evidence base for stem cell innovations.

9.3.1 Conceptual Frames for Stem Cell Innovation

Despite occurring in many fields of medicine, and often with minimal oversight, medical innovation is often difficult to distinguish from practice and research, and innovative SCBI are unlikely to be any different. The Belmont Report, which forms the basis of the Common Rule and is applied internationally in legislative instruments that regulate human subject research, defines medical practice as an intervention designed solely to benefit an individual patient and research as an activity that is designed to test hypotheses and produce generalisable knowledge (United States Health and Human Services 1979). According to the Report, innovation occurs when there is a significant departure from accepted practice or variation in the standard of care that is not the object of a formal research protocol. The procedure may be new or experimental but does not constitute research unless it is the object of a formal protocol.

The distinctions made in this Report have drawn sharp criticisms. While the Report recommends that 'major' innovations be subject to research at an early stage, it provides considerable latitude in what constitutes as a significant departure from the standard of care (Eaton and Kennedy 2007). The Report is also inadequate in its analysis of how innovation is applied in the context of patient care or how novel treatments are developed when there are no accepted therapies (Agich 2001). In addition, there are financial incentives in how clinicians are reimbursed, and innovators may have a personal financial stake in the intellectual property that is connected to an innovative therapy (Taylor 2010). These incentives may encourage clinicians to experiment with innovative procedures and adopt them as practice rather than engage in the often onerous and costly burdens of research.

As they currently read, the ICMS guidelines are likely to have such implications by not only encouraging members to experiment with autologous ASCs as a matter of practice without initiating formal clinical research but providing a cover of legitimacy for commercial purveyors of unproven SCBI. The treatment of innovation in the guidelines, as delineated from research and practice, is tacit, at best: it does not explicitly mention innovation nor does it make any clear distinctions between research and practice. The recommended investigational stages within the accreditation programme imply that some sort of research protocol would be required, but the various circumstances under which IRB approval would not be needed are confusing and, in some places, contradictory. The endorsements that allow physicians to charge patients to take part in research, as if providing a service, obscures the distinction between research and practice even further.

A similar approach has been adopted by the Texas Medical Board. Following Governor Rick Perry's intervention in 2011, the Board determined that the 'use of investigational agents constitutes the practice of medicine' and established standards that allow physicians to use stem cells in certain clinical settings pending approval from an IRB (Texas Medical Board 2012). According to these rules, which were adopted in July 2012, the use of stem cells shall be considered investigational 'unless they are used in the conduct of an FDA-approved protocol or until such time as they are approved by the FDA' (Texas Medical Board 2012). Such 'investigational' uses shall be permitted with the approval of an IRB but will exclude off-label uses of FDA-approved drugs and biologics, any HCT/Ps that are manufactured 'pursuant to Sections 351 and 361' of the PHS Act, or anything that has already been approved by an IRB (Texas Medical Board 2012). As federal law requires that all stem cells, including autologous ASCs, be manufactured pursuant to the PHS Act, it is unclear precisely which type of stem cells would fall under the purview of these standards. The circumstances in which they would be used without IRB approval are also unclear as the standards exclude products that are IRB-approved from the definition of an investigational agent.

Setting aside these possible drafting issues, the Board has been widely criticised for appearing to allow physicians to substitute FDA oversight with an IRB approval. For instance, Levine argues that stipulations for IRB approval (without regulatory oversight) should only apply to investigational agents that fall outside the scope of FDA regulation and/or would not require premarketing approval (Levine 2012). He also raises concerns over the lack of reporting required for the evaluation of safety and efficacy. While the rules stipulate that investigational agents, whatever they may be, should be administered as part of a 'systematic program competently designed, under accepted standards of scientific research' (Texas Medical Board 2012), there is no requirement for results to be submitted to the Board or published more broadly within the scientific and medical literature, and physicians may presumably charge for these interventions on a pay-for-service basis. These conditions blur the boundary between research and practice, and it is debatable as to whether the standards will provide physicians, let alone their patients, with clear guidance on when and how an innovative SCBI should be administered outside the context of formal trials.

The ISSCR guidelines provides some greater clarity by drawing on the distinctions made in the Belmont Report and explicitly laying out some circumstances that might justify administering stem cells to patients outside clinical trials and without IRB approval. Members of the ISSCR taskforce have reasoned that innovation with some types of stem cells could occur within the context of patient care rather than achieving epistemic goals (Taylor 2010, Hyun 2010). Such contexts might include prescribing an approved stem cell line off-label for other non-approved purposes or applying for 'compassionate use' of investigational agents that are being administered in registered clinical trials (Lindvall and Hyun 2009). Certain approaches also might not be fully amenable to clinical trial, such as those with a strong surgical component. In these circumstances, an 'unproven' SCBI could be administered to a small number of patients who have exhausted existing treatment options providing that the clinician has intentions of eventually investigating the intervention in a formal research protocol.

While the ISSCR guidance provides a far more sophisticated framework for stem cell innovation, further clarification is needed to specify what types of treatments would constitute as 'innovative', when they should be provided to patients and according to what standards of evidence, whether and when they should move into clinical trials, and who should oversee this process (Cohen and Cohen 2010a). The guidance also does not indicate how far outside the standard of care an unproven

SCBI should be before formal clinical research is initiated nor how such variations in the standard of care should be assessed. Importantly, while distinguishing 'legitimate' innovations from those provided to large numbers of patients on a fee-for-service basis, the ISSCR guidance does not specify how small patient population should be before clinicians should initiate clinical trials (Cohen and Cohen 2010b) nor who should pay for an unproven intervention if not the patient as neither public nor private healthcare systems are compelled to cover experimental procedures. The construct thus requires further development.

9.3.2 Oversight Frameworks for Stem Cell Innovations

In establishing oversight frameworks, some have looked towards the field of surgery, which claims to have a rich tradition of innovation (Riskin et al. 2006). In 2006, criteria were developed to help surgeons identify when a variation in standard practice should be considered as an innovation and when it should be incorporated into formal research protocols (Reitsma and Moreno 2006). According to these criteria, innovations may be categorised as either a routine variation in the standard of care made in response to a patient's individual needs, a significant variation that is intended to produce generalisable knowledge and requires IRB approval or a planned variation that is not aimed at testing a hypothesis but is to be administered to more than one patient in response to their needs. These criteria were subsequently adopted in the guidelines of the US Society of University Surgeons for the ethical development and application of surgical innovations (Biffl et al. 2008). They are meant to provide surgeons with the freedom to innovate with experimental procedures and techniques without the burdens of clinical research.

Comparisons with surgery have some utility for stem cell innovation due to the likelihood that many SCBI will involve a surgical component in either retrieving and/or transplanting cells. For example, an international team of surgeons and scientists have successfully transplanted two engineered tracheas seeded with autologous bone marrow-derived stem cells (Jungebluth et al. 2011, Macchiarini et al. 2008). Both procedures were performed as life saving measures in the context of clinical care and represented a significant departure from the standard of care after established treatment options had been exhausted. The highly experimental innovations, which involved the surgical removal of the patients' windpipe and transplantation of an engineered trachea, were not part of a formal clinical trial and were supported with only minimal preclinical animal data (Hollander 2010). Oversight was provided by local ethics committees, and, since both procedures also involved the use of engineered cells, special permission was obtained from the relevant authorities that regulate HCT/Ps in the various countries involved.

According to the 2008 amended version of the World Medical Association's Declaration of Helsinki, unproven medical interventions such as these may be administered to small patient populations where existing treatments are ineffective or non-existent. Article 35 of the Declaration also states that:

> In the treatment of a patient, where proven interventions do not exist or have been ineffective, the physician, after seeking expert advice, with informed consent from the patient or a legally authorized representative, may use an unproven intervention if in the physician's judgement it offers hope of saving life, re-establishing health or alleviating suffering. Where possible, this intervention should be made the object of research, designed to evaluate its safety and efficacy. In all cases, new information should be recorded and, where appropriate, made publicly available. (World Medical Association 2013)

These humanitarian provisions are reflected in the FDA's expanded access programme for investigational new drugs, as well as other regulatory regimes: for example, the 'hospital use' exemption of the Advanced Therapy Medicinal Products directive of the European Commission, which would have permitted the second trachea experiment to proceed. A major caveat when invoking these provisions, however, is that they are essentially reserved for one-off lifesaving interventions where no other comparable treatment options are available: *they do not justify administering an unestablished intervention to large numbers of patients outside the context of formal clinical research, and charging them for it*. Indeed, the statement strongly suggests that experimental procedures should preferably be tested within a research protocol and be subject to the approval of an ethical review committee. Ordinarily, research protocols that involve human subjects are reviewed and approved by an IRB, but if the intervention is not subject to research, an approval could be sought from a hospital-based ethics committee or some other sanctioned group that provides oversight of ethical issues that arise in the context of clinical practice or patient care.

This approach concurs with the ISSCR guidelines, which recommends oversight from an *appropriate body of experts*, but not necessarily an IRB. Patrick Taylor, who presided on the ISSCR taskforce, has argued that IRBs are unsuitable for this task because they are already overburdened and they lack the institutional influence needed to manage the clinical risks involved in testing an innovative procedure (Taylor 2010). Instead, he and other members of the taskforce have proposed new oversight bodies similar to the local innovation committees recommended by the US Society of University Surgeons for surgical innovations (Lindvall and Hyun 2009). These committees are meant to operate like an IRB but focus on quality control and patient safety and are aimed at giving surgeons the flexibility to innovate without further burdening the IRB system (Lindvall and Hyun 2009). An innovator would be directed to an IRB only after the committee believes that an innovation should be investigated under more rigorous conditions.

While appealing, innovation committees will likely be based within institutions that are appropriately resourced to provide administrative support and access to expertise in both stem cell science and the targeted medical condition. Clinicians who are not institutionally based or operate in countries that lack these resources could 'outsource' their review to the innovation committees of larger institutions.[6] However, the committee would not have the type of influence in clinical risk man-

---

[6]This outsourcing occurs occasionally with specialist oversight committees for stem cell research (i.e. ESCRO or SCRO committees).

agement that Taylor (2010) argues is lacking from the IRB structure in the first place. There are further questions over representation as committee members would have considerable influence over the classification of stem cell innovations as either routine variations or major departures from the standard of care, which will affect the degree of oversight and standards of evidence required of clinicians. Influence over these decisions not only raises possible conflicts of interest on the behalf of members who work in the field of stem cell research (Cohen and Cohen 2010a), but it also reinforces and legitimates the authority of certain types of knowledge and expertise over the translation of stem cell medicines, which arguably lies at the heart of the tensions in the two guidelines.

9.3.3 *Standards of Evidence in Stem Cell Innovation*

Both guidelines use framing strategies that normatively prescribes an evidence base that reflects the institutional interests of the organisation's membership. One presents a version that privileges the authority of science in medicine by inserting scientific knowledge and expertise in the collection, validation and oversight of evidence. This version aligns itself with the regulatory authorities and incorporates the relevant laws that regulate HCT/Ps and medical experimentation with human subjects. The other privileges the knowledge and experience of medical professionals operating within less regulated and market-centric sectors of the private healthcare industry in the USA. This version not only ignores existing laws and regulations that govern human subject research and commerce in biological drug products, but it fails to acknowledge the basic standards of evidence that should be required to establish an innovation with stem cells in clinical practice.

The ICMS guidelines prioritise case studies/series, which is unlikely to establish a persuasive evidence base to support the efficacy of cell cultures that are created for, and administered only once to, an individual patient. Placebo effects have been shown to be strong with somatic stem cell transplants, which, once tested in adequately controlled studies outside their current uses, have tended to produce mixed results: probably due, in part, to a lack of standardisation and characterisation of ASC cultures (Trounson et al. 2011). Until these basic standards of evidence are met, it is unlikely that insurers and other third-party payers will provide coverage for any SCBI that is administered outside the standard of care, especially those using cells that lawfully fall under the jurisdiction of drug regulators. For physicians that operate outside of public and private reimbursement systems, however, the need to establish these standards becomes less pertinent. Instead, the emphasis shifts towards establishing accepted norms within a community of professionals who can attest to the reasonableness of an intervention in medical malpractice suits.[7] While

---

[7] In the USA, medical torts laws fall within the jurisdiction of individual states although the general framework, as inherited from British common law, requires proof that substandard medical care resulted in an injury. For an overview of medical negligence law in the USA, see Bal (2008).

it is unclear what sort of legal protection the ICMS guidelines might provide if members are sued for medical negligence following an injury with ASCs, both the registry and the compassion use provisions are far more aligned with these interests than meeting standards set out within EBM.

The ISSCR guidelines, on the other hand, embody the dominant paradigm of translational research in EBM by prioritising evidence that is gathered using scientific methods in formal clinical trials. The current gold standard in EBM is multiple-site double-blinded RCTs (Timmermans and Berg 2003). However, medical innovations are often introduced into clinical settings, and accepted into practice, well before these standards are established (Denis et al. 2002). Indeed, EBM and its dominance in practice standards have been heavily criticised by scholars who have argued that the philosophical basis of EBM is epistemologically flawed for focusing too narrowly on the RCT (Cohen et al. 2004, Ashcroft 2004). Value judgements that are routinely used in clinical decision-making are systematically excluded from the RCT (Kerridge et al. 1998), while intrinsic biases are frequently ignored (Kelly and Moore 2012). The legitimacy of EBM has also been criticised for essentially replacing one system of authority (the clinician) with another (the clinician-scientist) in the evaluation, legitimation and dissemination of evidence (Upshur and Tracy 2004): these tensions are apparent in the two guidelines.

However, it is challenging to circumvent the need for rigorously tested evidence in medicine. This need was highlighted starkly with the use of autologous bone marrow stem cell transplant (BMT) with high-dose chemotherapy (HDC) for the treatment of breast cancer in the late 1980s/1990s (Rettig et al. 2007). Considered as an innovation in medical practice at the time, the use of BMT-HDC for breast cancer did not require any specific regulatory approvals (the chemotherapy drugs were already approved, and the isolated cells were not expanded in culture) but was accepted into practice following a number of small clinical studies indicating its effectiveness in reducing the size of tumours. However, it was not until double-blinded RCTs had concluded that it was no more effective than the standard treatment in reducing mortality that the procedure was abandoned. A confluence of commercial interests, patient advocacy and lawsuits against the third-party payers attributed to this inefficacious procedure being accepted into practice standards before reliable evidence of efficacy could be established in properly controlled trials.

Similar scenarios have been witnessed in the field of surgery where innovations in metal on metal hip replacements made of cobalt-chromium alloy, rather than ceramic or polyethylene, were recently found to have toxic effects in post-marketing studies after the early failure of thousands of implants (Cohen 2012). The assumed effectiveness of the widely practiced arthroscopic surgery for knee osteoarthritis has also been recently undermined in two RCTs (Mounsey and Ewigman 2009). To help manage the need for transparency and evidence in surgical innovation, the US Society of University Surgeons has recommended the establishment of registries that are meant to help provide early rationale for an innovative procedure and facilitate the sharing of information about the outcomes of experimental surgical procedures (Biffl et al. 2008). However, a similar registry previously established by the American College of Surgeons is no longer available. Surgeons are now advised that for innovations that 'depart in a significant way from standard or accepted practice, the innovation should be made the object of formal research at an early stage to determine if it is safe and effective',[8] although what counts as a 'significant' departure from the standard of care, and who makes these decisions, remains unclear. Indeed, recommendations made by the IDEAL consortium in the UK for more rigorous evaluation and oversight suggest that many of the ethical, regulatory and evidentiary issues surrounding innovative surgery are still unresolved (Johnson and Rogers 2012) Thus, it would be premature to establish a framework for stem cell innovation based on the field of surgery.

9.4   Concluding Remarks

This paper analysed the guidelines of two international organisations on the clinical translation of innovative SCBI and found two different perspectives on when it may be justifiable to introduce stem cell innovations into clinical settings without evidence from clinical trials. By framing stem cell innovation as either research or practice, the two guidelines each endorse different methodologies and standards of evidence and situate the authority to oversee the collection, validation and dissemination with either the scientific community or the medical profession. Differences in the level of oversight and evidence prescribed in each guideline are indicative of the contested authority of science and scientific expertise in the practice of medicine. Through these guidelines, both organisations are arguably staking out territory in the enormously promising and potentially lucrative field of stem cell medicine on behalf of two very powerful institutions.

From the above analysis, the ISSCR clearly has the most coherent set of guidelines that are not overtly aligned with the interests of commercial purveyors who are seeking to profit while operating outside internationally accepted practice standards. However, a strong case for removing the oversight of new and innovative SCBIs from the existing ethics review framework is lacking. While current systems are imperfect and in need of ongoing improvement, especially where there are vested interests in the oversight of an innovation, it is unclear that their replacements will be any more efficient or effective in providing adequate protections for patients while enabling investigation into promising new therapies. Nor has an exceptionalist approach to stem cell innovation been adequately justified. Until then, better enforcement of existent laws and regulations that govern both medical practice and clinical research may be of greater value than inventing new and potentially less transparent mechanisms for stem cell innovation.

Part of the problem may be that the concept of 'innovation' is an unstable construct that manifests in practice as unstructured and largely unsupervised medical experimentation. Indeed, the framing of innovation as clinical practice, and not medical experimentation, potentially creates a false dichotomy between the ethics

---

[8] See section V(A) of the American College of Surgeons (1997).

of patient care and human subject research (Sugarman and Sipp 2010). If modern medical practice is predicated on evidenced-based approaches, then any innovation that is aimed at providing new forms of clinical care must necessarily contain an epistemological component. An intervention that has epistemic goals, even if they are not the primary focus, must still fall, at least in part, within an experimental paradigm. Therefore, administering stem cells of any sort to patients outside of their currently accepted uses to more than (perhaps) a *very* small number of patients without an IRB-approved protocol may be ethically problematic. How small this number should be and under what specific circumstances such procedures could occur without IRB approval, and who should oversee this process, warrants further interrogation.

Acknowledgements This research was supported by the National Medical Research Council under the funding initiatives of the National Research Fund, Singapore. Many thanks go to Dr. Karen Maschke of the Hastings Centre for her mentorship and Professor Alastair Campbell and Dr. Benjamin Capps at the Centre for Biomedical Ethics, National University of Singapore, for their valuable insights.

References

Agich GJ (2001) Ethics and innovation in medicine. J Med Ethics 27:295–296

American College of Surgeons (1997) Statements on principles [Online]. Available: http://www.ama-assn.org/ama/pub/physician-resources/medical-ethics/about-ethics-group/ethics-resource-center/educational-resources/federation-repository-ethics-documents-online/american-college-surgeons.page. Accessed 28 Mar 2012

American Stem Cell Therapy Association (2009) Physician group opposes FDA's position on adult stem cells [Online]. Available: http://web.archive.org/web/20090329110142/http://www.stemcelldocs.org/Media_Room.html. Accessed 5 July 2012

Ashcroft RE (2004) Current epistemological problems in evidence based medicine. J Med Ethics 30:131–135

Bal BS (2008) An introduction to medical malpractice in the United States. Clin Orthop Relat Res 467:339–347

Biffl WL, Spain DA, Reitsma AM, Minter RM, Upperman J, Wilson M, Adams R, Goldman EB, Angelos P, Krummel T, Greenfield LJ (2008) Responsible development and application of surgical innovations: a position statement of the society of university surgeons. J Am Coll Surg 206:1204–1209

Caulfield T, McGuire A (2012) Athletes' use of unproven stem cell therapies: adding to inappropriate media hype? Mol Ther 20:1656–1658

Cohen AM, Stavri PZ, Hersh WR (2004) A categorization and analysis of the criticisms of evidence-based medicine. Int J Med Inform 73:35–43

Cohen CB, Cohen PJ (2010a) International stem cell tourism and the need for effective regulation: part I: stem cell tourism in Russia and India: clinical research, innovative treatment, or unproven hype? Kennedy Inst Ethics J 20:27–49

Cohen CB, Cohen PJ (2010b) International stem cell tourism and the need for effective regulation: part II: developing sound oversight measures and effective patient support. Kennedy Inst Ethics J 20:207–230

Cohen D (2012) How safe are metal-on-metal hip implants? Br Med J 344:e1410

Cyranoski D (2012) Stem-cell therapy takes off in Texas. Nature 483:13–14

Daley GQ, Hyun I, Lindvall O (2008) Mapping the road to the clinical translation of stem cells. Cell Stem Cell 2:139–140

Denis J-L, Hébert Y, Langley A, Lozeau D, Trottier L-H (2002) Explaining diffusion patterns for complex health care innovations. Health Care Manag Rev 27:60–73

Eaton ML, Kennedy D (2007) Innovation in medical technology: ethical issues and challenges. John Hopkins University Press, Baltimore, MD

Gieryn TF (1999) Cultural boundaries of science: credibility on the line. University of Chicago Press, Chicago

Greenhalgh C, Rogers M (2007) The value of intellectual property rights to firms and society. Oxf Rev Econ Policy 23:541–567

Guyatt GH, Sackett DL, Sinclair JC, Hayward R, Cook DJ, Cook RJ (1995) Users' guides to the medical literature. IX. A method for grading health care recommendations. Evidence-based medicine working group. J Am Med Assoc 274:1800–1804

Hollander AP (2010) A case study of experimental stem cell therapy and the risks of over-regulation. In: Capps B, Campbell A (eds) Contested cells: global perspectives on the stem cell debate. Imperial College Press, London

Hyun I (2008) Stem cells from skin cells: the ethical questions. Hastings Cent Rep 38:20–22

Hyun I (2010) Allowing innovative stem cell-based therapies outside of clinical trials: ethical and policy challenges. J Law Med Ethics 38:277–285

International Cellular Medicine Society (2009) Clinical Guidelines Version 1.0 [Online]. Adopted by Committee on May 13, 2009. Available: http://www.cellmedicinesociety.org/attachments/054_ICMS%20Clinical%20Guidelines.pdf. Accessed 7 July 2011

International Cellular Medicine Society (2010) Join the ICMS [Online]. Available: http://www.cellmedicinesociety.org/home/join. Accessed 5 July 2012

International Cellular Medicine Society (2012) ICMS Guidelines [Online]. Available: http://www.cellmedicinesociety.org/icms-guidelines. Accessed 5 July 2012

International Society for Stem Cell Research (2002) Mission statement [Online]. International Society for Stem Cell Research. Available: http://www.isscr.org/home/about-us/mission-statement. Accessed 27 Feb 2013

International Society for Stem Cell Research (2008) Guidelines for the clinical translation of stem cells [Online]. International Society for Stem Cell Research. Available: http://www.isscr.org/docs/default-source/clin-trans-guidelines/isscrglclinicaltrans.pdf. 27 Feb 2013

Johnson J, Rogers W (2012) Innovative surgery: the ethical challenges. J Med Ethics 38:9–12

Jungebluth P, Alici E, Baiguera S, Le Blanc K, Blomberg P, Bozóky B, Crowley C, Einarsson O, Grinnemo K-H, Gudbjartsson T, Le Guyader S, Henriksson G, Hermanson O, Juto JE, Leidner B, Lilja T, Liska J, Luedde T, Lundin V, Moll G, Nilsson B, Roderburg C, Strömblad S, Sutlu T, Teixeira AI, Watz E, Seifalian A, Macchiarini P (2011) Tracheobronchial transplantation with a stem-cell-seeded bioartificial nanocomposite: a proof-of-concept study. Lancet 378:1997–2004

Kelly MP, Moore T (2012) The judgement process in evidence-based medicine and health technology assessment. Soc Theory Health 10:1–9

Kerridge I, Lowe M, Henry D (1998) Ethics and evidence based medicine. Br Med J 316:1151–1153

Kiatpongsan S, Sipp D (2008) Offshore stem cell treatments. Nat Rep Stem Cells. doi:10.1038/stemcells.2008.151

Kiatpongsan S, Sipp D (2009) Monitoring and regulating offshore stem cell clinics. Science 323:1564–1565

Lau D, Ogbogu U, Taylor B, Stafinski T, Menon D, Caulfield T (2008) Stem cell clinics online: the direct-to-consumer portrayal of stem cell medicine. Cell Stem Cell 3:591–594

Levine AD (2012) Improving oversight of innovative medical interventions in Texas, USA. Regen Med 7:451–453

Lindvall O, Hyun I (2009) Medical innovation versus stem cell tourism. Science 324:1664–1665

Lysaght T, Campbell AV (2011) Regulating autologous adult stem cells: the FDA steps up. Cell Stem Cell 9:393–396

Lysaght T, Campbell AV (2013) Broadening the scope of debates around stem cell research. Bioethics 27:251–256

Macchiarini, P., Jungebluth, P., Go, T., Asnaghi, M. A., Rees, L. E., Cogan, T., A, Dodson, A., Martorell J, Bellini, S., Parnigotto, P. P., Dickinson, S. C., Hollander, A. P., Mantero, S., Conconi M. T., Birchall, M. A. 2008. Clinical transplantation of a tissue-engineered airway. Lancet, 372, 2023-2030.

Mounsey A, Ewigman B (2009) Arthroscopic surgery for knee osteoarthritis? Just say no. J Fam Pract 58:143–145

Power C, Rasko JE (2011) Promises and challenges of stem cell research for regenerative medicine. Ann Intern Med 155:706–713

Regenberg AC, Hutchinson LA, Schanker B, Mathews DJH (2009) Medicine on the fringe: stem cell-based interventions in advance of evidence. Stem Cells 27:2312–2319

Reitsma AM, Moreno JD (2006) Ethical guidelines for innovative surgery. University Publishing Group, Hagerstown, MD

Rettig RA, Jacobson PD, Farquhar CM, Aubry WM (2007) False hope: bone marrow transplantation for breast cancer. Oxford University Press, New York

Riskin DJ, Longaker MT, Gertner M, Krummel TM (2006) Innovation in surgery: a historical perspective. Ann Surg 244:686–693

Sackett DL, Rosenberg WMC, Gray JAM, Haynes RB, Richardson WS (1996) Evidence based medicine: what it is and what it isn't. Br Med J 312:71–72

Sugarman J, Sipp D (2010) Ethical aspects of stem cell-based clinical translation: research, innovation, and delivering unproven intervention. In: Hug K, Hermerén G (eds) Translational stem cell research: issues beyond the debate on the moral status of the human embryo. Springer, New York

Taylor PL (2010) Overseeing innovative therapy without mistaking it for research: a function-based model based on old truths, new capacities, and lessons from stem cells. J Law Med Ethics 38:286–302

Texas Medical Board (2012) Practice guidelines for the use of investigational agents [Online] Available: https://texreg.sos.state.tx.us/public/readtac$ext.ViewTAC?tac_view=4&ti=22&pt=9&ch=198&rl=Y. Accessed 7 October 2015

Timmermans S, Berg M (2003) The gold standard: the challenge of evidence-based medicine and standardization in health care. Temple University Press, Philadelphia

Trounson A, Thakar R, Lomax G, Gibbons D (2011) Clinical trials for stem cell therapies. BMC Med 9:52

United States Health and Human Services (1979) Ethical principles and guidelines for the protection of human subjects of research [Online], April 18, 1979: U.S. Health & Human Services. Available: http://www.hhs.gov/ohrp/humansubjects/guidance/belmont.html. Accessed 12 Oct 2015

Upshur RE, Tracy CS (2004) Legitimacy , authority , and hierarchy: critical challenges for evidence-based medicine. Brief Treat Crisis Interv 4:197–204

Wilson-Kovacs DM, Weber S, Hauskeller C (2010) Stem cells clinical trials for cardiac repair: regulation as practical accomplishment. Sociol Health Illn 32:89–105

World Medical Association (2013) Declaration of Helsinki: ethical principles for medical research involving human subjects [Online], WMA. Available: http://www.wma.net/en/30publications/10policies/b3/. Accessed 23 May 2011

Chapter 10
The Regulation of Clinical Stem Cell Research in China

Li Jiang

10.1 Introduction

In the money-oriented global market, stem cell research is in some commercial area lacking proper moral standard and legal guidance. In 2003, the Chinese government launched the Ethical Guideline for Human Embryonic Stem Cell (HESC) Research. With increased interest in this field of research, the Chinese government has ordered a halt to the provision of unapproved stem cell treatments and clinical trials from 2015. However, despite these emerging regulatory controls, stem cell therapy and clinical stem cell research are booming in China from the early 2000s until now. The deep root for this booming is the regulatory system for laboratory science and medical application in stem cell therapy developments.[1]

10.2 The Economic Motivation for Clinical Stem Cell Research

Many scientists believe that stem cell may cure a patchwork of unprecedented diseases, such as diabetes, cancer, Parkinson's, Alzheimer's and heart diseases (Robbins-Roth 2001). The boundless potential of stem cells motivates countries to rebuild their current regulatory frameworks, facilitate stem cell research and maintain a competitive position in the global health market. In the USA, statistics from the National Institutes of Health (NIH) showed that the federal government

---

[1] Parts of this chapter are based on previously published work: (1) Jiang (2016a), (2) Jiang (2016b). Material from these two works is reprinted with permission from Springer Press and Mary Ann Liebert Inc. Publishers.

L. Jiang (✉)
Kenneth Wang School of Law, Soochow University, Suzhou, Jiangsu, China
e-mail: jelly3636@hotmail.com

© Springer International Publishing AG 2017

P.V. Pham, A. Rosemann (eds.), *Safety, Ethics and Regulations*, Stem Cells in Clinical Applications, DOI 10.1007/978-3-319-59165-0_10

invested nearly 1.429 billion dollars in 2015 and 1.495 billion dollars in 2016 for stem cell research.[2] In Europe, 54 billion Euros were budgeted for HESC research from 2007 to 2013 (Watt 2006). In China, the Ministry of Science and Technology launched the 2500 million RMB key national research programme for stem cell and translational medicine from 2015.[3] Certain health markets for luxury treatments have already matured around the world, such as for PGD, IVF, cancer treatment and sex selection. The question is why such governments have invested so heavily in stem cell research.

Chinese government desires an international and transnational competitive advantage. Stem cell research is still at an early stage, and researchers are developing the basic technologies for this area. To secure a competitive position in future commercial development, China provided excellent state funding-oriented stem cell research networks.[4] The Chinese government believes that scientific research using stem cells is a worthwhile investment.

10.3   Stem Cell Research Funding in China

Generally speaking, the funding strategy was successful. The major funding of stem cell research in China is obtained from governmental organizations, ranging from the Ministry of Science and Technology and the National Natural Science Foundation to the Chinese Academy of Sciences. The 973 programmes[5] and the major scientific research project programmes[6] are the two main sources of stem cell research funding. The research field of funded programmes is within the popular areas of world stem cell research.[7] The supporting priorities depend on China's national Five-Year Plan for National Economic and Social Development. During the eleventh Five-Year Plan, 29 stem cell research projects were funded by the 973 programmes and the major scientific research project programmes (Chen et al. 2011). The money from these programmes exceeded 832 million RMB. Over 50 research centres throughout the country obtained sponsorship from these programmes. In the latest

---

[2] https://www.bioinformant.com/stem-cell-funding/. Accessed online 13 Nov 2016.

[3] http://www.most.gov.cn/tztg/201502/t20150226_118286.htm. Accessed online 13 Nov 2016; see also http://www.infzm.com/content/115819 Accessed 13 Nov 2016.

[4] Ibid.

[5] The 973 programmes, also called the national basic research programme, were established in June 1997 in order to promote creativity and the sustainable development of China. Stem cell research is one supporting priority project by the 973 programmes.

[6] The major scientific research project mainly sponsors four areas: protein research, research on quantum control, nanotechnology research and research on development and reproduction.

[7] The hot research area in the world stem cell research is the embryo differentiation and transplant, iPS, HESC, tumour stem cell, neural stem cell, regulatory network of stem cell, stem cell used in heart disease treatment and core blood stem cell. See ibid.

thirteenth Five-Year Plan, stem cell and regenerative medicine are still hot and heavily invested areas.[8]

China's stem cell research is conducted by experts, most of whom have either obtained an overseas university degree or have spent some time training overseas.[9] The China Global Expert Recruitment Programme is highly attractive with a variety of financial and research incentives.[10] Third, some results of the funded programme are considered to be pioneering research worldwide. For example, Chinese scientists were the first to verify the totipotency of induced pluripotent stem (iPS) cells (Zhao et al. 2009), as well as the first to find a way of generating the induced pluripotent cell (Esteban et al. 2009). The research funding programme launched by the Chinese government facilitated China to be a rising star in regenerative medicine.

10.4   Stem Cell Industry in China

The stem cell research development to some extent depends on economic progress. Although the Chinese economy has grown in recent years, there is still a tremendous gap between China and Western countries. With regard to stem cell research, the fundamental facilities in some laboratories such as those in Beijing or Shanghai are considered world class (Esteban et al. 2009). The environmental facilities and equipment of some laboratories are even envied by the world leading experts (Guotong 2007). Average Chinese laboratory facilities still lag behind those in developed countries. However, the stem cell industry in China, both with regard to technology and business models, is in a rapid development phrase, and this bodes well for future prosperity.

Focusing on therapy, stem cell research in China is in the rapid process of being transferred from basic scientific research to practicable diagnostic procedures. Shenzhen Beike (Beike) is one such company that has won world renown for its stem cell therapy. From the laboratory to hospital application, Beike's highly reputable therapy has attracted patients from all over the world to undergo treatment in China. With the benefit of the first special economic zone of China, Beike combined laboratories and hospitals to establish treatment centres.[11] As the president of Beike, Xiang Hu said, '[i]nitially, we only cooperated with laboratories and hospitals which offered a good standard of equipment, excellent environment and a high level team' (Yong 2011). In order to promote the interaction, 'Beike creatively launched a stem cell public technical service platform and constructed a stem cell clinical research network' (Yong 2011). So far, Beike has announced the world's largest

---

[8] http://www.tech-ce.com/news_detail.aspx?tid=2&id=15222. Accessed online 24 Sept 2016.

[9] Ibid.

[10] Ibid.

[11] The city of Shenzhen has benefited of the 'opening and reform' policy by the Chinese leader Deng Xiaoping. As the first 'special economic zone', Shenzhen attracted many foreign investments as well as tax deductions (see Song 2011).

clinical application security evaluation of allogeneic human umbilical cord blood-derived stem cells, as well as publishing the research data of effective treatment in systemic lupus erythematosus, hereditary ataxia and muscular dystrophy.[12]

Da An Gene is another high-tech enterprise oriented in molecular diagnostic techniques in China.[13] Da An Gene offers a variety of genetic testing services including the whole genome sequencing and the targeted sequencing, which is approved by the Chinese Food and Drug Administration (CFDA). It is noticeable that the biggest shareholder of Da An Gene is Sun Yat-sen University. Based on it, three research institutions of school level (Da An gene diagnostic centre, Sun Yat-sen University tissue matching centre and Sun Yat-sen University biotechnology research institute) are established. These platforms facilitate clinical biochemical tests, clinical immunology test, clinical haematology inspection, clinical genetic tests, clinical gene inspection and clinical stem cell research as they are carried out in university's laboratory and used for research purpose.

Even in the capital market, it is possible to find companies whose main business relays on the stem cell industry. As the only one in the Shanghai and Shenzhen market, Zhongyuan Union Stem Cell Bioengineering Corporation successfully operates three famous stem cell enterprises: Union Stem Cell Genetic Co. Ltd., Union East China Stem Cell Gene & Engineering Co. Ltd. and Heze Biotechnology Co. Ltd.. The company holds certain important patents such as umbilical cord tissue-derived mesenchymal seeded separation method, human umbilical cord mesenchymal stem cell antifibrotic injection and its preparation method, human adipose adult stem cell acquisition method and construction of the stem cell bank (Xiaoqin 2011). From the above, we can conclude that Chinese companies have already entered the downstream market of the stem cell industry.

10.5 The Legal Framework of Stem Cell Clinical Research in China

Awareness concerning regulatory issues associated with stem cell treatment is increasing in China, and various government responses have attempted to monitor these therapies. In 2003, the Ministry of Science and Technology and the Health Ministry jointly announced the Ethical Guideline for HESC Research (Guideline). The Guideline attempts to clarify the prohibitive issues for reproductive and regenerative therapies, such as researches on reproductive cloning and implanting embryos used for research into women's wombs. Although the Guideline regulates some prohibited areas of research, no clause in the Guideline stipulates the qualifications of the research institutions or the licencing and supervising mechanism of the research project (The Ethical Guideline for HESC research in China 2003.). The Guideline seems to be too vague to monitor stem cell therapy.

---

[12] The Beike Biotech website. http://beikebiotech.com/. Accessed 20 Nov 2015.

[13] The Da An Gene website. http://daan.joomcn.com/index.php?option=com_content&view=article&id=1&Itemid=113. Accessed 24 Sept 2016.

In 2007, the Health Ministry launched the Ethical Guideline for Biomedical Research Involving Humans (Ethical Guideline) (The Ethical Guideline for Biomedical research involving human 2016). The Ethical Guideline attempts to protect human dignity and the rights of research subjects. Although the Ethical Guideline states that clinical research shall be reviewed by an ethics committee before commencement, most studies are not examined by an ethics committee in practice because of the lack of a penalty clause in the Ethical Guideline (Esteban et al. 2009).

The authority seems now no longer to turn a blind eye to stem cell therapy anymore. In August 2015, the Administration Measure on Clinical Stem Cell Research was launched, which is the first normative document related to stem cell therapy in China. The normative document is one source of regulation resulted from administrative decision, which should not have the force of law but is the basis of administrative enforcement. The promulgation of this document has been of great significance to the promotion of the sound and orderly development of stem cell research, as well as the supervision of stem cell therapy. Generally, the regulatory requirements and procedures in Administrative Measure are similar to other countries, such as stem cell clinical trials using Good Clinical Practice of Pharmaceutical Products ('GCP'), the licencing of manufacturing using Good Manufacturing Practice ('GMP') and the adverse event reporting system (The Administrative Measure on Clinical Stem Cell Research in China 2015). The Administration Measure states that the research must comply with the principles of science, standardization, openness, ethics and sufficient protection of the rights of research subjects (The Administrative Measure on Clinical Stem Cell Research in China 2015). Remarkably, the Administration Measure specially stipulates the following three perspectives:

10.5.1   Clarifying the Qualifications of Institution Conducting Clinical Stem Cell Research

Article 7 of the Measure states that the applicants for carrying out clinical stem cell research must demonstrate that:

1. The institution shall be a grade III class A hospital and have relevant department involving with the clinical stem cell research.
2. The institution shall be qualified for clinical testing of the relevant drug based on the law.
3. The institution has a strong ability in medical treatment, research and teaching. The institution undertakes the major research project in stem cell area. Moreover, the research project shall be fully supported by a legitimate and sustainable fund.
4. The institution shall facilitate a complete stem cell quality management capacity, an entire clinical stem cell research quality management system and an independent stem cell research quality assurance office. The institution shall establish a qualified person responsibility system; record the whole preparation of stem cell drug, clinic research, quality and risk management; and keep the relevant documents

(including quality management manual, clinical research procedure, standard operation procedure and research record, etc.). The institution shall also establish a clinical stem cell research audit system, including the qualified internal examiner, internal examination and external examination.

5. The responsible person for clinical stem cell research project and the qualified person for drug quality, who must be a senior professional title with academic goodwill, shall be appointed by the chief deputy person of the institution. The leading researcher shall be trained about GCP and attain the relevant qualification. The institution shall fully facilitate human resources to conduct the accordingly clinical stem cell research, regulate and implement the training plan for clinical stem cell researcher and evaluate the results of training.
6. The institution shall establish academic committee and ethical committee comprised by high-level experts who have the capability to conduct clinical stem cell research.
7. The institution shall establish the system to deal with the risk of the clinical stem cell research and handle the adverse reactions and events (The Administrative Measure on Clinical Stem Cell Research in China 2015).

10.5.2   Establishing the Initial Review, the Record Filing Mechanism and the Research Project Reporting System

Article 17 of Measure stipulates that research and pharmaceutical products involving with stem cell shall comply with some requirements, including GCP and GMP (The Administrative Measure on Clinical Stem Cell Research in China 2015).

Article 19 of Measure regulated that the institution's academic committee is responsible for conducting a scientific review of record filing materials of the research application, especially in the following aspects:

(1) The necessity of conducting clinic stem cell research
(2) Whether the research project is scientifically appropriate
(3) The feasibility of research project
(4) The qualification of leading researchers and the training about clinical stem cell research
(5) The risk of research project and relevant measures for preventing it
(6) The quality control measure of preparing pharmaceutical products related to it

Article 20 further indicates that the institution's ethical committee conducts an independent ethical review of the project (The Administrative Measure on Clinical Stem Cell Research in China 2015).

Article 24 requires the project materials, which has been reviewed by institution's academic committee and ethical committee, should be jointly examined by provincial health administration and the food and drug administration and recorded by NHFPC and CFDA (The Administrative Measure on Clinical Stem Cell Research in China 2015).

10.5.3   Reporting Unsafe Events in Clinical Stem Cell Research

Article 34 requires institutions to promptly report to the national and provincial health administration in case of severe adverse reactions and accidents (The Administrative Measure on Clinical Stem Cell Research in China 2015).

Article 35 requires institutions to promptly report to the institution's academic committee and ethical committee in case of errors. The institution's academic and committee and ethical committee shall report these errors to the national and provincial health administration (The Administrative Measure on Clinical Stem Cell Research in China 2015).

Article 36 requires that milestone of progress of research project shall be reviewed by the institution's academic committee and ethical committee. The results of examination shall be reported to the national and the provincial health administration (The Administrative Measure on Clinical Stem Cell Research in China 2015).

10.6   The Remaining Unsolved Problems of Stem Cell Clinical Research Regulation in China

The new administrative measure seems to outline requirements for clinical stem cell research (Cyranoski 2015). Researchers who want to carry out stem cell-related research must register and seek approval by the CFDA. The administrative measure also shows that the Chinese government aims to regulate stem cell clinical research in line with the international documents, such as CIOMS/WHO International Ethical Guidelines on Biomedical Research Involving Human Subjects and the Helsinki Declaration. Indeed, the Administrative Measure has strengthened ethical and scientific review of clinical stem cell research, by requiring informed consent and using clinical grade stem cell. However, in principle, the Administrative Measure still needs more work in the following aspects.

10.6.1   The Lacking of a Liability Clause

The liability and penalties for violating the rules are not yet clear. Although the Ministry of Health wishes to supervise stem cell research through this Administrative Measure, it contains neither the clause about liability of persons involving clinical stem cell research nor the clause related to liability of persons monitoring clinical stem cell research. Without specific liability clauses, some institutions, hospitals and clinics might continue conducting uncertified stem cell therapy. The review, the record filing mechanism and the research reporting system might not be well performed. The Administrative Measure would receive little attention from clinical researchers due to the lack of liability clauses. A similar concern is from the implementation of the Guideline for ethical review of biomedical research issued by

Ministry of Health in 2007.[14] According to this Guideline, the ethical review must be taken before the commencement of the clinical trial.[15] However, many stem cell therapies were conducted without ethical review by the ethical committee. One important reason for that is the vague liability clause of the Guideline. Under this Guideline, the principal personnel of these non-reviewed stem cell therapies were not penalized (The Ethical Guideline for Biomedical research involving human 2016). Therefore, the lack of a liability clause will strongly influence the implementation of Administrative Measure.

10.6.2   The Lacking of a Traceability System

A system allowing complete traceability of the product and its starting materials is essential to monitor the safety of stem cell therapy.[16] The traceability of stem cells and products developed from them is critical for the notification of serious adverse reactions and events. In addition, the traceability requirements have significant effect on the rate at the donated embryos and other materials in stem cell therapy (Corrigan et al. 2006). However, the Administrative Measure did not mention traceability system. In the absence of traceability mechanisms, crucial details of a product might not be provided to the patient. Traceability also plays a more general role in providing safe, high-quality and efficient clinical stem cell application. Verifiable quality assurances are delivered through traceability. The introduction of a traceability system is central to the risk management of stem cell therapies. Most importantly, a traceability system can bolster liability incentives for institutions to practice its due diligence—conduct an act with a certain standard of care. Whether for statutory or civil liability, traceability is a key element of proof. Not participating in the traceability system might release the liability burden of researchers, investigators and supervisors involved in stem cell therapy. Without a liability and traceability system, the effectiveness of Administrative Measure cannot be guaranteed.

10.6.3   The Lacking of Expert Responsible Authorities

In general, if stem cell regulation wants to be effective, it should be enforced by a special agency/committee, delegated by government, with expertise in the field of study. In China, the first Administrative Measure was jointly launched by NHFPC, which is affiliated to the Ministry of Health, and CFDA. However, the Administrative Measure has not certified who is the responsible authority, the Ministry of Health or CFDA. And the Administrative Measure does not require establishing an expert committee focused on stem cell therapy. Without the expert responsible authority,

---

[14] The Guideline for Ethical Review of Biomedical Research 2007 in China.

[15] Ibid.

[16] EC Regulation No 1394/2007 on Advanced Therapy Medicinal Products and Amending Directive 2001/83/EC and Regulation (EC) No 726/2004, [2007] O.J.L324/121.

the demand of reviewing, recording and reporting clinical stem cell research might not be satisfied, and the quality and safety control of clinical stem cell research might not be assured.

10.6.4 Non-Applicable to Military Hospital

Military hospitals are beyond the regulatory regimes of Administrative Measures despite that the Administrative Measures are to establish a rigorous system to supervise stem cell therapy. In China, military hospitals are the main research institutions of stem cell therapies. If you search stem cell therapy online, hundreds of military hospitals providing stem cell therapy would be in the result list, including People's Liberation Army Hospitals, Air Force Hospitals and Armed Police Force Hospitals. According to a survey by Dr. Dominique McMahon from the University of Toronto, 36% of all stem cell therapies carried out in China are performed by military hospitals (The Ethical Guideline for Biomedical research involving human 2016). Military hospitals are affiliated to the Health Division of the General Logistics Department in an administrative hierarchy. The General Logistics Department is in equal footing to the Ministry of Health. Therefore, military hospital is not the institution governed by the Ministry of Health. Since the Administrative Measure is a normative document issued by the Ministry of Health, surely military hospitals are not bound to comply with the Administrative Measure.

10.7 Stem Cell Industry in China

Regulating clinical stem cell research, which has been one of the most ethically controversial technologies of the present age, is an unenviable task for China. The ambit of the Administrative Measure is wide. The Chinese government hopes to rein the clinical stem cell research with the enforcement of the new order. However, the legal framework of clinical stem cell therapy leaves a door open for untested and unverified stem cell therapies. And it is still an open question whether the combination of the Ministry of Health and CFDA is capable of enforcing the Administrative Measures on hospitals.

References

Chen T, Jiang H, Qian W (2011) The funding mode of stem cell research in China. China Basic Sci 4:31

Corrigan O, Liddell K, Mcmillan J, Stewart A Wallace S (2006) Ethical legal and social issues in stem cell research and therapy. Cambridge Genetic Knowledge Park, p 16

Cyranoski D (2015) China announces stem cell rules-long awaited regulations may curb rogue clinics offering unproven treatments. Nature. http://www.nature.com/news/china-lannounces-stem-cell-rules-1.18252. Accessed online 25 Sept 2016

Esteban MA et al (2009) Vitamin C enhances the generation of mouse and human induced pluripotent stem cells. Cell Stem Cell 6:71

Guotong X (2007) The challenge and opportunity of stem cell research in China. Front Sci 4:36

Jiang L (2016a) Regulating human embryonic stem cell in China: a comparative study on human embryonic stem cell's patentability and morality in US and EU. Springer, Heidelberg and New York

Jiang L (2016b) Will the first administrative measure on clinical stem cell research end the wild east scientific feast? Biotechnol Law Rep 35(1):27–35

Robbins-Roth C (2001) From alchemy to IPO: the business of biotechnology. Perseus Publishing, p 91

Song P (2011) The proliferation of stem cell therapies in post-Mao China: problematizing ethical regulation. New Genet Soc 30:141

The Administrative Measure on Clinical Stem Cell Research in China. (2015). Available at http://www.moh.gov.cn/qjjys/s3581/201508/28635ef99c5743e294f45e8b29c72309.shtml. Last visited 8 Sept 2015

The Ethical Guideline for Biomedical research involving human. (2016). Available at http://www.moh.gov.cn/mohbgt/pw10702/200804/18816.shtml. Last visited 9 Sept 2015

The Ethical Guideline for HESC research in China. (2003). Available at http://www.cncbd.org.cn/News/Detail/3376. Last visited 8 Sept 2015

Watt N (2006) EUROPE reaches deal on stem cell research. The Guardian, London

Xiaoqin R (2011) The patents granted to Heze Biotechnology. Shanghai Stock Newspaper

Yong Y (2011) Beike biotech: win the respect and appraise by the stem cell frontier technology. The Chinese Economic, http://district.ce.cn/zg/201106/02/t20110602_22458434.shtml. Accessed 20 Nov 2013

Zhao XY, Li W, Lv Z, Liu L, Tong M, Hai T, Hao J, Guo CL, Ma QW, Wang L, Zeng F, Zhou Q (2009) iPS cells produce viable mice through tetraploid complementation. Nature 461:7260

Chapter 11
Governing the Stem Cell Sector in India

Shashank S. Tiwari, Paul Martin, and Sujatha Raman

11.1 Introduction

India is a key player in the stem cell sector with significant government investment and research activities including the creation of new embryonic cell lines and publication of a number of scientific papers (DBT Annual Report 2015–2016; Inamdar et al. 2009; Pandey and Desai 2016; Sharma 2009). While these efforts have been commended nationally and in the international community, significant concerns began to emerge from the mid-2000s over unproven stem cell treatments being offered in clinics with apparently little by way of regulatory oversight (Jayaraman 2005; Lander et al. 2008). In 2014, the Indian government began to respond to these

Parts of this chapter were previously published in the following article: Tiwari SS, Raman S (2014) Governing stem cell therapy in India: regulatory vacuum or jurisdictional ambiguity? New Genetics and Society, 33(4), 413–433. Material from that article is reprinted with permission from Taylor & Francis.

S.S. Tiwari
Individualized Treatment Technology Laboratories LLC,
1601 12th Avenue South, Birmingham, AL 35205, USA

Institute for Science and Society (ISS), School of Sociology and Social Policy, University of Nottingham, Nottingham NG7 2RD, UK
e-mail: shashank17t@gmail.com

P. Martin
Department of Sociological Studies, University of Sheffield, Sheffield, S10 2TU, UK
e-mail: paul.martin@sheffield.ac.uk

S. Raman (✉)
Institute for Science and Society (ISS), School of Sociology and Social Policy, University of Nottingham, Nottingham NG7 2RD, UK

Science and Technology Studies (STS), University of Nottingham,
Nottingham NG7 2RD, UK
e-mail: sujatha.raman@nottingham.ac.uk

© Springer International Publishing AG 2017
P.V. Pham, A. Rosemann (eds.), *Safety, Ethics and Regulations*, Stem Cells in Clinical Applications, DOI 10.1007/978-3-319-59165-0_11

concerns by announcing legal changes that would, in theory, outlaw stem cell therapies given the absence of clinical trial evidence on their safety and efficacy (CDSCO 2014). In a further step, a new Drugs and Cosmetics (Amendment) Bill 2015 has been drafted to regulate clinical trials and medical devices and to classify the clinical use of stem cells under the category of drugs.

In this chapter, we first review the background to these developments drawing from social scientific studies of the stem cell sector in India. We then consider the challenges and prospects for governing this sector in practice, building on concepts of governance in the study of law, politics and biomedicine which highlight the need to understand how regulatory objects are constituted and different possible modes by which an emerging technology might be actually governed. This helps clarify why stem cell activities, more specifically stem cell-based therapies, have been historically difficult to regulate and the implications for the implementation of recent laws (CDSCO 2014; ICMR-DBT 2013; MoH&FW 2014).

In India, strictly research-based activities in stem cells have been relatively uncontroversial; rather, it has been clinical activities which have been objects of national and international concern. It is reported that many clinics in India make false claims about the efficacy of a wide range of stem cell treatments and, in some cases, have offer fake declarations of approval from governing bodies (Pandya 2008; Sipp 2009). The reported use of embryonic stem cell therapies by Nutech Mediworld in New Delhi attracted widespread condemnation in the early days (Cohen and Cohen 2010; Khullar 2009; Padma 2006). Other claims related to the clinical use of adult stem cells have also been controversial (Pandya 2008; Rosemann and Chisainthop 2016). A rise in so-called stem cell tourism with patients from the West travelling to India to be treated has also been highlighted in international commentary (Cohen and Cohen 2010). Such uses of stem cells are taken to be 'experimental' in that the regimes in question are yet to be proven treatments established as such through a recognised framework of clinical trials. It has therefore been argued that stem cell development in India operate in a 'governance vacuum' (Salter 2008).

Concerns about regulatory shortcomings around stem cell activities are not unique to India. For example, studies also highlight examples of unproven/untested stem cell treatments being advertised and offered in a range of countries including the USA (see Rosemann and Chiasinthop 2016; Turner and Knoepfler 2016). While we focus in this chapter on India, this case should be understood within the wider global political economy of stem cell activity.

Early studies on the stem cell sector in India concluded that the governance vacuum in India was a result of the lack of *statutory* regulation of stem cell activities (Cohen and Cohen 2010; Glasner 2009; Patra and Sleeboom-Faulkner 2009, 2010; Salter et al. 2007; Salter 2008; Sleeboom-Faulkner and Patra 2011). In 2007, the Indian Council of Medical Research (ICMR) and the Department of Biotechnology (DBT) had jointly issued a set of *Guidelines for Stem Cell Research and Therapy* (ICMR-DBT 2007). This 76-page document specified general ethical principles for research and processes for formal committee approval of stem cell activities and for their periodic review/monitoring. In terms of procedures and underlying norms, the content was in line with mainstream bioethics. The guidelines stipulated that clinical use of stem cells was not permitted and that any use of stem cells in clinical contexts (with the exception of already standardised uses of bone marrow transplantation and epithelial therapies for corneal disorders) must be part of a clinical trial conducted after approval by a committee set up to oversee stem cell activity and the Drug Controller General of India (DCGI) who sits within the Central Drugs Standard Control Organization (CDSCO). However, since these guidelines lacked statutory backing, many scholars concluded that the way forward would be to give them legislative weight. Rather than start from this assumption, we work from a more fundamental set of questions. How has the governance problem around stem cells been framed in India? What are the possible pathways for governing stem cell activity, including but not restricted to statutory guidelines? What does the effort to debate and govern stem cell therapy conceal as well as reveal about India's engagement with biomedicine?

We begin by summarising key insights from social scientific studies of how law and governance work in practice, especially in relation to technologies associated with the life sciences. We then examine the significance of how the 'object of governance' is constituted in the Indian stem cell case, paying specific attention to boundaries of jurisdiction between different regulatory agencies in the domains of biomedicine, biomedical research and stem cells. Insofar as stem cell treatments are offered in a clinical context, it should in theory be possible to govern them through the regime governing medical practice and practitioners, a point that those focusing on stem cell-specific guidelines fail to consider. We follow this up by examining the range of forms of regulation available (in principle) for addressing the governance vacuum around stem cells.

Second, the focus on a statutory gap as the underlying cause of legitimacy problems seems to be based on the assumption that laws, once enacted, automatically coerce people to behave in the ways intended by their designers. Yet, scholars in sociolegal studies have long highlighted the limits of a purely 'top-down' approach to understanding the nature and abilities of state intervention and of law itself (e.g. Kagan et al. 2003; May 2005). Issues of meaning, discretion and judgement remain important in the domain of more formal and codified laws. In democratic societies, when law 'works'—or when it is seen to work—it is through a process that unfolds and becomes enacted *through* society rather than being imposed *on* it. This in turn also means that nonstatutory guidelines do not necessarily have to produce disorder or ethical transgressions as social or professional norms sometimes produce law-like behaviour (Jasanoff 2011). Multiple approaches to governing health care in India are now emerging beyond those represented by statutory laws alone (Peters and Muraleedharan 2008). In the main part of the chapter, we follow this insight by investigating different forms of governance (Pierre and Peters 2000) of medical practice in India and their relevance to stem cell clinical activity.

11.2 Law-in-Practice and the Making of Governance

In this section, we distil key questions from scholarship on law, governance and the life sciences and technologies that help us identify the issues facing regulators of new developments such as stem cells and consider the different ways in which the stem cell sector in India might be governed.

First, how does a particular activity such as stem cell treatment, in our case, come to be seen as risky or needing regulation? A central problem for governance of research and development in biotechnology and the life sciences in all countries is the tension between the imperative to promote new technologies and the imperative to regulate them. In the case of biomedical research in the West, the tension was transformed into a productive one for scientists with regulation seen as a way of managing reputational risk (Dixon-Woods and Ashcroft 2008) and, in that respect, enabling rather than only constraining research. But how this tension plays out in different contexts is a matter to investigate. In India, reputational concerns are indeed regularly cited as a reason for needing statutory regulation.

Second, when new objects of governance are made visible as needing intervention, how are they constituted and ordered? For example, both human and agricultural biotechnologies have been constituted as a series of *products* in the USA, thus allowing them to be regulated by existing frameworks of contract law for market transactions with any grievances handled through the courts (Jasanoff 2011). By contrast, countries like Britain handled technologies such as those relating to surrogacy through existing frameworks in family law. Different ways of constituting the object lent themselves to different spheres of jurisdiction, the political and cultural legitimacy of which allowed governments to manage controversies around new technologies. Research on lawmaking therefore highlights the value of understanding how legal or regulatory questions are framed in the first place and how this matters for jurisdictional boundaries. This is a key issue for our case as framing the object in question as stem cell 'research' did not have the intended success.

Third, mechanisms of governance are not simply embedded in the state. Professional and political norms of practice and judgement may acquire law-like qualities despite never having been formally articulated as such (Jasanoff 2011); this means that stem cell practices could, in theory, be governed from 'below' and not just from 'above'. Pierre and Peters (2000) distinguish between four modes of governance: a hierarchical, state-led model of command-and-control regulation, a market-based mode, a community-based mode and a distributed or network mode consisting of actors from different institutions and social groups. In the stem cell case, we could therefore explore alternatives to statutory regulation as well as ask how different modes *interact* in practice. For example, civil society groups might put pressure on the professions, industries or the state in order to hold them to account in terms of their role in governing. Professions and industries might call for more regulation by the state if they are concerned about reputational risk; equally, they might also put pressure on the state to relax regulatory constraints. We will explore these dynamics later in the chapter.

Fourth and related to the above, a key question is what happens to statutory laws once they come into existence. Laws cannot by themselves compel people to act in expected ways. This point allows us to consider how the state, professions or other communities, industries and civil society interact in the process of implementing formal laws and policies.

Finally, bringing in a civil society perspective into the study of law opens up the possibility of fundamentally rethinking the terms on which the regulatory problem at stake has been framed in the first place. Salter et al. (2016) argue that given growing demand for stem cell treatments, more attention must be paid to the health consumer market rather than only efforts to regulate supply. By contrast, Kim (2014) highlights civil society activists in South Korea who have argued for stricter controls on stem cells in order to rein in a capitalist-developmental drive towards biotechnology and promote alternative pathways in the public interest. Given the inequalities entailed in the politics of life (Raman & Tutton 2010), we therefore might ask why a governance vacuum matters and to whom.

11.3 Methods

To understand the making and interpretation of law-in-practice, it is essential to consider how stakeholders 'on the ground' perceive the key issues—in this case, difficulties around stem cell governance and prospects for their remediation. Hence, a qualitative study of documents in different media (news and opinion, scientific literature, policy reports) and interviews with key stakeholders was undertaken.

Semi-structured interviews were conducted after ethical approval from the University of Nottingham, UK, during June 2010–January 2011, and again during September–October 2011, in various cities in India including New Delhi, Mumbai, Pune, Chennai, Bangalore, Hyderabad, Tirupati, Kolkata and Chandigarh where most of the research and clinical activities in stem cells are being carried out. Locations were identified on the basis of a mapping exercise in 2010 using documents available on the internet. Twenty-seven interviews (five scientists, 11 clinicians, seven firms' representatives and four policymakers) were conducted by the first author, lasting between 45 min to an hour (with one exception, where the interview finished in 15 min). The majority of the interviews were recorded with the informants' permission and transcribed. However, in three cases, informants were not comfortable with the prospect of being recorded; hence, notes were taken and subsequently written up.

Documents included news items on stem cell activities published in leading newspapers available on the Internet (*The Times of India*, *The Hindu*, *The Indian Express*), science magazines (*BioSpectrum India*), official documents related to stem cell research and medical governance published by government bodies and articles published in international journals on stem cells in India (e.g. *Nature*, *Science*). There is a lively debate in Indian newspapers and journals (especially the *Indian Journal of Medical Ethics*) on the state of medical ethics in the country and wider issues of negligence. These articles also provided key insights into how professionals and commentators in India perceive the issues that are being discussed elsewhere in the international media and journal literature.

11.4 Constituting Stem Cell Research as the Object of Governance

As news emerged of unproven stem cell treatments being offered in India, respondents identified a 'vacuum' in governance which they traced to two key factors: the nonstatutory status of the 2007 ICMR-DBT guidelines and an overall fragmentation of regulatory authority (Patra and Sleeboom-Faulkner 2009; Salter et al. 2007). The Indian Council of Medical Research (ICMR) is part of the Ministry of Health and Family Welfare, while the Department of Biotechnology (DBT) is in the Ministry of Science and Technology. Yet, such a structure does not necessarily have to fail as it might represent an effective way of combining forces in complex situations calling for multiple sources of expertise. Following Jasanoff's (2011) injunction to consider how biotechnology is ordered, we ask how stem cell therapy has been constituted as a regulatory object. Framing the question this way sheds light on the terms in which the 'problem' to be regulated is made visible and jurisdictional boundaries drawn, which, in turn, allows us to consider if these boundaries might be defined differently.

In the early days, the Indian government was keen to promote Nutech Mediworld's work with the then health secretary quoted in 2005 as saying that 'sometimes, scientific knowledge cannot wait for bureaucratic apparatus' (Mudur 2005). However, fears about reputational risk (Dixon-Woods and Ashcroft 2008) began to emerge and stimulated a regulatory response. Many Indian scientists and clinicians expressed concerns to journalists about unwarranted claims made by Nutech Mediworld and others (e.g. Life Line Hospital, Chennai; All India Institute of Medical Sciences, New Delhi) regarding successful treatments based on stem cells (see Mudur 2005). A few took to journals in science and in medical ethics to criticise these claims and the lack of an effective response from government (e.g. Jayaraman 2005; Padma 2006; Pandya 2008). Many urged the ICMR to take a firmer stance and 'mandate' medical ethics (Mudur 2005).

Given their keenness to secure investor confidence in biomedical research, it is not surprising that the Indian government responded quickly to these developments with the 2007 ICMR-DBT guidelines that were modelled on established Western bioethical frameworks. However, critics argued that these needed to have legislative force (e.g. Pandya 2008), a point that was also evident during interviews conducted with key players. One scientist (Scientist 1) working in a government-funded research laboratory who was interviewed pointed out that malpractice would only stop when the Indian Parliament backed up guidelines with legislation. He also argued that those who violated the guidelines should be punished. A private medical practitioner (Clinician 1), who himself offered experimental stem cell therapy for muscular dystrophy, lamented that everyone was free to provide stem cell treatments and that guidelines were ineffective since violations went unchecked.

This underlines the point made by others (Patra and Sleeboom-Faulkner 2010; Salter 2008) that guidelines cannot compel action in the way that laws potentially can. However, while the development of guidelines as a response to controversy was seemingly straightforward, the question of jurisdictional authority over their implementation has been more complicated. What was missed in this debate was the fact that neither the ICMR nor the DBT had a *legislative* remit over medical research. Interviewees in government and industry pointed out that the ICMR funds research and provides advice, while the DBT is an agency for funding (rather than regulating) preclinical and clinical R&D. Also, the DBT has no remit over activities taking place outside government-funded R&D programmes (Policymaker 2). Its stem cell task force and committees oversee the DBT's own research activities, but these do not cover clinical trials.

The Drug Controller General of India (DCGI) which is frequently characterised as the 'Indian FDA' already had a mandate to regulate clinical trials and would have been the obvious candidate to extend its remit to stem cells. Only the DCGI has the authority to regulate their activities, an industry representative was quick to point out (Firms Representative 2). However, the DCGI had no experts of its own who were able to evaluate stem cell proposals, according to a policymaker. Also, in these early days, it appeared that the DCGI was uncertain about the reach of its powers which may be due to the fact that it is only nominally similar to the FDA with a remit primarily related to drug approvals (Sunder Rajan 2007). 'Our FDA is not that strong', noted a policymaker (Policymaker 2). Clinician 1 quoted above noted that the Drug Controller of India likewise acknowledged limits to their office's remit over stem cell treatments.

This suggests a fundamental jurisdictional ambiguity with even the relevant agency unsure of what falls under its regulatory scope. Referring to the wider landscape of medical law, a policymaker (3) explained that those offering stem cell therapies did not interpret their activities as 'research' but rather as 'treatment' which they had a right to give as doctors. The ICMR-DBT guidelines did not seem to apply to them, in this view. Here, we begin to see that the jurisdictional difficulties around identifying who has authority to regulate stem cell 'research' have arisen partly from the ambiguous boundary between research and therapy. For those offering stem cell therapy, guidelines pertaining to 'research' did not appear to have meaning since they saw themselves as treating patients rather than using them as research subjects for publishable studies (see also Bharadwaj 2014).

Comparing the 2007 guidelines with the new guidelines published in 2013, the most dramatic change related to the very title. In 2007, the document was labelled *Guidelines for Stem Cell Research and Therapy* (emphasis added). By 2013, this read *Guidelines for Stem Cell Research*. The writers of the foreword to the 2013 document drew attention to the change, explaining that this was done to avoid confusion over the fact that stem cell therapy is not allowed in the first place; hence, there can be no guidelines to govern it! The 2013 guidelines reiterated the point that any clinical use of stem cells must be part of an authorised clinical trial, a point that was already present in the 2007 version but did not have meaning for those carrying out the activity in question.

To summarise, constituting the regulatory object as 'research' enabled the ICMR to bring its expertise in stem cells and bioethics to bear on the problem. However, since the ICMR could only provide advice, the DCGI's statutory powers were highlighted as the answer to the problem of ICMR's guidelines being ignored in practice. However, insofar as unproven therapy was being provided in clinical settings *outside* recognised clinical trials, bringing this activity under DCGI's remit proved challenging in the first instance.

11.4.1 Reconstituting Stem Cell Governance

So far we have highlighted the jurisdictional ambiguities that challenge one-dimensional accounts of a governance vacuum in stem cell research. Yet, jurisdictional boundaries and the objects of regulation can be open to reconstitution as was evident at the time of fieldwork and confirmed by recent developments in stem cell governance. In light of the persistent controversy over stem cell therapies, the story of what the 'Indian FDA' can or cannot do was slowly being opened up to alternative interpretations in interviews conducted in 2010–2011. For example, one policymaker (1) observed that initially the DCGI thought that stem cells were not 'biological entities' relevant to their purview, but later, they realised that it could be classified under the category of biological products which they had authority to regulate. As clinicians began to offer marketable stem cell products, they were opening themselves up to scrutiny by the DCGI under its existing remit which (unlike the DBT) covers both public *and* private activities.

In 2012, the DCGI constituted a special division for stem cells in response to criticisms that it did not have any internal evaluation mechanism (*BioSpectrum India*, April 30, 2012). In practice, DCGI's reliance on the ICMR is likely to continue as ICMR's Director General is also the Chair of the new division, though as one interviewee noted, both agencies are part of the same ministry and this co-working need not be construed as a problem (Policymaker 2). In 2014, in addition to the publication of revised guidelines mentioned above from the ICMR-DBT, the DCGI announced that it would modify the Drugs and Cosmetics Act to treat 'stem cells and cell-based products' as new drugs (CDSCO 2014). In a subsequent initiative, the Ministry of Health and Family Welfare introduced the draft Drugs and Cosmetics (Amendment) Bill 2015 for the same reason.

In sum, jurisdictional ambiguities over the governance of stem cell therapy seem to have finally been resolved with the ICMR-DBT revising their guidelines and the DCGI extending their statutory remit to stem cells. Statutory laws are in place to address earlier criticisms of guidelines. Yet, if the meaning of law is determined in practice, legal amendments are insufficient in themselves to draw a line under the challenges of stem cell governance. While the DCGI has addressed the ambiguity over who has legal jurisdiction over stem cell uses, we still need to know how medical law works on the ground in order to make sense of the practical implications of DCGI's efforts. In the next section, we explore ways in which clinical medicine is regulated in India and their prospects for contributing to the governance of stem cell therapy.

11.5   Enacting Stem Cell Governance Through Regulation of Clinical Practice?

Even well-ordered statutory laws require mechanisms for enforcement. One interviewee (Policymaker 3) implicitly raised the possibility of potential violations altogether going unnoticed—in the absence of complaints actually being lodged, there was little that regulators could do. In this section, we consider two possible routes by which recent stem cell laws may—or may not—be enacted in practice, first, through professional self-regulation and second, through the broader edifice of statutory law governing clinical practice. We examine the difference, if any, that law makes through its interaction with the medical profession and the courts.

We adapt Pierre and Peters' (2000) four-part typology of governance to a simpler distinction between *centralised* ('hierarchical' models) and *decentralised* forms (more distributed models). This also allows us to consider decentralised approaches which derive from legislation (e.g. consumer legislation, professional guidelines such as those issued by ICMR-DBT on stem cells) as well as the activities of those formally 'external' to the state (e.g. civil society) but which might still involve putting pressure on governments to act in particular ways. Decentralised modes of governance do not mean the state is absent.

International and national guidelines stipulate that clinical uses of stem cells must be as part of a clinical trial for research conducted under established regulatory protocols. If, as we have argued, stem cell therapy escaped the regulatory net due to its location in a health-care market rather than research per se, one option might be self-regulation through ethical codes of conduct that cover clinical practice. However self-regulation is complicated by the fact that professionals will inevitably have their own views about how regulation should or should not proceed. For example, many leading Indian stem cell clinicians have requested the government of India and the current Prime Minister Mr. Modi to reconsider the proposal of classifying autologous stem cells under the category of drugs (Stem Cell Society of India 2015). We therefore need to consider how professional norms are themselves shaped, including the role of the state in regulating the medical profession and providing an overarching structure within which medical providers govern themselves.

Sanctioned by the Indian Medical Council Act of 1956, the Medical Council of India (MCI) is the primary regulatory body for maintaining uniform standards of medical education and certifying medical qualifications. Medical practitioners register through state-level councils overseen by the MCI. In 2002, the MCI introduced the *Indian Medical Council (Professional Conduct, Etiquette and Ethics) Regulations of 2002* to cover codes of conduct for practitioners, which again operate through state councils. However, violations of the code have been noted and the code itself challenged as impractical in a highly market-driven health-care sector. For instance, some clinicians and corporate hospitals advertise their medical services through media interviews or hoardings at public places although the medical code considers advertising to be unethical (Balasubramanian 2008). It is only recently that MCI started taking action on medical service advertisements (Perappadan 2014). Overall, the MCI, at both central and state levels, is perceived to be ineffective in monitoring codes of conduct with critics charging that 'they have not bothered to exercise the powers given to check unethical medical practice' (Pandya 2007, p. 2). The MCI and state medical councils have also been plagued with corruption charges over the years (Pandya 2007; *The Times of India*, April 24, 2010). The government of India policy think-tank, the National Institution for Transforming India (NITI Aayog), has recently called for replacing the tainted MCI with a new National Medical Commission, and a bill along these lines has since been drafted (Bhargava 2016).

In addition to state-sanctioned councils, the Indian Medical Association (IMA) is the main professional body for doctors. Its website highlights that '[IMA] looks after the interest of doctors as well as the well-being of the community at large'.[1] However, in a stinging critique, one doctor charges the IMA with behaving 'as an interest group pushing the special interests of doctors instead of society as a whole' and altogether failing to contribute to policy on improving health indices in India (Thomas 2011, p. 2). Dr. Thomas also takes the MCI to task for failing to provide leadership on ethics education for doctors (Thomas 2011).

Madhiwalla (2011, p. 3) argues that the medical profession in India has not traditionally faced the type of public scrutiny that medicine received in the West owing to its origins as a sector built 'by both the colonial and the independent Indian state as the vehicle of modernity and welfare'. An interest in bioethics emerged in the 1980s from controversy over the role of medicine in the 1984 Bhopal disaster and earlier, in sterilisation programmes introduced during the 1975 Emergency. However, it remained a niche interest with few roots in professional education.

Others emphasise the need for guidelines to be backed up by threat of sanctions. Commenting on the lack of efficacy of the MCI's code of conduct, one doctor said, 'it is important to have ethical guidelines. But the profession should enforce them. We need to develop mechanisms so that a variety of transgressions are regulated and penalised' (Dr. K. Reddy quoted in Jain 2010). Once again, there are strong hopes pinned on statutory law. However, the question is how violations of laws or professional codes become visible in the first place. Who notices if something goes wrong? We turn to this question below.

At present, medical ethics violations are dealt with indirectly under various sections of the Indian Penal Code which defines criminal acts and related punish-

---

[1] http://www.ima-india.org/IMA.html.

ments (Dhar 2010). Section 304-A of the Code deals with complaints against medical practitioners for alleged medical negligence (Nayak 2004) which includes violations of medical ethics (Dhar 2010). The civil Law of Torts is considered to be among the most significant for governing medical malpractice as it has been successfully applied in many cases (Peters and Muraleedharan 2008). It applies to all health professionals, whether in the public or the private sector. This law also covers circumstances when a clinician treats a patient without informed consent (Nandimath 2009). The Indian Contract Act of 1872 provides legal protection to agreements between the parties but has hardly been used for health issues in India (Peters and Muraleedharan 2008).

Taken together, these legal avenues appear to offer some statutory weight for governing medical practices including unproven stem cell treatments. So, if such treatments were offered despite recent legal amendments, these cases could, in theory, be pursued by underpinning legislation such as the Indian Penal Code. However, the social meaning (Jasanoff 2011) of any of these laws as they have been applied in the medical sector is problematic given the way in which they have been tended to be interpreted in the courts and entrenched delays in completing court cases (BBC, October 24, 2013; *The Times of India*, March 6, 2010). According to one Indian Supreme Court order, the opinion of an expert or panel of doctors is necessary to begin a case (Kamath 2010). It is also alleged that courts have tended to favour medical providers in their rulings (Peters and Muraleedharan 2008).

For these reasons, Peters and Muraleedharan (2008) suggest that focusing on enforcement of legal mechanisms is insufficient since 'the limited ability to enforce civil and criminal laws in India is well known' (Peters and Muraleedharan 2008, p. 2137). Hence they call for approaches focusing on the capacity of consumers to raise complaints through alternative forums. This then opens up the possibility of making sense of stem cell governance through a wider perspective offered by investigations of the relationship between law, medicine and civil society, a question to which we now turn.

11.6 Stem Cell Governance Through Civil Society

If neither statutory laws nor professional self-regulation is sufficient for governance, we need to ask how law may be supported or given meaning through its embedding in civil society. Second, a civil society perspective also reopens the very question around which the notion of a 'governance vacuum' in the Indian stem cell sector has emerged. We consider each of these issues in turn.

One way to deal with the limits of the court system is to develop alternative mechanisms of enforcing medical laws. Here, the Consumer Protection Act (CPA) of 1986 is potentially relevant as it is meant to protect the interests of consumers from poor-quality products/services and provide quicker responses to grievances by circumventing the delays of court cases (Peters and Muraleedharan 2008). Cases are brought to consumer forums which do not require court fees. Medical services were included in this act in the year 1995 after a Supreme Court ruling that 'patients aggrieved by deficiencies in medical services rendered for payment can claim damages under the act' (Mudur 1995, p. 1385). However, complainants still do have costs and delays beyond the stipulated 3-month limit remain a problem. Unsurprisingly, most health cases tend to be brought by wealthier and educated families (Peters and Muraleedharan 2008).

Also, the CPA covers only private clinicians who offer paid services. As with the Indian Penal Code, the CPA requires expert advice from other doctors for a case of alleged malpractice or negligence to be brought (Joshi 2011). This act also does not cover clinicians working in public hospitals. In theory, this gap should not be relevant to the case of stem cell treatments which are primarily offered in private practice for a fee, while services in government-run public hospitals are, for the most part, free. However, the link between private and public health care is more blurred in practice with public-sector doctors doing private practice and sometimes referring their patients to private facilities for certain services. This may also account for the fact that the private sector accounts for 80% of health-care services in India (Peters and Muraleedharan 2008), despite high levels of poverty.

If public-sector medicine does become relevant to the governance of stem cell therapy through interfaces with the private, the gap in the CPA could, in principle, be addressed through right-to-know laws (Jasanoff 1988) which allow individuals to access information held by public authorities. India introduced a Right to Information Act in 2005 which has been used over the years by activists to seek information on clinical trials and publicise ethical violations (Paliwal 2011). Mere exposure of violations does not guarantee that perpetrators will be held to account. For example, in one case in Madhya Pradesh, government doctors were found to have made millions of dollars through their role in corrupt clinical trials though this resulted in a mere $100 fine (Yee 2012). Still, it is worth asking if such outcomes might be transformed in future through civil society activism.

The role of organised civil society action around biomedicine is similarly becoming stronger; however, the capacity to impact on 'high-tech' biomedicine is more complex. Bhattacharya et al. (2008) claim that the exploitation of poor people during clinical trials and surrogacy or issues related to the trade in organs and human tissues have gone unnoticed, failing to create a mass movement. Controversial HPV vaccine trials did, however, spark organised action and led to the trials being suspended (Sarojini et al. 2010). Following protests in February 2011 around clinical trials involving victims of the Bhopal gas disaster (Rajalakshmi 2012), the government of India was compelled to take action against clinicians who were involved, though health activists subsequently criticised the penalty imposed as a mere token (Yee 2012). Such episodes helped trigger the government effort to amend and strengthen the 1940 Drugs and Cosmetics Act with new powers to punish violations. So, while health activism does have a history in India (Madhiwalla 2011), it is only now increasing in national visibility.

In sum, the ability to enact stem cell laws through civil society mechanisms remains limited, unless patients perceive their rights to have been violated and have the wherewithal to follow through. More significantly, looking at stem cell governance through a civil society lens allows us open up broader questions as Kim (2014) has done with reference to South Korea. A key question for future work might be to explore opportunities for activists to reframe the concern about a 'governance vacuum' around unproven therapies to a social justice concern about the very development of stem cell biomedicine in a context of radical social and health-care inequalities. If stem cell therapies were to be developed in a way that might benefit a wider spectrum of people, questions about governance would look very different.

11.7 Conclusion

We began this chapter by asking why has it been difficult to govern stem cell treatments offered in India and the prospects for this vacuum in governance (Sleeboom-Faulkner and Patra 2008; Salter 2008) to be remedied. Many have argued that the answer is to create statutory/legal backing for stem cell research guidelines developed by the two major agencies in the sector, the ICMR and the DBT. We have shown that this diagnosis of a statutory gap is inadequate since the construction of law and the boundaries of regulatory objects need attention as do the ways in which laws and law-like behaviour work in practice through social and institutional interactions (Jasanoff 1988; 2011). Indeed, the statutory gap in Indian stem cell governance was recently addressed with changes announced in 2014–2015 to the Drug Controller of India's (DCGI) legal remit and a revised set of guidelines produced by the ICMR-DBT guidelines at the same time. But questions still remain over the capacity to enact law-in-practice and enforce the new laws.

Our analysis highlighted a key jurisdictional ambiguity around stem cell therapy. The ICMR-DBT guidelines were framed in terms of stem cell research, but research implies the conduct of clinical trials. Until recently, stem cell therapies escaped the regulatory net of the Drug Controller General of India (DCGI) as they did not take place under the auspices of a trial, sitting primarily in a private market for clinical services. Patients may have taken the risk—or opportunity—of 'therapeutic consumption' (Sunder Rajan 2007) without a system of governmental protection backed up by the ability to enforce sanctions. But this indicates that the reputational controversy around Indian stem cell activities did not affect clinical practice even though it damaged the cause of research and eventually led to the recent amendments. The ability to enact these laws in practice depends on their interactions with the medical profession (specifically, mechanisms such as codes of conduct issued by the Medical Council of India), the wider edifice of health-care governance (specifically, statutory and quasi-statutory options available through the Indian Penal Code or the 1986 Consumer Protection Act) and civil society activism. However, we found that these too are problematic for a number of reasons. If legal violations go unnoticed, a de facto 'governance vacuum' would still persist. Medical negligence cases may be on the rise in India but hardly represent a viable option for the majority, not least for the costs they impose on individuals and on the health-care system as a whole. Civil society activism around health is becoming more visible, but this is necessarily centred on remedying the serious inequalities of health-care access in a country where the commercial sector accounts for 80% of health-care services rather than violations in stem cell treatments per se.

In the end, we need to acknowledge that stem cell treatments are primarily offered to those who can afford them—or who find the means to afford them. This then means asking not only whose rights are potentially being violated by unethical/unregulated treatments, but what the rise of such commercial treatments means for others' rights to health care. It also means attending to the question of who can currently afford to participate in such a market and who is effectively excluded at the outset from future markets for better-regulated/certified forms of stem cell treatment. The vacuum around stem cell activity in India—be it a vacuum in governance or in bioethical behaviour—is more problematic than a simple failure to adequately enforce guidelines through statutory or nonstatutory means. Rather, the vacuum encompasses multiple inequalities in the politics of life (Raman and Tutton 2010) that shape the governance and delivery of health care which need to be placed centre stage in such debates over biomedical research governance.

Acknowledgements This chapter draws from and reworks material from a paper previously published by the authors in the journal *New Genetics and Society* (DOI:10.1080/14636778.2014.9702 69). The fieldwork reported here was supported by a Wellcome Trust Developing Countries Doctoral Studentship (grant number: WT087867MA) awarded to Shashank Tiwari. The Trust is not responsible for views expressed in this paper. The authors would like to thank Pranav Desai for support and feedback during the course of the project.

References

Balasubramanian S (2008) KMC Questions Ethics of Healthcare Ads. CNN-IBN, March 03. Available via http://ibnlive.in.com/news/kmc-questions-ethics-of-healthcare-advertisements/60370-3.html. Accessed 20 Sept 2016

Bharadwaj A (2014) Experimental subjectification: the pursuit of human embryonic stem cells in India. Ethnos 79:84–107

Bhargava Y (2016) States approve proposal to replace Medical Council of India. The Hindu, 9 Sept 2016. Available via http://www.thehindu.com/news/national/states-approve-proposal-to-replace-medical-council-of-india/article9087055.ece. Accessed 5 Oct 2016

Bhattacharya N, Proctor C, Hodges S (2008) Medical Garbage and the Limits of Global Governance in Contemporary Tamil Nadu, India. GARNET Working Paper No: 43/08

CDSCO (2014) F.No.X-11026/65/13-BD: directorate general of health services central drugs standard control organization biological division, February 18. Available via http://www.cdsco.nic.in/writereaddata/Guidance%20Document%20For%20Regulatory%20Approvals%20of%20Stem%20Cells%20and%20Cell%20Based%20Products.pdf. Accessed 22 Sept 2016

Cohen CB, Cohen PJ (2010) International stem cell tourism and the need for effective regulation: part 1: stem cell tourism in Russia and India: clinical research, innovative treatment, or unproven hype? Kennedy Inst Ethics J 20(1):27–49

DBT Annual Report (2015–2016) Department of Biotechnology, Government of India

Dhar A (2010) Medical ethics violation to be made punishable offence. The Hindu, August 01. Available via http://www.thehindu.com/news/national/medical-ethics-violation-to-be-made-punishable-offence/article544767.ece. Accessed 20 Sept 2016

Dixon-Woods M, Ashcroft RE (2008) Regulation and the social licence for medical research. Med Health Care Philos 11(4):381–391

Glasner P (2009) Cellular division: social and political complexity in Indian stem cell research. New Genet Soc 28(3):283–296

ICMR-DBT (2007) Guidelines for stem cell research and therapy. Indian Council of Medical Research

ICMR-DBT (2013) Guidelines for stem cell research. Indian Council of Medical Research

Inamdar MS, Venu P, Srinivas MS et al (2009) Derivation and characterization of two sibling human embryonic stem cell lines from discarded grade ill embryos. Stem Cell Dev 18(3):423–433

Jain K (2010) Doctors question impact of code of conduct without punishment. Business Standard, 2 Jan 2010. Available via http://www.businessstandard. com/article/economy-policy/doctors-question-impact-of-code-of-conduct-without-punishment-110010200009_1.html Accessed 29 May 2017

Jasanoff S (1988) The Bhopal disaster and the right to know. Soc Sci Med 27(10):1113–1123

Jasanoff S (2011) Introduction: rewriting life, reframing rights. In: Jasanoff S (ed) Reframing rights: bioconstitutionalism in the genetic age. MIT Press, Cambridge, MA, pp 1–27

Jayaraman KS (2005) Indian regulations fail to monitor growing stem-cell use in clinics. Nature 434:259

Joshi S (2011) No place to be sick. Tehelka, February 05. Available via http://www.tehelka.com/no-place-to-be-sick/. Accessed 5 Oct 2016

Kagan RA, Gunningham N, Thornton D (2003) Explaining corporate environmental performance: how does regulation matter? Law Soc Rev 37(1):51–90

Kamath MS (2010) Court has killed all medical negligence. DNA India, February 22. http://www.dnaindia.com/mumbai/report-court-has-killed-all-medical-negligence-cases-1350940

Khullar M (2009) Unfettered by regulation, India pulls ahead on stem cell treatments. Global Post, October 09. Available via http://www.globalpost.com/dispatch/india/091009/unfettered-regulation-india-pulls-ahead-stem-cell-treatments. Accessed 30 Sept 2016

Kim S-H (2014) The politics of human embryonic stem research in South Korea: contesting national sociotechnical imaginaries. Sci Cult 23(3):293–319

Lander B, Thorsteinsdottir H, Singer PA et al (2008) Harnessing stem cells for health needs in India. Cell Stem Cell 3(1):11–15

Madhiwalla N (2011) The NBC and the bioethics movement in India. Indian J Med Ethics 8(1):3–4

May PJ (2005) Compliance motivations: perspectives of farmers, homebuilders, and marine facilities. Law Policy 27(2):317–347

MoH&FW (2014) The drugs and cosmetics (amendment) bill 2015. Ministry of Health & Human Welfare, Government of India

Mudur G (1995) Indian doctors may be tried in consumer courts. BMJ 311(7017):1385

Mudur GS (2005) Stem cell march, minus checks: lack of research rules allows doctors to do as they like. The Telegraph, November 17. Available via http://www.telegraphindia.com/1051117/asp/nation/story_5487047.asp. Accessed 23 Sept 2016

Nandimath OV (2009) Consent and medical treatment: the legal paradigm in India. Indian J Urol 25(3):343–347

Nayak RK (2004) Medical negligence, patient's safety and the law. Reg Health Forum 8(2):15–24

Padma TV (2006) Unchecked by guidelines, indian stem cell scientists rush ahead. Nat Med 12(1):4

Paliwal A (2011) Ethics on trial. Down to Earth, June 30. Available via http://www.downtoearth.org.in/content/ethics-trial. Accessed Oct 2016

Pandey A, Desai PN (2016) Emerging structure of stem cell research in India: an analysis of publication output, 1990–2014. J Sci Res 5(1):49–61

Pandya S (2007) Medical ethics in India Today: concerns and possible solutions. Paper Presented at the National Seminar on Bioethics, Thane, January 24–25

Pandya SK (2008) Stem cell transplantation in India: tall claims, questionable ethics. Indian J Med Ethics 5:16–18

Patra PK, Sleeboom-Faulkner M (2009) Bionetworking: experimental stem cell therapy and patient recruitment in India. Anthropol Med 16(2):147–163

Patra PK, Sleeboom-Faulkner M (2010) Bionetworking: between guidelines and practice in stem cell therapy enterprises in India. SCRIPTed 7(2):295–310

Perappadan BS (2014) MCI backs probe into ads by doctors in violation of ethics. The Hindu, 20 September. Available via http://www.thehindu.com/news/cities/Delhi/mci-backs-probe-into-ads-by-doctors-in-violation-of-ethics/article6429237.ece. Accessed 5 Oct 2016

Peters DH, Muraleedharan VR (2008) Regulating India's health services: to what end? What future? Soc Sci Med 66:2133–2144

Pierre J, Peters BG (2000) Governance, Politics and the State, Macmillan Press Ltd

Rajalakshmi TK (2012) Criminal trials. Frontline 29(2): January 28–February 10. Available via http://www.frontline.in/navigation/?type=static&page=flonnet&rdurl=fl2902/stories/20120210290203300.htm. Accessed 24 Sept 2016

Raman S, Tutton R (2010) Life, science, and biopower. Sci Technol Hum Values 35(5):711–734

Rosemann A, Chiasinthop N (2016) The pluralization of the international: resistance and alter-standardization in regenerative stem cell medicine. Soc Stud Sci 46(1):112–139

Salter B (2008) Governing stem cell science in China and India: emerging economies and the global politics of innovation. New Genet Soc 27(2):145–159

Salter B, Cooper M, Dickins A, Cardo V (2007) Stem cell science in India: emerging economies and the politics of globalization. Regen Med 2:75–89

Salter B, Zhou Y, Datta S (2016) Governing new global health-care markets: the case of stem cell treatments. New Polit Econ. doi:10.1080/13563467.2016.1198757

Sarojini NB, Srinivasan S, Madhavi Y et al (2010) The HPV vaccine: science, ethics and regulation. Econ Polit Weekly 45(48):27–34

Sharma A (2009) Stem cell research and policy in India: current scenario and future perspective. J Stem Cells 4(2):133–140

Sipp D (2009) The rocky road to regulation. Nat Rep Stem Cells. September 23

Sleeboom-Faulkner M, Patra PK (2008) The bioethical vacuum: national policies on human embryonic stem cell research in India and China. J Int Biotechnol Law 5(6):221–234

Sleeboom-Faulkner M, Patra PK (2011) Experimental stem cell therapy: biohierarchies and bionetworking in Japan and India. Soc Stud Sci 41(5):645–666

Stem Cell Society of India (2015) Letter to the Drugs Controller General of India, 5 Feb 2015. Available at http://stemcellsocietyofindia.com/assets/scs-activities-aboutregulations. pdf Accessed 29 May 2017

Sunder Rajan K (2007) Experimental values: Indian clinical trials and surplus health. New Left Rev 45:67–88

Thomas G (2011) Medical council of India and the Indian medical associations: uneasy relations. Indian J Med Ethics 8(1):2

Turner L, Knoepfler P (2016) Selling stem cells in the USA: assessing the direct-to-consumer industry. Cell Stem Cell 19(2):154–157

Yee A (2012) Regulation failing to keep up with India's trails boom. Lancet 379:397–398

Chapter 12
Patenting Human Embryonic Stem Cells in the European Union Context: An Updated Analysis of a Complex Issue

Iñigo de Miguel Beriain

12.1 Introduction

Is it possible to patent human embryonic stem cells (hESC)? This question plays a key role in the debate on the future of regenerative medicine in so far as patentability is extremely important to ensure the development of a concrete biotechnology. As Brian Salter wrote some years ago,

> Without IPR, and in particular patent protection, emerging markets would find it difficult (or more difficult) to develop since the tangible product has yet to appear and economic value is embedded in the potential application of the knowledge. This problem is particularly acute in high-tech and research based Small to Medium Enterprises (SMEs) for whom their IPR is their main asset (Salter 2007).

However, the access to hESC patents has been largely denied to enterprises and researchers in the EU context. This has been for a number of different reasons that have contributed to darken the debate and, sometimes, to arrive at rulings that contradicted the state of the art in biology, such as the prohibition to patent hESC lines produced, thanks to a parthenogenesis process.

The aim of this chapter is to discuss the current juridical debate on the patentability of the hESC in the European context via an updated analysis of the European Patent Office (EPO onwards) and the Court of Justice of the European Union (CJEU onwards) jurisprudence and the normative framework created by both the European Patent Convention (EPC onwards) and the EU Directive on biotechnological inventions (Directive 98/44/EC of the European Parliament and of the Council of 6 July 1998 on the legal protection of biotechnological inventions). This will include a final reflection on the recent changes introduced by Case C-364/13 (International Stem Cell Corporation v Comptroller General of Patents, Designs and Trademarks)

I. de Miguel Beriain (✉)
Department of Public Law, University of the Basque Country,
Lehendakari Aguirre 135. Zubiria Etxea, Bilbao, 48015 Bizkaia, Spain
e-mail: inigo.demiguelb@ehu.eus

© Springer International Publishing AG 2017
P.V. Pham, A. Rosemann (eds.), *Safety, Ethics and Regulations*, Stem Cells in Clinical Applications, DOI 10.1007/978-3-319-59165-0_12

and the probable consequences it might bring to the patentability of hESC according to the attitudes adopted by the EPO in 2016.

12.2 Patenting hESC: The Normative Framework

What are the normative limits to the patentability of hESC? One can hardly answer this complicated question without mentioning the so-called ordre public clause, which was included in Article 53(a) of the European Patent Convention (Convention on the Grant of European Patents (EPC) of 5 October 1973, as revised by the Act revising Article 63 EPC of 17 December 1991 and the Act revising the EPC of 29 November 2000). It states: http://www.epo.org/law-practice/legal-texts/html/epc/1973/e/ar53.html—FOOTNOTE-29.

> Exceptions to patentability
> European patents shall not be granted in respect of:
> (a) inventions the publication or exploitation of which would be contrary to "ordre public" or morality, provided that the exploitation shall not be deemed to be so contrary merely because it is prohibited by law or regulation in some or all of the Contracting States.

For a long time, this clause was rarely used. Nevertheless, this situation changed dramatically after the approval of the EU Directive on biotechnological inventions, which included an article (Art. 6) that reads:

> 1. Inventions shall be considered unpatentable where their commercial exploitation would be contrary to ordre public or morality; however, exploitation shall not be deemed to be so contrary merely because it is prohibited by law or regulation.
> 2. On the basis of paragraph 1, the following, in particular, shall be considered unpatentable:
> … (c) uses of human embryos for industrial or commercial purposes;

Of course, the approval of the Directive introduced no compulsory changes in the EPO framework, in so far as EU rules are not legally binding on the EPO, which is not part of the EU at all. However, its influence was immediately felt. In the aftermath of the Directive, a new rule (Rule 28) was incorporated to the EPC's implementation regulations, stating that "Under Article 53(a), European patents shall not be granted in respect of biotechnological inventions which, in particular, concern the following: … (c) uses of human embryos for industrial or commercial purposes". Keeping this normative framework present, the Opposition Division of the EPO decided, in the "Edinburgh Patent" case (European Patent No. EP 0695351, with the title "Isolation, selection and propagation of animal transgenic stem cells"), to adopt a broad interpretation of the exceptions to patentability included in the EPC. As a consequence, it excluded from patentability both (a) processes that involve the extraction of stem cells from a human blastocyst (and therefore directly entail the destruction of the human embryo) and (b) claims relying on already-established hESC cell lines (Porter et al. 2006).

This ruling did not deny the idea that hESC could be patentable *as such*. The EPO, indeed, considered that, in theory, they could be patentable in so far as there was a scientific consensus on that hESC should not be considered as embryos because they alone cannot give rise to a full-grown organism (Condic et al. 2009; Condic 2014). However, the EPO concluded that, in practice, they could not be patentable due to the fact that their production involved the destruction of a human embryo at any stage, a practice that should be considered the forbidden commercial use of human embryos. This idea was, on its behalf, based on two assumed beliefs: the impossibility of obtaining hESC from an embryo without destroying that embryo and the impossibility of creating hESC from sources other than a human embryo.

These arguments provoked a general rejection of all patent claims involving hESC for a time in the EU context, a situation that generated a lively debate in the arenas of industry and academia (Porter et al. 2006). However, both ideas have been successfully challenged in the following 10 years. As a consequence, the EPO has adequately turned its position on this issue. Indeed, it has created a new paradigm by recognising (explicitly or not) the lack of consistency of both premises according to the updated knowledge. As a final result, hESC has been progressively allowed in the European context. In the next section, this normative U-turn will be exposed, and its present and future consequences described.

12.3 The WARF Case and the Possibility of Obtaining hESC Without Destroying Human Embryos

The first of the two premises to be challenged was the one stating that it was not possible to obtain hESC without destroying human embryos, an assumption that led to the delivery of the first patents on hESC. This turn was made in the context of the "primate embryonic stem cells" patent, filed by the Wisconsin Alumni Research Foundation [WARF], which claimed cell cultures including primate embryonic stem cells (EP0770125, application number 196903521.1). This claim was initially rejected by the Examining Division of the EPO in 2004, on the basis that the method disclosed made use of human embryos for industrial or commercial purposes. The applicant, however, decided to appeal to the Technical Board of Appeal of the EPO, which referred the question to the Enlarged Board of Appeal [EBoA]. The Board reached a decision in 2008, ruling that claims directed to products which, at the filing date, could be prepared *exclusively* by a method *necessarily* involving the destruction of human embryos are not patent eligible, even if the said method is not part of the claims (Davey et al. 2015).

This statement opened the gate to the patentability because of the concrete use of the word *necessarily*. Indeed, by using it, the EBoA emphasised that the moral clause only concerned inventions obtained by any methods involving the destruction of human embryos. On the other hand, it would not be applicable to general inventions relating to human stem cells or human stem cell cultures (Zhu 2011) as such. Therefore, it recognised that where it is possible to create hESC without destroying human embryos, these hESC could be patented. As a consequence, following the WARF decision, a European patent involving hESCs could be issued if its date of filing was after 10 January 2008, because it was considered that a technology able to produce this result (creating hESC without destroying embryos) had been disclosed at that moment (Chung et al. 2008). The invention, as disclosed in the application, did not necessarily have to be obtained with cells generated according to non-destructive embryo technology. However, the applicant had to be able to demonstrate that the invention could be reproduced, at the date of filing, using such cells (if the examiner had any doubt, an objection was raised, even if the application was filed after 10 January 2008). This demonstration did not, however, necessarily have to feature in the description of the application (Faure Andre 2014).

In any case, it was clear from the very beginning that the impact of this new decision depended on the definition of "human embryo" under the Directive on Biotechnology (Davey et al. 2015), an issue which was left blank by the ruling on purpose. Indeed, the decision did not adhere to any position regarding the human embryo definition. Therefore, it was not possible to determine whether Rule 28(c) should be applied only to inventions involving human embryos produced by fertilisation or also to those created by other techniques, such as, for instance, somatic cell nuclear transfer (cloning), which was already available at that moment. However, this was precisely the issue that was about to be faced by the European Court of Justice.

12.4 Oliver Brüstle Vs. Greenpeace and the New EPO Position

In 1997, a German scientist, Dr. Oliver Brüstle, submitted a patent concerning the production of neural precursor cells used for the treatment of neurological diseases made from human embryonic stem cells. At the request of Greenpeace, the federal tribunal of patents (Bundespatentgericht) ruled that Dr. Brüstle's patent was invalid since it concerned processes which allowed precursor cells to be obtained from human embryonic stem cells (Puppinck 2013). Dr. Brüstle decided to appeal the decision to the Federal Court of Justice (Bundesgerichtshof). This Court, on its side, requested a preliminary ruling from the Court of Justice of the European Union (CJEU onwards) on the interpretation of the Parliament and the Council's Directive 98/44/CE, from 6 July 1998, regarding the legal protection of biotechnological inventions, in order to define what should be understood as a human embryo in the context of the Directive.

Thus, the CJEU finally undertook to define the term "human embryo" in Article 6(2)(c) of the Directive in a ruling (Case C-34/10, Oliver Brüstle vs. Greenpeace e.V., 18 October 2011). The nucleus of this ruling is contained in point 39, which states that "any human ovum after fertilisation, any non-fertilised human ovum into which the cell nucleus from a mature human cell has been transplanted, and any non-fertilised human ovum whose division and further development have been stimulated by parthenogenesis constitute a "human embryo"". The reasoning behind this conclusion was showed in points 35 and 36. Point 35 stated that "any human ovum must, as soon as fertilised, be regarded as a 'human embryo' within the meaning and for the purposes of the application of Article 6(2)(c) of the Directive, since that fertilisation is such as to commence the process of development of a human being", a consideration that should "also apply to a non-fertilised human ovum into which the cell nucleus from a mature human cell has been transplanted and a non-fertilised human ovum whose division and further development have been stimulated by parthenogenesis" (36).

This position was due to a scientific belief that was considered an undeniable fact by the Court: "although those organisms have not, strictly speaking, been the object of fertilisation, due to the effect of the technique used to obtain them they are, as is apparent from the written observations presented to the Court, capable of commencing the process of development of a human being just as an embryo created by fertilisation of an ovum can do so" (36). Therefore, the reasoning reinforced one of the two main beliefs that served as a basis to the "Edinburgh Patent" case ruling, as previously mentioned: the impossibility of creating hESC other than from a human embryo.

Moreover, the Brüstle vs. Greenpeace ruling went a step further than the Edinburgh Patent case. Together with this extremely extensive definition of the embryo, the Court adopted a similar position regarding the range of the use of the embryos in the process, by stating that the *use of human embryos exclusion* applied not only to those inventions which directly involved in the destruction of an embryo but also to those requiring the use of an hESC which was originally derived through the destruction of an embryo. This meant that any methods or products which involved as their base material cells taken from an established hESC line which was originally obtained through the destruction of a human embryo, no matter how long ago the hESC line was established, were excluded from patentability (Rigby 2015). Accordingly, it also stated that a claim would be excluded from patentability even if it involved the use of an existing hESC line where the destruction of a human embryo for preparing the existing hESC line occurred a long time before implementation of the invention (Young-In et al. 2016). Finally, the CJEU detailed that the ban on the patentability of human embryos for industrial and commercial purposes also covered the use of embryos for scientific research, since the granting of a patent for an invention involved, in principle, such types of uses.

The adoption of these criteria created an important (even if not lasting) difference between the CJEU and the EPO, which was made clear by U. Storz: "while the WARF decision has often been interpreted in such way that (i) patent applications which relate to inventions made after the underlying hES cell lines became available, and (ii) patent applications related to the production of stem cells, or stem cells as such, which describe at least one alternative way to produce the said cells (i.e., not related to, or involving, hES cells), are both patentable, such bypass is no longer possible in the understanding of the ECJ" (Storz 2013). However, differences did not last for long. As previously mentioned, an ECJ ruling cannot directly influence the EPO, since it is not part of the EU, but it usually happens in practice. Indeed, it was as soon as 3 November 2011 when the latter's EPO president Benoît Battistelli stated in his weblog that "if the judges rule in favour of a restrictive interpretation of biotech patentability provisions, the EPO will immediately implement it". Therefore, it was not surprising to see that 3 years after the Brüstle vs. Greenpeace ruling, on 4 February 2014, the Technical Board of Appeal of the EPO (EPO. Boards of Appeal, *Technion Research and Development Foundation Ltd.*, T2221/10. 4 February 2014) tightened the argument held in the WARF case, explicitly noting that CJEU rulings were persuasive. In fact, the Board repeated the CJEU argument, detailing that hESC would not be patentable if they had been obtained from the destruction of human embryos, no matter when such destruction might take place (Mahalatchimy et al. 2015). Moreover, it clarified that inventions which made use of publicly available hESC lines initially obtained by methods involving the destruction of human embryos were excluded from patentability. Finally, it decided that the use of commercially available hESC lines was not sufficient to meet the requirements of Article 53(a) EPC if, at the "relevant date", these cell lines could only be obtained by destroying an embryo. This was particularly important, in so far as, between 2008 and 2012, the EPO granted patents for inventions involving such cell lines, as their use did not momentarily require the involvement of a human embryo (Faure Andre 2014).

12.5 A New Turn: Judgement C-364/13

Oliver Brüstle vs. Greenpeace received wide criticism due to its extensive interpretation of the moral clause, its construction of the concept of human beings, and the consequences that these facts might bring to the EU biotechnological industry. However, it was not formally defied until 2014. The history reads like this: at the beginning of the 2010s, a private firm called International Stem Cell Corporation devised methods for provoking human eggs to divide into clusters of pluripotent stem cells, known as parthenotes, and methods for turning such cells into corneal cells (Hitchcock 2014). Soon afterwards, it made a claim to register patents on them in the UK. The hearing officer of the UK Intellectual Property Office, acting for the comptroller, refused to register those applications by the decision of 16 August 2012. The reason provided was quite simple: as far as the inventions disclosed in the applications for registration related to unfertilised human ova which were however "capable of commencing the process of development of a human being just as an embryo created by fertilisation of an ovum can do so", they should be considered human embryos, within the meaning of paragraph 36 of the judgement in Brüstle.

This decision made perfect sense, according to the Brüstle ruling, but, in any case, the ISCO decided to bring an appeal against it before the High Court of Justice (England and Wales), Chancery Division (Patents Court), claiming that it was unclear whether the expression "capable of commencing the process of development of a human being" referred to an entity that could in fact develop into a human being or something that could start the process of becoming a human being but was unable to complete that process. At the same time, it amended its applications for registration to exclude the prospect of the use of any method aimed, through additional genetic manipulation, at overcoming the inability of a parthenote to develop into a human being, in order to facilitate the Court to make a decision in its interest.

The High Court of Justice (England and Wales), Chancery Division (Patents Court), felt quite dubious regarding the decision to be made, in so far as it considered that the balance between respect to human dignity and promotion of biological research development could hardly be maintained while banning the patents on parthenotes. Therefore, it finally decided to refer a simply and clear question to the Court of Justice for a preliminary ruling: "are unfertilised human ova whose division and further development have been stimulated by parthenogenesis, and which, in contrast to fertilised ova, contain only pluripotent cells and are incapable of developing into human beings, included in the term 'human embryos' in Article 6(2)(c) of Directive 98/44 ...? ".

On 17 July 2014, the general advocate of the case, P. Cruz Villalón, redacted an opinion on the issue that mostly accepted the arguments of International Stem Cell Corporation. This was only the preliminary to the Court's ruling, which was finally made on 18 December 2014 (Case C-364/13 International Stem Cell Corporation vs. Comptroller General of Patents, Designs and Trademarks). Mainly addressing the opinion formulated by its advocate general, it corrected the ruling made by Brüstle vs. Greenpeace by stating that

> Article 6(2)(c) of Directive 98/44 must be interpreted as meaning that an unfertilised human ovum whose division and further development have been stimulated by parthenogenesis does not constitute a 'human embryo', within the meaning of that provision, if, in the light of current scientific knowledge, that ovum does not, in itself, have the inherent capacity of developing into a human being, this being a matter for the national court to determine.

At first glance, it could be concluded that the new ruling had simply corrected a scientific mistake included in Brüstle. Indeed, in a very polite way, the Court stated that "according to current scientific knowledge, a human parthenote, due to the effect of the technique used to obtain it, is not as such capable of commencing the process of development which leads to a human being" (point 33). Of course, the same scientific knowledge was available at the time that Brüstle was ruled, but it probably seemed less rude to avoid mentioning the gross scientific mistake that the Court had made on that previous occasion. That mistake was made on the grounds that these eggs *are* "capable of commencing the process of the development of a human being just as an embryo created by fertilisation of an ovum can", a fact which is not certain at all, as scientists repeatedly mentioned in the aftermath of the ruling (Green 2011).

However, if we only concentrate on that part, we would be missing the fact that the really groundbreaking statement included in the ruling referred to the notion of what an embryo is. Indeed, the most astonishing statement made by the Court was that "in order to be classified as a 'human embryo', a non-fertilised human ovum must necessarily have the inherent capacity of developing into a human being" (point 28) and "consequently, where a non-fertilised human ovum does not fulfil that condition, the mere fact that that organism commences a process of development is not sufficient for it to be regarded as a 'human embryo', within the meaning and for the purposes of the application of Directive 98/44" (point 29). Of course, this was in huge contrast to the Court's previous ruling in Brüstle, where it held that an ovum would constitute a "human embryo" if it were "capable of commencing the process of development of a human being" instead of "capable to develop into a human being", as this new decision stated (Dannreuther 2014). This subtle difference of criteria involves an enormous change in scientific terms, in so far as the capability to develop into a human being involves a much higher exigency than the capability to commence the process of development. Therefore, some of the entities that should be considered as human embryos according to Brüstle vs. Greenpeace should no longer be named as such in the terms settled by Case C-364/13. This ruling, thus, involved a decisive U-turn in the jurisprudence of the High Court, not just because it permitted patents on human parthenotes but for the reasons that justified the decision: their lack of the necessary capability to develop into a human being. In doing so, the Court recognised the possibility to create hESC from a source different to human embryos.

By adopting this new criterion, Case C-364/13 has brought an important consequence: claims on patents on hESC obtained from parthenotes are now considered acceptable because they are not embryos, and, thus, their use does not contradict the moral clause restriction. Furthermore, it seems reasonable to foresee that the EPO might start accepting applications relating to hESCs filed after 5 June 2003. This is due to a simple fact: even if judgement C-364/13 does not indicate the date from which the skilled person would have been able to generate hESCs from parthenotes, it is precisely that moment when the EPO considers that the skilled person would have been able to generate parthenotes and derive hESCs from them because it was that point at which WO03046141, a patent application including an efficient methodology to reach this purpose, was disclosed.

However, changes might reach a much deeper dimension, depending on the consequences that the new criterion to define a human embryo involves in practice. Indeed, it does not seem at all illogical to consider that if a parthenote must not be considered an embryo because it does not hold the inherent capacity to develop into a human being, then *any* egg which shares this lack does not constitute a human embryo, regardless of the method it was created by (De Miguel Beriain 2014). This opens up a number of extremely interesting questions (Rigby 2015): Can it therefore be concluded that using non-viable embryos with chromosomal abnormalities to isolate hESCs (Alikani and Munné 2005) is outside the scope of the prohibition? A further question is whether a blastocyst that has been rated as being of poor quality (and that is therefore deemed by IVF experts to be unsuitable for implantation) has or not have the capacity to develop into a human being in the terms expressed by the EPO (Mitalipova et al. 2003). Moreover, let us think that our latest discoveries (Fragouli et al. 2015; Diez-Juan et al. 2015) about the correlation between mitochondrial DNA (mtDNA) and development potential are confirmed. Where this is the case, we will soon be perfectly able to distinguish which fertilised eggs are not capable of being implanted due to a fatal mtDNA composition. Should we then conclude that hESC extracted from these entities could be patented, because they could not be considered human embryos according to the judgement C-364/13 criterion? If this were the case, we could think about a revolutionary change in the framework, as hESC lines obtained from these defective fertilised eggs would be 100% similar to embryonic stem cell lines (blastocysts are totally equal) and they could be obtained in a much easier and cheaper way than the parthenogenesis process, because IVF could probably provide us with thousands of embryo-like structures every year, even for free, if the users were willing to donate these entities to science.

Of course, all of these modifications will only happen if the EPO decides to assimilate the new CJEU criterion (and, moreover, if its interpretation adopts the idea that entities possessing the same capabilities should be considered morally equivalent, no matter the way they were created). At present, this change in practice does not appear to have been publicised in the *Official Journal of the EPO* or reflected in the changes to the EPO official guidelines (Young-In et al. 2016; Rigby 2015), but there are good reasons to think so. First of all, it is necessary to highlight that whenever any difference between the positions of the EPO and the CJEU exists, the EPO deliberately surrenders its position in favour of that of the CJEU (Storz 2013). Moreover, there are some evidences to suggest that this case will not make an exception: a recent examination report in respect of EPO application No.13186524.8 concluded, with reference to the judgement C-364/13, that the invention was not excluded from patentability under the EPO's morality provisions. Thus, it seems reasonable to anticipate that it should be easier to protect technologies using human parthenotes in the future in Europe, including those involving their destruction (Faure Andre 2014). If further decisions confirm this initial impression, the future will know a totally different normative framework, a framework that might work much better with the necessities of the industry and also with the requirements of our common moral basements, which recommend an adequate protection of human embryos'—but only real human embryos' (De Miguel Beriain 2014; Meskus and De Miguel 2013)—life.

Acknowledgements I would like to take this opportunity to acknowledge the support from the Spanish Ministry of Economy, awards DER 2013-41462-R and DER2015-68212-R.

References

Alikani M, Munné S (2005) Nonviable human pre-implantation embryos as a source of stem cells for research and potential therapy. Stem Cell Rev Rep 1(4):337–343

Chung Y et al (2008) Human embryonic stem cell lines generated without embryo destruction. Cell Stem Cell 2:113–117

Condic ML (2014) Totipotency: what it is and what it is not. Stem Cells Dev 23(8):796–812. doi:10.1089/scd.2013.0364

Condic ML, Lee P, George RG (2009) Ontological and ethical implications of direct nuclear reprogramming: response to Magill and Neaves. Kennedy Inst Ethics J 19:33–40

Dannreuther A (2014) The CJEU clarifies when stem cells can be patented in Europe, Friday, 19 December 2014, EU Law Analysis Expert insight into EU law developments. http://eulawanalysis.blogspot.com/2014/12/the-cjeu-clarifies-when-stem-cells-can.html. Accessed 21 Sept 2016

Davey S, Davey N, Gu Q et al (2015) Interfacing of science, medicine and law: the stem cell patent controversy in the United States and the European Union. Front Cell Dev Biol 3:71. doi:10.3389/fcell.2015.00071

De Miguel Beriain I (2014) What is a human embryo? A new and extremely difficult to ensamble piece in the bioethics puzzle. Croat Med J 55(6):669–671

Diez-Juan A, Rubio C, Marin C, Martinez S, Al-Asmar N, Riboldi M, Diaz-Gimeno P, Valbuena D, Simon C (2015) Mitochondrial DNA content as a viability score in human euploid embryos: less is better. Fertil Steril 104(3):534–541

Faure Andre G (2014) Human embryonic stem cell patentability in Europe and the United States, Paris. https://www.google.es/url?sa=t&rct=j&q=&esrc=s&source=web&cd=2&cad=rja&uact=8&ved=0ahUKEwit-6bWnaPPAhUBFxQKHS9MDD0QFgglMAE&url=http%3A%2F%2Fwww.regimbeau.eu%2FREGIMBEAU%2FGST%2FCOM%2FPUBLICATIONS%2F2015-04-hESCs-GFA_EN.pdf&usg=AFQjCNFMrelIBoCpPLv-iUsXPGSH1KMTUQ&sig2=R3P0LL8cxfDWLPAYZfnDUg. Accessed 21 Sept 2016

Fragouli E, Spath K, Alfarawati S, Kaper F, Craig A, Michel CE, Kokocinski F, Cohen J, Munne S, Wells D (2015) Altered levels of mitochondrial DNA are associated with female age, aneuploidy, and provide an independent measure of embryonic implantation potential. PLoS Genet 11:e1005241

Green JBA (2011) Patenting: European stem-cell ruling is misleading. Nature 479:41. doi:10.1038/479041a

Hitchcock J (2014) Brüstle in the Dock: ISCC v Comptroller General of Patents, Opinion of AG Cruz Villalón, Bionews. http://www.bionews.org.uk/page_442731.asp. Accessed 21 Sept 2016

Mahalatchimy A, Rial-Sebbag E, Duguet A-M, Taboulet F, Cambon-Thomsen A (2015) The impact of European embryonic stem cell patent decisions on research strategies. Nat Biotechnol 33(1):41–43

Meskus M, De Miguel I (2013) Embryo-like features of induced pluripotent stem cells defy legal and ethical boundaries. Croat Med J 54(6):589–591

Mitalipova M et al (2003) Human embryonic stem cell lines derived from discarded embryos. Stem Cell 21:521–526

Porter G, Denning C, Plomer A, Sinden J, Torremans P (2006) The patentability of human embryonic stem cells in Europe. Nat Biotechnol 24:653–655. doi:10.1038/nbt0606-653

Puppinck G (2013) Synthetic analysis of the ECJ Case C-34/10 Oliver Brüstle v Greenpeace e.V. and its ethical consequences. http://www.oneofus.eu/docs/france/SyntheticanalysisoftheECJcaseofBrustlevGreenpeaceanditsethicalconsequences.pdf. Accessed 21 Sept 2016

Rigby B (2015) Another U-turn on stem cell patents in Europe. Available via Dehns, http://www.dehns.com/site/information/industry_news_and_articles/another_u-turn_on_stem_cell_patents_in_europe.html. Accessed 21 Sept 2016

Salter B (2007) State strategies and speculative innovation in regenerative medicine: the global politics of uncertain futures. Working Paper 20, July Available at: https://www.google.es/url?sa=t&rct=j&q=&esrc=s&source=web&cd=1&cad=rja&uact=8&ved=0ahUKEwjrjYHXzqzPAhWBkBQKHe6GCK8QFggeMAA&url=http%3A%2F%2Fwww.kcl.ac.uk%2Fsspp%2Fdepartments%2Fpoliticaleconomy%2Fresearch%2Fbiopolitics%2Fpublications%2Fworkingpapers%2Fwp20.pdf&usg=AFQjCNEaoOEltwJlP8D9k0RrV4ANLgn-Mg&sig2=gjjmYuzk07SOirp20dpeQ. Accessed 21 Sept 2016

Storz U (2013) The limits of patentability: stem cells. In: Hübel A, Storz U, Hüttermann A (eds) Limits of patentability. Plant sciences, stem cells and nucleic acids. Springer, Amsterdam, pp 9–32

Young-In TK, Sheperd M, Zvesper T (2016) Patentability of human embryonic stem cells, 1. https://www.marks-clerk.com/Home/Knowledge-News/Articles/Human-embryonic-stem-cells-at-the-EPO.aspx#.V-TuBeZImzl. Accessed 21 Sept 2016

Zhu H (2011) A comparative study on human embryonic stem cell patent law in the United States, the European Patent Organization, and China. https://www.google.es/url?sa=t&rct=j&q=&esrc=s&source=web&cd=1&cad=rja&uact=8&ved=0ahUKEwja-sK7sK_PAhXMmBoKHV6kAJIQFggeMAA&url=https%3A%2F%2Fkuscholarworks.ku.edu%2Fbitstream%2Fhandle%2F1808%2F11472%2FZhu_ku_0099D_12182_DATA_1.pdf%3Bsequence%3D1&usg=AFQjCNH2AHExGRmu7JhNDooUmB90bqd1dQ&sig2=e307P56ECa85i8ek_MQt1g. Accessed 21 Sept 2016

Chapter 13
The Regulatory Situation for Clinical Stem Cell Research in China

Achim Rosemann, Margaret Sleeboom-Faulkner, Xinqing Zhang, Suli Sui, and Adrian Ely

13.1 Introduction

This chapter provides an overview of the regulatory landscape of clinical stem cell research in the People's Republic of China. It provides, first, an overview of the regulation for the donation of human gametes and embryos and for their use in basic and clinical research. The chapter will then in Part II present an overview of all regulations and laws that were issued in China since the1990s to govern human subjects research. While these regulations do not mention stem cell research as a distinct regulatory category, these rules affect clinical stem cell research horizontally.

The final part of the chapter introduces the stepwise formation of a regulatory approach especially for clinical stem cell applications. This process was initiated in the mid-2000s and is ongoing. The chapter ends with a conclusion that discusses open questions and the implications of China's current regulation of clinical stem cell research for both domestic researchers and international clinical collaborations.[1]

---

[1] This chapter is based on a working paper and two publications that the authors of this chapter have published in 2013 and 2015:

A. Rosemann (✉)
Centre for Education Studies, Faculty of Social Sciences, B1.20, University of Warwick, Coventry CV4 7AL, UK
e-mail: A.Rosemann@warwick.ac.uk

M. Sleeboom-Faulkner
Centre for Bionetworking, Department of Anthropology, School of Global Studies, Arts C 209, University of Sussex, Brighton BN1 9SJ, UK

X. Zhang • S. Sui
School of Social Sciences and Humanities, Peking Union Medical College, #9, Dongdan, 3d Alley, Beijing 100730, China

A. Ely
Science Policy Research Unit (SPRU), University of Sussex, Jubilee Building 371, Brighton BN1 9SJ, UK

© Springer International Publishing AG 2017

P.V. Pham, A. Rosemann (eds.), *Safety, Ethics and Regulations*, Stem Cells in Clinical Applications, DOI 10.1007/978-3-319-59165-0_13

13.2 The Regulation for the Donation and Use of Human Gametes and Embryos

Two regulations have been issued in China to govern the donation of human gametes and embryos for research purposes. Both of these regulations focus exclusively on human embryonic stem cell research. The first concerns the sourcing of human embryos and oocytes in the context of IVF clinics. The second addresses the specific conditions under which human embryos can be produced and used research and clinical application. No regulation exists currently that addresses (a) somatic cell nuclear transfer techniques for research purposes (i.e., "therapeutic cloning"), (b) basic or preclinical research with iPS cells, and (c) research with human-animal hybrids.

13.2.1 The Ethics Guiding Principles for Assisted Reproductive Technology [2001–2003]

The "Ethics Guiding Principles for Assisted Reproductive Technology" have been issued by the Ministry of Health (MOH) between 2001 and 2003.[2] These guidelines regulate the donation and transfer of human embryos and gametes for reproductive purposes and research. This document addresses stem cell research, by stipulating that all ART institutions must set up ethics committees and that these committees must approve applications of human embryos to be donated for research (Hu et al. 2011; Cure 2009). The regulation affects the donation, circulation, and use of human embryos, gametes, and fetal tissue for research in four additional ways: (1) by stating that the buying and selling of human ova, sperm, embryos, or fetal tissues is prohibited; (2) by restricting the use of embryos for research to supernumerary

---

1. Rosemann A, Zhang X, Sui S, Su Y, Ely, A (2013) Country report: Stem cell research in China. Working Paper, Centre for Bionetworking, University of Sussex. (This working paper forms the basis for Sects. 13.1, 13.2, and 13.3 of this chapter.)
2. Rosemann A (2013) Medical innovation and national experimental pluralism: Insights from clinical stem cell research and applications in China. BioSocieties 8(1):58–74. (This paper forms the basis for Sects. 13.4, 13.4.1, 13.4.2, and 13.4.3 of this chapter.) (Palgrave MacMillan provides the right to the authors to use this information in this chapter.)
3. Rosemann A, Sleeboom-Faulkner M (2015) New regulation for clinical stem cell research in China–expected impact and challenges for implementation. Regen Med Doi:10.2217/rme.15.80. (This paper forms the basis for Sects. 13.4.4 and 13.5 of this chapter.) (This article has been published under a Creative Commons CC-BY license, and we are allowed to use published this information in this chapter.)

[2] Please note that the Chinese Ministry of Health (MOH) was in 2012 renamed to the National Health and Family Planning Commission (NHFPC). In this article we use both of these terms: for regulatory documents that were issued before 2012, we use the term Ministry of Health (MOH), and for regulatory documents that were issued after 2012, we use the term National Health and Family Planning Commission (NHFPC).

embryos in the context of an IVF treatment and by explicitly prohibiting the creation of IVF embryos for research only; (3) by specifying that embryos and gametes must be voluntarily donated, on the basis of informed consent; and (4) by forbidding hormonal super-stimulation, to harvest a larger number of oocytes. This regulation is backed up by punitive measures: IVF clinics or ART centers can lose their license if they violate these guidelines (Cure 2009).

13.2.2 The Ethics Guiding Principles for hESC Research [2003]

The derivation of human embryonic stem cells (hESC) and the use of these cells for research are regulated with the "Ethics Guiding Principles for hESC research" (人胚胎干细胞研究伦理指导原则). Regulation occurs at a national level through ministerial guidelines, joint issued in 2003 by the Ministry of Health (MOH) and the Ministry of Science and Technology (MOST). The core aspects of this regulation are that (1) embryos are not allowed to be used for the derivation of hESC after 14 days post-conception and (2) embryos used for research cannot be implanted in human beings (prohibition of human reproductive cloning). The principles demand, furthermore, that institutions that are involved in hESC form an ethics committee that details regulatory rules and exact conditions under which research can be conducted. These principles have been criticized by members of the National Ethics Committee of the MOH in China because (a) they do not introduce a registration or licensing system of research institutes or clinics that conduct hESC research, (b) they are not backed up by law, and (c) no clear control pathways for the principles are provided. Plans and efforts to revise this regulation, from the side of the MOH National Ethics Committee, are ongoing (Zhai 2007).

13.3 The Regulation of Human Subjects Research in China

Before introducing the regulations, laws, and institutions that play a role in the governance of clinical stem cell research and applications in Sect. 13.4, we will first of all provide an outline of the regulations and laws that were issued since the 1990s to govern human subjects research in China. As will become clear in Sect. 13.4 of this chapter, even though these regulations do not mention stem cell research, they play a role in the governance of clinical stem cell research and applications in China at a horizontal level.

13.3.1 The Foundation of a National-Level Ethics Committee [1998]

An important step for the governance of human subjects research in China was the foundation of the Ministry of Health's "ethics committee on biomedical research involving human subjects" in 1998. The committee was renamed in 2000 as "Medical Ethics Expert Committee." Following a reform in 2007, the committee comprises 17 members from a multidisciplinary background. These members have been instrumental in the formation of many of the regulatory documents discussed in this chapter.

13.3.2 The Regulation on Ethical Review of Biomedical Research Involving Human Subjects [2007]

The "Regulation on ethical review of biomedical research involving human subjects".
(涉及人的生物医学研究伦理审查办法 [试行]) addresses the formation of institutional review committees, and the procedures through which ethics review of human subjects research shall be conducted. A first document on the regulation of ethical review procedures was drafted in 1998. However, due to internal controversies, the document was not endorsed (Hu et al. 2011). A revised version of the regulation was then issued by the MOH in 2007. According to this regulation, all forms of research and experimental clinical interventions that involve human subjects require review by an ethics committee at the level of research institutes and hospitals. The regulation specifies information on the procedures and criteria for ethics committee review, the structure of the committees as well as details on informed consent procedures. This regulation has been of significance to clinical stem cell research and applications in China, because it requires mandatory EC approval at the level of medical institutions.

13.3.3 Regulation on the Governance of Medical Institutions [1994]

The "Regulation on the governance of medical institutions" was issued by the State Council in 1994. It clarifies that informed consent is required for the conduct of surgical operations, special investigations, participation in clinical studies as well as experimental medical interventions. The regulation introduces performance rules for medical institutions such as registration procedures, required qualifications of medical staff, and institutional safeguards that shall prevent the misuse of patients. An example is a clause that specifies that that approval documents for treatments (which were provided by the Chinese health authority to a particular hospital) cannot be "inherited," if the owner or name of a hospital changes. This regulation is of relevancy for clinical SC research, in particular with regard to the governance of clinics that offer experimental for-profit interventions with stem cells. The Jilin Silicon Valley Hospital, for instance, was criticized by the media on the basis of this regulation. The reason is that the hospital had changed its name and proprietor but used the same approval license for its cellular treatments (issued by a local and a provincial health bureau) in order to attract patients. The Chinese health authorities withdrew the license of the hospital shortly after this discovery.

13.3.4 The Drug Clinical Trial Regulations Law on Practicing Doctors [1999]

This regulation protects patients by stating that doctors who violate a patient's privacy or who conduct experimental medical interventions without informed consent will be legally persecuted. Even though cases where this regulation has been applied in the context of stem cell-based forms of clinical intervention, this regulation offers a legal instrument could be applied to providers of experimental stem cell treatments, if these fail to sufficiently inform patients of medical risks or make exaggerated treatment claims (Cure 2009). In other words, based on this regulation, patients could sue doctors who offered experimental stem cell treatments based on fraudulent claims.

13.3.5 The Drug Administration Law [2001, amended in 2015]

The Drug Administration Law was issued by the National People's Congress and is implemented through the Chinese Ministry of Health (which since 2012 was renamed to the "National Health and Family Planning Commission") and the Chinese State Food and Drug Administration (SFDA) (which since 2012 was renamed to the China Food and Drug Administration).[3] The law has been amended in 2015 and covers the use of pharmaceutical products in research as well as routine clinical applications following market approval. It clarifies that GCP and GMP standards must be followed (Cure 2009). This regulation is of relevancy for the development of stem cell-based medicinal products once the Chinese regulators have issued

---

[3] Please note that the Chinese State Food and Drug Administration (SFDA) was in 2012 renamed to the China Food and Drug Administration (CFDA). In this article we use both of these terms: for regulatory documents that were issued before 2012, we use the term State Food and Drug Administration (SFDA), and for regulatory documents that were issued after 2012, we use the term China Food and Drug Administration (CFDA).

a regulation for stem cell research that specifies the exact conditions for market approval under the authority of the China Food and Drug Administration.

13.3.6 The SFDA Good Clinical Practice Standards [1999, 2003]

The State Food and Drug Administration (SFDA) good clinical practice standards (药物临床试验质量管理规范) were issued in a first version in 1999 and in a second more complete version in 2003. Both versions were drafted and issued by the SFDA. The SFDA GCP standards specify procedures for clinical trials in the context of market authorization pipeline and for the accreditation of medical institutions that take part in drug trials. The regulation requires that each hospital that conducts clinical trials acquire GCP certification (Cure 2009). An interesting feature is that GCP certification is based on examinations of high-level clinical staff (heads of department). They emphasize strict informed consent and review by ethics committees and include provisions on how IRBs should be composed and be organized. The Chinese GCP standards draw actively on the ICH GCP standards (Cure 2009). These SFDA GCP standards are of relevancy to hospitals that conduct clinical trials with SC in the context of an SFDA registered IND application.

13.3.7 The SFDA Guidance for Human Somatic Cell Therapy Research and Quality Control of the Products [2003]

The Guidance for Human Somatic Cell Therapy Research and Quality Control of the Products (人体细胞治疗研究和制剂质量控制技术指导原则) forms an important precursory regulation for the clinical use of cells and stem cells. It was issued by the Chinese SFDA in 2003 and is still valid. The guidance addresses fundamental issues of therapy research with somatic cells. It focuses on aspects such as the collection, isolation, and verification of somatic cells, the kind of (medical and personal) information that is required from cell donors, directives on the use and documentation of specific types of culture mediums, and a broad range of specifications on quality control. Quality control encompasses measures for both preclinical research and production, culture, and storage protocols of somatic cellular products in the context of clinical research. While this guidance is still valid at the time of writing, many of the biological characteristics and particularities of stem cell research are not—or only insufficiently—addressed in this document. This is one of the reasons why the Chinese SFDA and MOH in the late 2000s decided that a new and more comprehensive regulation for the regulation of clinical stem cell research is necessary.

13.4 The Evolving Regulatory Approach for Clinical Stem Cell Research

In the subsequent paragraphs, we provide a detailed overview of the evolving regulatory framework for clinical stem cell research and applications in China. At the moment of writing, the process of developing a regulatory framework that can be used for both clinical testing and market approval of stem cell treatments is still ongoing. Since 2009 a range of regulatory documents have been issued by the Chinese health authorities that have affected clinical research and experimental for-profit interventions in various ways. The most recent and fargoing regulatory measure has been provided in 2015.

13.4.1 The Management Measures for the Clinical Use of Medical Technologies [2009]

On May 1, 2009, the MOH promulgated the "Management Measures for the Clinical Use of Medical Technologies" [医疗技术临床应用管理办法], a regulation that classified a range of new medical technologies and procedures into three categories. Stem cell transplant technology was grouped into category III, which included technologies considered as risky, ethically controversial and in need of clinical verification (Chen 2009). To implement the regulation, the MOH assigned five institutions (Chen 2009, p. 271); among them are the Chinese Medical Association, the Chinese Hospital Association, and the Chinese Doctors Association. According to an associate of the MOH in Beijing, clinics that used SC transplantation technology were summoned to register at these institutions. These organizations in turn were assigned to grant licenses on the basis of newly formed assessment criteria and review and inspection committees. In practice, this regulation has not yet been implemented for SC transplantation technologies. As stated by a senior SC scientist, who as a member of the Chinese Doctors Association was involved in the formulation of review criteria, there were widespread disagreements among experts of the assigned five institutions, over the precise characteristics of these criteria, over feasible implementation pathways, as well as the extent to which the situation should be controlled.

13.4.2 Notification on Self-Evaluation and Self-Correction Work Regarding the Development of Clinical Stem Cell Clinical Research and Applications [2012]

On January 6, 2012, the National Health and Family Planning Commission (NHFPC, the former MOH) issued a regulatory document called "Notification on Self-Evaluation and Self-Correction Work regarding the Development of Clinical Stem

Cell Clinical Research and Applications" [关于开展干细胞临床研究和应用自查自纠工作的通知].[4] With this document, the NHFPC initiated a 1-year phase that was announced to be followed by a more comprehensive regulatory approach at a later point. In the January 2012 document, four subsequent stages of this forthcoming approach were announced: self-evaluation (*zicha*), self-correction (zijiu), recertification (*chongxin renzheng*), and standardized management (*guifan guanli*). The initial 1-year phase that is set out in the 2012 document, however, addresses only the first two of these stages: self-evaluation and self-correction. Self-evaluation of the hospitals that carry out SC-based clinical research and applications shall occur in the following way. First, clinics are required to fill in the "Self-Evaluation Form for Inquiry into Conditions of Stem Cell Clinical Research and Applications."[5] In this form, clinics are asked to report truthfully on previously and currently developed kinds of clinical research and applications with stem cells. Information is requested on (1) types of cells and forms of cell-processing, (2) the disease types for which cells have been used, (3) forms of ethics and regulatory approval mechanisms, (4) informed consent procedures, (5) information on risks and experienced problems, (6) sources of funding and patient fees, (7) number of patients experimentally treated, and (8) publications or summarizing reports from clinical trials or other types of clinical studies. Second, this information is evaluated by province-level MOH workgroups, which are coordinated by the "Stem Cell Clinical Research and Application Standardization and Rectification Work and Leadership Group," cofounded by the MOH and SFDA in Beijing (paragraph 2). The task of these province-level workgroups is to appraise the incoming data, to produce summarizing reports to Beijing (paragraph 4), and, during later stages, to play an active role in the implementation and enforcement of the regulation (paragraph 2).

Self-correction means that all institutes that have not yet received approval, either by the MOH or the SFDA, must stop clinical stem cell research or application activities until approval has been obtained. Institutes that continue to carry out unauthorized clinical research or applications have been announced to be targeted as focal points for rectification (paragraph 2). On the other hand, clinical trials for stem cell products that have obtained approval by the SFDA are expected to act in strict accordance with the requirements set out by the SFDA and in compliance with the Chinese GCP standards (paragraph 2). The document has announced that no registration applications will be accepted by the MOH or the SFDA until July 1, 2012 (paragraph 2). Information on how applications for registration will be handled, however, has not been provided in the text. Uncertainty also remains as to how non-compliance will be dealt with, and which role the MOH and its province-level workgroups will play in this. It is not clear, furthermore, whether military hospitals (which operate under the command of the Health Department of the Army General Logistics Department) will be subjected to the same review and approval procedures as state hospitals or whether a different regulatory approach shall apply.

---

[4] http://www.moh.gov.cn/publicfiles/business/htmlfiles/mohkjjys/s3582/201201/53890.htm.

[5] This document has been put on the MOH website. http://61.49.18.65/publicfiles///business/cms-resources/mohkjjys/cmsrsdocument/doc13829.docx. Translations of these two documents can be requested from the author of this article per email.

13.4.3 The March 2013 Announcement of Three Interrelated Draft Regulations

On March 7, 2013, the MOH published online three regulatory documents on clinical stem cell research for public feedback and commentary. These documents introduced the underlying rationale of the planned regulatory framework, an overview of its basic structure, central regulatory instruments, and planned implementation structure. In contrast to initial media reports (Zornoza 2013), these documents did not yet constitute regulatory draft documents themselves, and they had no legal authority. This consultation formed the first publicly visible step toward regulatory intervention of clinical stem cell research and applications in China, since the abovementioned January 6, 2012, notification. It represents the first move toward realization of the third and the fourth of the four regulatory phases that the MOH announced in its 2012 notification (as introduced above): "recertification" [chongxin renzheng] and "standardized management" [guifan guanli], following, from phases I and II, "self-evaluation" [zicha] and "self-correction" [zijiu], which were initiated in the course of 2012.

The most important announcements in these documents were as follows. The approval of stem cell-based therapeutic applications must be based on phase I, II, and III clinical trials. These trials must be approved by hospital internal ethics committees and joint expert committees of the NHFPC and CFDA. Clinical stem cell trials must follow from solid preclinical evidence that documents the safety and therapeutic potential of a candidate therapy in animal models. Clinical or corporate sponsors of these trials will not be allowed to charge patients for participation in these trials. The stem cell collection, purification, amplification, certification, packaging, storage, and transport of the stem cells that shall be used for clinical trials must occur in accordance with good laboratory practice (GLP) and good manufacturing practice (GMP) standards. Another announcement was that violation of these requirements would be subject to punishment procedures and legal persecution under the existing drug management law.

13.4.4 The Regulation for Clinical Stem Cell Research and the Stem Cell Preparations Quality Control and Preclinical Research Guidelines [2015]

Following the public consultation in 2013, another 2 years went by until a formal regulation for clinical stem cell research was published. On August 22, 2015, the Chinese National Health and Family Planning Commission (NHFPC, the former Ministry of Health, MOH) and the China Food and Drug Administration (CFDA, the former SFDA) have issued two interrelated documents: (1) the regulation for clinical stem cell research and (2) the Stem Cell Preparations Quality Control and

Pre-clinical Research Guidelines. These documents form the long-awaited follow-up from the regulatory announcement that was issued in March 2013. The August 2015 "regulation for clinical stem cell research" presents itself as a "trial" or "interim" (试行) regulation. This is not unusual. In China, regulation usually starts out as a draft (草案) or trial regulation (试行). A "trial" regulation should be regarded as valid as formal regulation, but it is flexible enough to leave space for change. The document announces the central elements of a regulatory foundation for the clinical translation of stem cell-based medicinal products and procedures. What does China's future regulation for clinical stem cell trials look like? What challenges can be expected with regard to its implementation? And what impacts will the regulation have for domestic researchers, clinics, and corporations in China and at an international level?

13.4.4.1 Overview of the "Trial" (or "Interim") Regulation

The trial regulation applies to the clinical use of human autologous and allogeneic stem cells that are manipulated in vitro, with the exception of the routine transplantations of hematopoietic stem cells and of clinical trials that use stem cells that are affirmed as pharmaceutical products. Stem cell treatments have to pass through methodical clinical studies and follow from systematic preclinical evidence. These trials must comply with the Chinese "Quality Control Standards for Clinical Drug Trials" (the Chinese good clinical practice [GCP] standards), which has guided the approval of new drugs by the China Food and Drug Administration (CFDA) since 2007. Furthermore, first-in-human clinical trials must be based on systematic evidence of preclinical research proving the therapeutic value and safety of a candidate treatment in appropriate animal models.

The standards and technical procedures for the collection, manufacturing, and storage of stem cells for clinical use are laid down in the "Stem Cell Preparations Quality Control and Preclinical Research Guidelines," a supplementary document published by the CFDA, which also specifies the required criteria for safety and efficacy assessment in the context of preclinical studies. Only level 3 hospitals—the highest-ranked hospital category in China—are permitted to conduct stem cell clinical trials. To qualify, such hospitals must have established institutions for research, health care, and teaching and be in possession of the relevant professional qualifications. Hospitals must have ethics and academic committees capable of dealing adequately with adverse effects and preventing high-risk applications. Moreover, hospitals are required to establish stem cell preparation facilities that are compliant with international GMP standards.

Investigators applying for stem cell clinical trials must do so at provincial branches of the NHFPC and CFDA and register the trials online at the Chinese Medicine Registry and Management System. The NHFPC and CFDA will jointly review the projects at a provincial level with the help of specifically formed expert committees. These committees do not only review incoming applications but also will conduct on-site verification and evaluation of academic institutions, ethics committees, and project management. If a clinical trial application is accepted, phase I of the trial can go ahead. Clinical trial progress reports must be submitted to the authorities on a regular basis, and after each phase investigators need to report the research results to the provincial agencies. Based on these reports, decisions are made about progression to the next phase and ultimately about routine clinical application.

The regulation seeks to protect the interests of patients in the following ways. First of all, clinical investigators may not charge money for patients taking part in clinical studies, and hospitals are not allowed to advertise stem cell trials as treatments. Hospitals are required to fully inform patients of the potential risks of the research involved and to arrange insurance coverage for human subjects for projects involving a high level of risk. In case of emergency, life-saving facilities need to be in place. Moreover, serious adverse events must be reported to the hospital ethics committee and the provincial health authorities and will result in the immediate halt of the research project and withdrawal of approval for the application of the stem cell therapy concerned.

Stem cell clinical trials must be conducted in accordance with the "2007 Interim Regulation on the Review of Biomedical Research Involving Human Subjects" of the MOH (now NHFPC) and the "Drug Administration Law," issued by the MOH in 2001 (and amended in 2015). Clinical trials using human embryonic stem cells must harvest and process the cells in line with the "Guiding Principles for the Ethics for Human Embryonic Stem Cell Research," a joint regulation issued in 2003 by the Ministry of Science and Technology and the MOH. With the new trial regulation, stem cell-based treatments are no longer regulated as class III medical technology in accordance with the 2009 regulation for clinical stem cell applications [4], which indicates that the former regulation is no longer valid.

Medical institutions and staff who violate regulatory provisions are directly held responsible in accordance with specifically designed penal procedures. The provincial branches of the NHFPC and CFDA have the authority to suspend stem cell trials and to punish investigators and staff in line with appropriate laws and regulations.

13.4.4.2 Commentary

China has invested heavily into stem cell medicine in recent years. This has resulted in a growing body of publications and the development of new candidate therapies (Song 2011). Simultaneously, due to a permissive regulatory environment for clinical stem cell applications, the country has witnessed the mushrooming of commercial stem cell clinics. Between 2002 and 2012, China became a global hub for the sale of unproven clinical for-profit interventions (Rosemann 2013). A first attempt to control this situation was undertaken in 2009 in the context of a new regulation for medical technologies (Sui and Sleeboom-Faulkner 2015). However, because of disagreements within the health authorities on feasible implementation pathways, this regulation was never enforced for stem cell research, and the number of unproven stem cell interventions was widely reported to grow (McMahon 2014; Sui and Sleeboom-Faulkner 2015). In 2012 the MOH undertook a renewed regulatory effort by introducing a notification, which stipulated that all medical institutions without prior approval from the MOH or the CFDA must stop clinical stem cell procedures. This notification had limited effect, mainly on state-supported scientific institutions. An article in *Nature* reported that 3 months after the ban, numerous clinics in China were continuing their services (Cyranoski 2012). Then, in March 2013, the NHFPC published three interrelated draft regulations for public comments. These documents announced stringent controls on experimental stem cell interventions and emphasized clinical translation through systematic clinical trials overseen by the Chinese health authorities.

Elements of the 2013 regulation have now been incorporated in the regulatory documents published in August 2015. The 2015 "trial" regulation indicates an important step toward the improved governance and review of stem cell clinical research and applications in China. With the enforcement of systematic clinical studies required to comply with scientific principles, standardization, transparency, and the improved protection of research subjects, the CFDA and NHFPC have established a framework intended to cater the needs of researchers in China and internationally. The regulation rejects the use of unproven experimental for-profit interventions with stem cells (Song 2011), while introducing a clear strategy toward more responsible forms of clinical translation. The prohibition to advertise unproven stem cell treatment and charging patients for taking part in experimental studies alone could potentially result in the permanent halt of experimental for-profit interventions in a large number of hospitals that have profited from unclear regulations for years (McMahon 2014). Institutions that work under the publicized rules can be expected to raise methodological standards, improve the validity of research data, and subject patients to less risk.

However, the actual impact of the regulation depends on its enforcement and implementation. By sharing administrative duties for review and certification of clinical stem cell research and applications between provincial NHFPC and CFDA branches and by training specialist staff and expert committees to operate at the provincial level, China's health authorities create a regulatory infrastructure that promises to hit its target. The document's grounding in the country's "Drug Administration Law" and the backing of its stipulations by punitive measures reinforce this impression. Implementation, nonetheless, can be expected to be a difficult and gradual process, with several challenges along the way. A first challenge will be to train sufficient numbers of staff and to recruit well-qualified experts for independent review, so that incoming applications can be dealt with in a reliable and simultaneously efficient way. A further challenge concerns the geographical size of China, the country's large number of medical institutions, and the lucrative business opportunities that have evolved in the stem cell field in recent years (McMahon and Thorsteinsdóttir 2010). In the light of the well-established national and international networks of for-profit stem cell therapy providers in China (Sui and Sleeboom-Faulkner 2015), it will be difficult to control for-profit stem cell clinics. The problem of implementing the regulation to established institutions that seek to approve stem cell clinical trials is different from that of controlling stem cell clinics. While the new trial regulation delegitimizes unapproved for-profit stem cell interventions and provides a legal basis to close down such clinics, it does not provide concrete details on how the enforcement of such controls might occur. While the Chinese authorities in the last years have sporadically clamped down on for-profit stem cell clinics (Sui and Sleeboom-Faulkner 2015), it is unclear whether the resources, administrative infrastructure, and the political will can be mobilized to counter these clinics on a large scale and on a nationwide level. Enforcement of the regulation in the context of level 3 hospitals, on the other hand, can be expected to be successful: China's elite stem cell researchers have long since demanded a kind of regulation that can legitimize their research and resulting clinical applications. It remains to be seen, however, how tightly oversight procedures for clinical stem cell applications will be organized and whether the number of staff and available resources will be sufficient to assure dependable implementation.

Moreover, variation can be expected in the interpretation of regulation and policies among the provinces. Will these divergent interpretations thwart homogenous implementation? Despite possible variation across provinces, it is clear in the trial regulation that all research and commercial activities fall under the responsibility of the main units of the NHFPC and the CFDA in Beijing, which prohibit unauthorized for-profit interventions at the national level. Exemptions from the national standard at the provincial level (which has proven a hindrance for the effective regulation of autologous stem cell treatments in the USA (Knoepfler 2014)) are not possible.

It is also not clear to what extent the regulation affects practices in army and police hospitals, which have their own regulatory bodies and where much of the commercial stem cell activities have been located in recent years (Yuan et al. 2012). Much will depend on the political prioritization of tackling all experimental stem cell therapy providers, ranging from small for-profit providers to powerful military organizations.

The promise of greater dependability of approval procedures for the clinical development of stem cell treatments and greater compatibility with international procedures should be a relief to many stem cell scientists in China. The absence of a functioning regulatory framework for clinical stem cell research for many years has deprived researchers and R&D companies of the possibility to apply for the official registration of newly developed candidate treatments (Rosemann 2013). It has also limited the opportunity for building international clinical research collaborations (Zhang 2012). By introducing systematic approval procedures for stem cell clinical trials, the forthcoming regulation will strengthen domestic innovation trajectories, facilitate collaborations with foreign researchers, and also allow for joint applications for the approval of candidate therapies at drug regulatory authorities in China and in other countries.

The trial regulation's commitment to systematic preclinical studies, clinical trials, reliable quality controls, the Chinese GCP standards, GMP, and external review by independent expert committees promises to create congruence with both the benchmarks set out in the "Guidelines for the Clinical Translation of Stem Cells" of the International Society for Stem Cell Research (ISSCR 2008) and the standards for clinical stem cell research handled by the US Food and Drug Administration (FDA) and the European Medicines Agency (EMA).

13.5 Conclusion and Final Questions

The new trial regulation provides a basis to define experimental for profit interventions with stem cells in China as illegal and to investigate and punish stem cell clinics that operate outside the supervision of the NHFCP and the CFDA. The focus of this new regulation, however, is exclusively on the governance of clinical research. It does not stipulate any details on how the transition from clinical trials to routine clinical use and market approval shall be handled. This leaves many questions to be answered that will be crucial for corporations and international collaborations that strive for the joint application of stem cell treatments at drug regulatory authorities in China and other countries. Because information on marketing conditions is absent in the publicized regulation, it is extremely difficult to discuss its implications for international collaborations. A possible explanation to the lack of information on market approval in the current regulation is that no agreement on this point has been reached yet between involved stakeholders.

Unclear is also what procedures will be handled for the clinical use of stem cells that are affirmed as pharmaceutical products and also what criteria the NHFCP and the CFDA handle in order to define pharmaceutical stem cell products. Clearly designated subcategories of different types of stem cell interventions have not yet been published. However, such definitions will be of crucial importance to determine the relevant regulatory authority (the NHFCP, the CFDA, or different subunits). The fact that the CFDA is closely involved in the drafting and implementation of this regulation suggests that at least some stem cell-based applications will be classified as medicinal products. No matter how, the fact that this important point remains undefined suggests that harmonization with regulatory agencies in the USA, Europe, and other highly developed countries is still a long way off. These uncertainties might cause confusion for biotech companies, especially those that produce stem cells as quantifiable batch products from a single cell line, as, for instance, Geron has done with its human embryonic stem cell product (Knoepfler 2014). Another question is what type of clinical studies the NHFCP and CFDA require to allow the go-ahead from clinic to the market and routine use. While in a former, now invalid draft of the new regulation that was issued for public consultation in 2013, it was stated that systematic controlled phase I–III trials would be required (Xinhua 2013), the current regulation only speaks of clinical trials that shall be conducted according to scientific principles. Do China's health regulators leave this question deliberately open, so as to have the flexibility to follow the current Japanese model rather than the USA or EU model, which allows for conditional and time-limited market approval after successful clinical studies with relatively small number of patients (Azuma 2015)? Another issue that remains unclear is whether the new regulation in China leaves space for the conduct of experimental clinical interventions with stem cells outside of the format of the clinical trial (for instance, as a "last resort" treatment in individual patients after all existing interventions have failed) and how these forms of clinical experimentation will be reviewed and approved. A further question is how the regulation will impact the affordability of stem cell trials. The requirement of systematic preclinical research, the availability of GMP laboratories, and clinical translation through systematic clinical trials will significantly increase the costs of clinical translation. Accordingly, the introduction of the new regulations may have drawbacks for less well-endowed research institutes (Sui and Sleeboom-Faulkner 2015). With increased costs and a system that allows clinical studies solely in qualified tier three hospitals, only a limited number of investigators and research institutions will be enabled to conduct clinical stem cell trials. The resulting unequal access to financial resources may redefine opportunities to clinical innovations in the stem cell field. It remains to be seen whether this new situation will reignite a new brain drain to the private sector or abroad.

Acknowledgments This article has benefited from research support provided by the European Research Council (ERC, 283219) and the Economic and Social Science Research Council (ESRC, ES/I018107/1). Due to ethical concerns, supporting data cannot be made openly available.

References

Azuma K (2015) Regulatory landscape of regenerative medicine in Japan. Curr Stem Cell Rep 1(11):118–128

Chen H (2009) Stem cell governance in China: from bench to bedside? New Genet Soc 28(3):267–282

CURE (2009) Medical research council: CURE committee report. URL: http://www.mrc.ac.uk/Utilities/Documentrecord/index.htm?d=MRC006303. Accessed 22 Dec 2015

Cyranoski D (2012) China's stem cell rules go unheeded. Nature 484(7393):149–150

Hu QL, Liu M, Wei Z (2011) Introduction of ethics committees in China. Humboldt University, Charite Campus, SIGENET Health China Week, Berlin

ISSCR (2008) Guidelines for the clinical translation of stem cells. http://www.isscr.org/docs/guidelines/isscrglclinicaltrans.pdf. Accessed 22 Dec 2015

Knoepfler PS (2014) From bench to FDA to bedside: US regulatory trends for new stem cell therapies. Adv Drug Deliv Rev 82:192–196

McMahon D, Thorsteinsdóttir H (2010) Regulations are needed for stem cell tourism: insights from China. Am J Bioeth 10(5):34–36

McMahon DS (2014) The global industry for unproven stem cell interventions and stem cell tourism. Tissue Eng Regen Med 11(1):1–9

Rosemann A (2013) Medical innovation and national experimental pluralism: insights from clinical stem cell research and applications in China. BioSocieties 8(1):58–74

Song P (2011) The proliferation of stem cell therapies in post-Mao China: problematizing ethical regulation. New Genet Soc 30(2):141–153

Sui S, Sleeboom-Faulkner M (2015) Governance of stem cell research and its clinical translation in China: an example of profit-oriented bionetworking. East Asian Sci Technol Soc 9:397–412. doi:10.1215/18752160-3316753

Xinhua (2013) China to further regulate stem cell clinical experiments. http://news.xinhuanet.com/english/sci/2013-03/07/c_132216784.htm. Accessed 22 Dec 2015

Yuan W, Sipp D, Wang ZZ, Deng H, Pei D, Zhou Q, Cheng T (2012) Stem cell science on the rise in China. Cell Stem Cell 10(1):12–15
Zhai XM (2007) Challenges and governance/regulatory responses in china. In conference paper, BIONET workshop, Shanghai
Zhang JY (2012) The cosmopolitanization of science: stem cell governance in China. Palgrave MacMillan, London
Zornoza L (2013) China proposes stem cell clinical trial rules. Website of Regulatory Focus, http://www.raps.org/focus-online/news/news-article-view/article/2983/china-proposes-stem-cell-clinical-trial-rules.aspx. Accessed 22 Dec 2015

Chapter 14
Contested Tissues: The Donation of Oocytes and Embryos in the IVF-Stem Cell Interface in China

Achim Rosemann

14.1   Introduction

Human embryonic stem cell research has been a rapidly growing area within the life sciences since their discovery in 1998 (Thomson et al. 1998). A fundamental prerequisite for human embryonic stem cell (hESC) research is the sourcing of human embryos. The supply of embryos is made possible by the in vitro fertilization (IVF) of oocytes as part of infertility treatments, a process that routinely involves the creation of larger numbers of embryos. Depending on national regulations, "surplus" or "spare" embryos that IVF patients do not plan to use for reproductive purposes can be donated for stem cell research. In the laboratory, the inner cell mass of the donated embryos is isolated, modified, and cultured to colonies of hESC. These "spare" embryos are typically donated through a system of voluntary gifting, which is based on informed consent. The IVF clinic is thus the pivotal point in a triangular relationship that links the stem cell laboratory with the donors of the "biological raw material" on which stem cell economies rely: women and couples undergoing IVF treatment. Sarah Franklin has called this relational space the "IVF-stem cell interface" (Franklin 2006, p. 86). The need of hESC for stem cell research has confronted IVF patients with new choices and moral dilemmas, and it has led to a conflict of interests between the needs of patients and the requirements of the research lab. IVF clinicians are between these two interests and faced with the difficult task to balance the professional codes of the clinic with the demands of research (Svendsen and Koch 2008, p. 94). This situation is particularly demanding when IVF clinicians are also stem cell researchers or when a stem cell lab is part of (or part of the same institution as) an IVF clinic, which is often the case.

A. Rosemann (✉)
Centre for Education Studies, Faculty of Social Sciences, University of Warwick,
Room B1.20, Coventry CV4 7AL, UK
e-mail: A.Rosemann@warwick.ac.uk

© Springer International Publishing AG 2017
P.V. Pham, A. Rosemann (eds.), *Safety, Ethics and Regulations*, Stem Cells in
Clinical Applications, DOI 10.1007/978-3-319-59165-0_14

During 3 months of ethnographic field research in China, conducted between 2008 and 2009, I followed the pathway of human reproductive tissues (human gametes and embryos) on their way out of the human body into the IVF clinic and then from the clinic into the stem cell research laboratory. As I have illustrated elsewhere (Rosemann 2011, 2016, under review), this journey is paralleled by the systematic reconceptualization of the value, the status, and the meanings ascribed to these tissues. While initially intimately entangled with the physical, emotional, and social realities of their biological originators, in the course of the donation process, these bonds are gradually disconnected, and the telos and biological significance of these tissues is redefined (Waldby and Mitchell 2006). These changes enable and legitimize the destruction of human embryos and their transformation to hESC lines and use for scientific research. These changes enable, furthermore, the integration of these lines into new relational networks and systems of exchange, including their exchange as a commodity (Waldby and Mitchell 2006).

This chapter focuses on value conceptions of embryos and donation practices in the context of the IVF clinic, with a particular focus on the enactment of informed consent procedures. A question that I ask in this respect is what ideas are communicated to potential embryo donors, so that the donation of their spare embryos for hESC research appears reasonable and justifiable. The structure of this paper is as follows. I will first explore the role and meaning of informed consent procedures and the attitudes on informed consent among IVF clinicians from five IVF clinics in China. I will then examine actual donation practices. I will focus in particular on the more problematic aspects of embryo donation, which involve the provision of false facts and forms of rhetoric deception. I will then discuss these findings with regard to the regulatory context in which hESC are donated, banked, and distributed to researchers. I will draw, in this regard, on a comparison with the UK and in particular the role of the UK stem cell bank. First, however, I will provide some information on the methodology of this study.

14.2 Methods

The data presented in this chapter were gathered in China in February and March 2008 as well as August 2009. In this time, I visited five IVF clinics and six stem cell centers in three cities in Southeast and Central China. Research methods comprised the analyses of documents, semi-structured open-ended interviews with 15 stem cell scientists, 15 IVF clinicians, and 15 embryo donors, as well as a quantitative survey study whose findings are presented elsewhere (Rosemann 2010; Rosemann 2016, under review). The names of researchers and clinicians that are cited in this article have been anonymized.

14.3   Informed Consent: Purposes and Attitudes

Informed consent (IC) is commonly portrayed as an indispensable bioethical principle that safeguards autonomy and protects the patients' right to take a fully informed decision (Corrigan 2003, p. 768). An alternative way of defining the role and functions of IC procedures has been proposed by Waldby and Mitchell (2006). Instead of looking at informed consent as a mechanism to protect patient rights, they analyze it as a mechanism for the protection of the interests of clinicians and researchers. The signing of a consent form, the authors suggest, relocates the rights of ownership from donors to researchers. This step encompasses the renouncement of all legal claims to the results, benefits, and profits that might be derived through the use of the donated embryo (Waldby and Mitchell 2006, pp. 71–73). This view was fully confirmed in the context of my research. As the following two quotations show, clinicians and researchers experience informed consent as a crucial safeguard for the protection of their activities and interests.

> Senior IVF clinician 1: For donation [after a treatment] patients must come back to here, personally and sign the consent form. Otherwise, several years later, a patient might say: "Where is my embryo?" No, this is impossible!
> I: That means the consent form is also a kind of a contract, a legal document that protects the clinic?
> Senior IVF clinician 1: Yes. It is very important … for both doctors and the patients. [It] is very important to protect the doctor! Otherwise it is terrible. It may be a disaster to a center.
> Senior IVF clinician 2: All details are literally stated there. Once a patient has signed a paper, for the researchers or doctors it is very safe, very safe.
> I: Very safe, why precisely?
> Senior IVF clinician 2: We have a paper here, a signature. We have a consent paper signed by the patient; of course, it is safe now for the technician to use the embryo.

Importantly, from the perspective of the donor, signing a consent form means they abstain from all rights to access to both medical and financial benefits that might be achieved through the act of donation:

> Senior IVF clinician 3: [It is specified here that] patients cannot achieve any kind of economic goods from a [developed] treatment.
> Senior IVF clinician and stem cell researcher 1: If you sign the informed consent form, you have to agree that you really have no rights to the medical benefits of the research. I know that because I am [besides my work as clinician] also involved in a research programme. It is said there that you will not get any benefits from the things we do. In my personal opinion I think that is not fair.

In my interviews, most of the IVF clinicians and researchers with whom I spoke expressed a well-developed responsibility awareness of the needs and concerns of patients. In particular, senior IVF clinicians endorsed the requirement to stick to ethical principles such as informed consent and the right of IVF patients for a voluntary and autonomous decision:

Senior IVF clinician 4: They [the patients] have the right ... their behavior is totally voluntary, not under any pressure from the researchers or the doctors and [...] no matter how they will decide, their clinical treatment will not be influenced at all.

Senior IVF clinician 5: We have to explain to them, we have to offer different options, and then the patients make a decision by themselves.

Senior IVF clinician and stem cell researcher 2: We have to protect the rights of patients, so that they receive all the embryos they need for a successful pregnancy. Then we have to give information to donors about our research [...]. We have to inform patients, also if we want their low-quality embryos.

Many more examples could have been provided here. It is important to note, however, that the overwhelming majority of these statements come from clinicians and stem cell researchers in senior organizational functions who would only sporadically conduct informed consent procedures themselves. It is not surprising that at the level of actual practice, a more varied picture emerged.

14.4   Informed Consent and Actual Donation Practices

For persons who undergo infertility treatment, the in vitro generated embryo signifies a source of profound hope and value. After the diagnosis and yearlong experience of infertility, the creation of these embryos constitutes an important source of "reproductive capital." The IVF embryo, in other words, embodies the promise to render a long-cherished but recurrently frustrated wish to have one's own child into the realm of the possible. However, whatever the initial ideas among infertility patients on their embryos may have been, in the course of the IVF treatment, these conceptions and feelings are subjected to considerable changes. IVF patients are exposed to new forms of expertise and explanations, and the embryos that are created in the laboratory are subjected to rigorous testing of their quality, morphology, and reproductive viability. In other words, in the context of their treatments, patients learn to think about the characteristics and value of their embryos through the technical categories and quality parameters of the IVF space. This process can give rise to disappointments and a significant "culture shock" among laypeople. This restructuring of ideas, attitudes, and mental images of patients' embryos does clearly facilitate the attempts of clinicians or stem cell researchers to motivate IVF patients to donate their spare embryos for research. An important reason for this is that IVF embryos are evaluated and categorized along parameters of reproductive potentiality, which means that some embryos are defined as being of lower reproductive value than others. Another reason is that the explanations and quality categories of the IVF clinic enable the overcoming of alternative understandings of life, value, ethics, and sociality, such as defined by common sense, tradition, or religion. While I have discussed some of these points elsewhere (Rosemann 2016, under review), I shall provide a number of examples that offer insights into the ways in which the donation of embryonic tissues is carried out in the IVF clinics I visited.

14.4.1 Exploring Clinical Practices: Disparities Between the Real and Projected Value of hESC

A key finding from my research was that there exist some fundamental contradictions between the value descriptions of donated embryos as conveyed by IVF clinicians to IVF patients in the context of the donation procedure and the ways in which potential benefits from hESC can be used in the future. These disparities between "projected" and "real" value refer to a critical power imbalance between embryo donors and IVF clinicians as well as stem cell researchers. This situation opens up fundamental questions regarding benefit sharing and justice (see also Dickenson 2006; Waldby and Mitchell 2006; Sleeboom-Faulkner 2014).

Communication practices and the ways in which the donation process of human embryos for hESC research were conducted varied considerably between but also within the different clinics I visited. The majority of IVF clinicians with whom I spoke expressed a well-developed responsibility awareness of the needs and the moral dilemmas of IVF patients. These clinicians and researchers stated that they do their utmost in informing patients, in answering their questions, and in offering time to discuss donation with friends or family. However, more problematic aspects could be observed. Some IVF clinicians would carry out IC procedures in a less mindful manner and in ways that violated the interests of patients and their right to receive complete and unbiased information. As the following example shows, ideas such as "the right to be informed" were sometimes handled in superficial and rather unsatisfactory ways:

> I: When you ask patients to donate their low-quality embryos for hESC research, what information do you provide to them?
>
> Junior IVF clinician 1: Information? (Laughs) … Not much information. Just these words written down on this paper here, not much more information. [She points to a multiple-purpose informed consent form that lies on the table in front of her, which has to be filled in and signed before the onset of the treatment; most of the issues that were covered here refer to the risks and procedures of the infertility treatment itself; the donation of low-quality embryos for hESC research was only one issue among many, and dealt with in two sentences. It is specified that the donated embryos shall be used for research and that they will be destroyed in the process and cannot be reclaimed.]
>
> I: But the woman [an IVF patient] we spoke to this morning, she didn't know anything about stem cell research. Don't you have to explain it to her?
>
> Junior IVF clinician 1: It is just a brief, a brief explanation. It doesn't have many details.
>
> I: But does that happen often that a woman is asked to sign a form and she does not really know what for? I mean, for what kind of research the embryos are given away?
>
> Junior IVF clinician 1: But they … most of the patients do not care about what research we are doing. They just focus on … if they can get successfully pregnant (laughs). [...] You know, most of them just don't have any questions about it. They just go over it. They agree or disagree and then talk about other things. They don't focus on this … this is not their focus.

Younger clinical staff members, in particular, appear to carry out informed consent procedures in careless and sometimes highly irresponsible ways. Occasionally, the conversations that accompanied the IC process appear to be characterized by the calculated handling of silence, that is, the facilitation of "choice" through strategic games of information concealment and disclosure. In some cases, the conversations with patients also included elements of overt deception and the making of untenable promises:

> I: How many percent of patients want to provide their embryos after you have talked to them?
>
> Junior IVF clinician 2: Mm, maybe 75%.
>
> I: Oh, that is a lot!
>
> Junior IVF clinician 3: Yes but that is because we encourage (*guli*) them, we persuade (*shuofu*) them.
>
> I: How do you do this?
>
> Junior IVF clinician 3 (laughs and points to Junior IVF clinician 2): She is good at this (laughs again). She is doing this very well, to persuade patients.
>
> Junior IVF clinician 2: I tell them that it is useful for scientists and useful for mankind, in the future, probably ... And, OK, I will make sure that the donated embryos will not be given to other people, so that they know they will not have another baby.
>
> I: And what else do you tell. How do you try to persuade a patient so that she really ...
>
> Junior IVF clinician 2: If patients come to our hospital their purpose is to have a baby, they do not care too much about the remaining embryos. [...] I tell them that the stem cells [derived from their donated embryo] can maybe be used for their children, in case they have a disease that can be cured in the future.
>
> Junior IVF clinician 3: If their child has leukaemia, for example, maybe our research would help to cure these diseases. Maybe the patients, if they hear this, they think it is better for their child [if they donate], so many times they will agree.

Similar tactics of leading patients to believe that stem cells derived from their donated embryos might directly benefit the future health of the donors or the donor's child could also be observed in another clinic. As a clinician in a senior position told me, occasionally he would tell patients the following:

> Senior IVF clinician 7: If in the coming days, there will be the necessity that you use the stem cell line [established from your embryo] for medical purposes, we will check whether your embryos have become a cell line, what and where the cell line is today, and whether it is possible to use this line for you. If in the future there is a technique, you are the first to use this technique. You have the privilege to use the stem cells.

I do not want to preclude that such promises are—at least partly—based on good intentions or at least on a genuinely optimistic attitude toward the medical potential of stem cell research. However, it is obvious that these conversations contain elements of deception and manipulation. In addition to this, from a legal perspective, such claims remain unsupported. In the consent forms that patients sign, it is unmistakably specified that with the act of donation, the donor gives up all future rights on the embryo, including any claim to get access to future therapies or economic profits derived from the research for which donated embryos have been used.

14.5   How is the Use of hESC Lines Regulated? China in Comparison with the UK

hESC cell lines created in China are distributed in relatively complex and difficult to oversee networks within, but also across, national borders. To understand how the distribution of hESC can be used for the extraction of specific forms of value (and how these forms of values could benefit embryo donors and citizens), it is necessary to know more about the regulatory conditions under which these exchanges occur. I have decided, in this regard, to briefly compare the regulatory situation in China with the situation in the UK. The reason for this is that the UK is, to my knowledge, the only country in the world in which distribution of hESC lines occurs under the centralized and legally binding control of a centralized institution: the UK Stem Cell Bank (UK-SCB). In contrast to the UK, where the distribution of hESC lines occurs entirely under the supervision and rules of the UK-SCB, the movement of hESC lines in China occurs in a more open and, in a regulatory sense, also a less stringent system. However, let us first look at the situation in the UK. In the UK, the transfer of hESC lines is permitted only after the completion of a wide range of meticulously prescribed check-up procedures, which range from informed consent protocols to standardized assessment procedures for cell characterization and quality control. The UK-SCB steering committee plays a crucial role here, as it checks the license, qualifications, reputation, research objectives, and capacities of applicant centers. In case of requests from centers abroad, the committee still evaluates the legality of the proposed research project in the acquiring country (Stephens et al. 2008). A further responsibility of the committee is to negotiate with applicant centers the precise conditions and terms of use of the attained cell materials. Agreed conditions must fully comply with the UK-SCB's code of practice. Transgression is punishable in law (Warrell 2009).

This situation differs significantly from China, where individual research institutes manage the distribution of hESC samples, and regional government branches carry out the necessary controls. Furthermore, a huge difference exists in China regarding the transfer of hESC materials within and across national borders. While transfers of hESC samples within China seem to occur on the basis of the institutes' internal approval procedures, a considerably more complex regulatory picture emerges in the case of transfers of hESC samples abroad. Here, two basic requirements must be met. The first is to obtain approval from the Chinese Inspection and Quarantine Bureau, which handles an online registration system and which has specified the conditions that apply to the transfer of human tissue in the "Work Norms for the Health Quarantine Examination and Approval of the Entry/Exit of Special (Biological) Items," a nationally binding memorandum issued in 2006. No distinct set of specifications, however, exists for the transfer of hESC samples in this document, which fall under the same category as blood, bone marrow, cord blood, and other tissue commonly used for medical purposes. Documentation requirements for this category include a range of standard operating procedures for the identification of cell identity, quality, and the presence of microbial contaminants and biohazards. Further requirements include a description of research purposes and potential risks.

The second requirement is the setting up of a Material Transfer Agreement (MTA), a document that has to be signed by the Chinese Human Genetic Resources Control Office (HGCO). The MTA specifies the conditions and terms of use of exchanged tissue as negotiated and agreed upon between the exchange partners. Besides issues related to intellectual property and benefit sharing, the document must include a technical description of the research and a risk assessment and safety evaluation form. The HGCO checks also the license and qualifications of the tissue recipient abroad. Once the MTA has been authorized, a local branch of the Inspection and Quarantine Bureau issues a final approval document (Warrell 2009).

A key difference from the UK is that neither the HGCO nor the Inspection and Quarantine Bureau carries out controls of issues relevant to the ethical oversight of the transfer of hESC samples, such as the documentation of appropriate informed consent. Furthermore, while research purposes and related risks, together with the license, reputation, and capacities of tissue recipients, are assessed in case of international transfers, in domestic transfers such controls are performed by individual research institutes. As I will show now, these differences have implications with regard to the benefits that can be achieved by researchers, embryo donors, and the wider citizenry.

14.6   Benefits for the National Community?

A widely expressed claim that IVF clinicians communicated to potential embryo donors was that the donation and use of embryos for research would contribute to the improvement of the future health of fellow citizens. As one of the IVF clinicians with whom I spoke put it: "We tell our patients that *the whole society* may benefit from the donation of their embryos in the future" (emphasis mine). Donation is portrayed here as an act of altruism and solidarity. Present-day citizens help, through their donation, the well-being of future citizens, a selfless expression of help that might also benefit oneself and one's family. While this may be true for more wealthy citizens of high-income countries with comprehensive health care systems, these representations provide donors with a biased and rather one-dimensional picture of the benefits that can be extracted by the donation and usage of embryos. Three points deserve attention in this regard.

First, many of the hESC cell lines that have been created over the past years have been distributed to research laboratories all over the world. However, the dispersion of places in which research is carried out implies also the spatial dispersion of therapeutic applications. As I have shown elsewhere (Rosemann 2011), the distribution of stem cell lines across borders is likely to result in the extraction of benefits that are shared beyond the relations of national citizenship, events that considerably contradict with the perceptions of most embryo donors in China, who are left in the belief that research findings contribute, in the first place, to the national health community. A justified concern in this regard is that access to these developing therapeutic possibilities will be highly selective, across national borders as well as within. Unlike blood, for example, whose donation benefits people regardless of their socioeconomic status, the donation of embryos for the labor-, technology- and capital-intensive stem cell research is likely to benefit exclusively the more wealthy segments of national populations. In low-income countries such as China, that means that larger parts of the population might be excluded from access to these therapies, including the majority of people who had donated their embryos for research.

Second, the one-sided focus on the communication of the health value of stem cell research neglects all other forms of value that are hoped to be extracted on the basis of hESC research, among which the accrual of profits by pharmaceutical companies and the biotech industry, the realization of political ambitions as well as financial gains, and increases in status for individual scientists and research centers.

Third, the communication of the future value of hESC research in terms of "benefits for the national citizenry" neglects to account for the concrete forms of value and benefit that the derivation, use, distribution, and circulation of hESC lines has for scientific user communities in the present. As I have shown in another publication (Rosemann 2011), the use and exchange of hESC lines between different research labs is producing various forms of value for researchers and research centers. These are not elusive forms of "future value" but tangible forms of "present-day value" that range from career benefits to the attainment of workforce, to augmented numbers of publications, to the initiation of national and international research collaborations, and to the initiation of sustained chains of exchange with other researchers and corporations. While these processes in themselves are in general positive and they are as they should be, it is nonetheless striking that these benefits remain completely unspecified in the everyday practice of embryo donation.

14.7   Conclusions

In this chapter I provided insights into the communication processes through which the donation of embryos is facilitated in IVF clinics in China. A varied picture emerged here that offered insights into practices that unfolded between the poles of a strong commitment to professional care and high responsibility awareness, on the one hand, and forms of deception and untenable promises, on the other hand. On average, differences in attitudes and practices could be noted between clinicians and researchers in lower and higher professional positions. These gaps between promoted principles of good practice at the top level and actual practices at the level of the bedside indicate a lack of adequate ethical training of clinical staff or researchers. The chapter clarified, furthermore, that there exists a fundamental gap between the ways in which the value of the donated embryo is communicated to donors and the forms of value that are extracted by hESC researchers and corporations. In the IVF clinics I visited, the donation of embryos for hESC research was framed primarily as an act of solidarity, which exemplified a selfless expression of support from citizen to citizen. However, in the light of the complexities and stumbling blocks of present-day systems of human tissue circulation and the concrete forms of benefit and profit that the derivation and possession of stem cell lines creates for user communities in the present, such claims are misleading. Cross-culturally informed types of bioethics recognize that there are variations in the ways in which social phenomena and processes are categorized and problematized. However, from my understanding, some of the practices I encountered did clearly transgress the (admittedly difficult to define) borderlands of mutual respect and the positive recognition of difference. The observed ways in which patients were misled by some clinicians are intolerable according to Chinese guidelines for embryo donation and according to international standards, as provided, for instance, by the International Society for Stem Cell Research.

Acknowledgments This article has benefited from research support provided by the European Research Council (ERC, 283219) and the Economic and Social Science Research Council (ESRC, ES/I018107/1). Due to ethical concerns, supporting data cannot be made openly available.

References

Corrigan O (2003) Empty Ethics: the problem with informed consent. Sociology of Health and Illness 25(3):768–792

Dickenson D (2006) The lady vanishes: what's missing from the stem cell debate. Biotech Inq 3:43–54

Franklin S (2006) The IVF-Stem cell Interface. International Journal of Surgery 4(4):86–90

Rosemann A (2010) The IVF-stem cell interface in China: ontologies, value-conceptions and donation practices of 'Chinese' forms of unborn human life. In: Doering O (ed) Life sciences in translation: a Sino-European dialogue on ethical governance of the life sciences. BIONET, London, 168–178

Rosemann A (2011) Modalities of value, exchange, solidarity: exploring the social life of stem cells in China. New Genet Soc 30(2):181–192

Rosemann A, Luo HY (under review) Attitudes and perceptions on hESC research and embryo donation among IVF patients, hESC researchers and students in China. Journal of Bioethical Inquiry

Sleeboom-Faulkner M (2014) Global morality and life science practices in Asia: assemblages of life. Palgrave MacMillan, London

Stephens N, Atkinson P, Glasner P (2008) The UK stem cell bank: securing the past, validating the present, protecting the future. Science as Culture 17(1):43–56

Svendsen MN, Koch L (2008) Unpacking the "spare embryo": facilitating stem cell research in a moral landscape. Social Studies of Science 38(1):93–110

Thomson JA, Itskovitz-Eldor J, Shapiro SS, Waknitz MA, Swiergiel JJ, Marshall VS, Jones JM (1998) Embryonic stem cell lines derived from human blastocysts. Science 282(5391):1145–1147

Waldby C, Mitchell R (2006) Tissue economies: blood, organs, and cell lines in late capitalism. Duke University Press, Durham and London

Warrell D (2009) Cure committee report: China–UK research ethics. UK Medical Research Council. http://www.mrc.ac.uk/Utillities/Documentrecord/index.htm?d=MRC006303. Retrieved 24 Feb 2016

Chapter 15
Challenges to International Stem Cell Clinical Trials in Countries with Diverging Regulations

Achim Rosemann

15.1 Introduction

International collaborations in stem cell medicine have the potential to increase the speed of clinical translation and to maximize access to new therapies in multiple countries (Martell et al. 2010). With phase II and III trials on the rise, the significance of multi-country stem cell trials is growing. At present, however, the conduct of international stem cell trials is hampered by a high level of regulatory heterogeneity across countries and the absence of internationally harmonized governance frameworks (Bubela et al. 2014). Even though drug regulatory authorities in the USA, the European Union, and Canada have now initiated collaborations that focus on the convergence of regulatory procedures for cellular therapy products, globally harmonized regulatory procedures are far off (Arcidiacono et al. 2012; Blasimme 2013). Regulatory procedures through which the safety and efficacy of stem cell-based treatments are determined vary widely and confront clinical trial sponsors with contrasting demands. In this chapter we show that this high level of regulatory variation and the lack of internationally binding standards for clinical stem cell research present a huge challenge to multi-country clinical trial collaborations. Four types of challenges will be highlighted. First is the need to inquire into and interact with regulatory procedures and law in multiple countries. Second, the interaction with medical authorities in multiple countries is resulting in a very high level of organizational complexity. Third is the challenge of time delays and unexpected costs that result from unclear, emerging, or changing regulatory arrangements in different countries. Fourth is that the high level of regulatory variation across countries necessitates far-reaching forms of scientific self-governance, training, and procedural adjustments in participating clinical trial sites. The chapter is organized in four parts. Part I

A. Rosemann (✉)
Centre for Education Studies, Faculty of Social Sciences, University of Warwick,
Room B1.20, Coventry CV4 7AL, UK
e-mail: A.Rosemann@warwick.ac.uk

© Springer International Publishing AG 2017
P.V. Pham, A. Rosemann (eds.), *Safety, Ethics and Regulations*, Stem Cells in
Clinical Applications, DOI 10.1007/978-3-319-59165-0_15

provides a brief overview of recent regulatory developments, to illustrate the types of differences that clinical trial sponsors and investigators can expect. Part II introduces some of the main practical and organizational challenges that arise from these differences. Part III illustrates the abovementioned challenges through a case study of an international clinical trial infrastructure that has been active across the contexts of China, Hong Kong, Taiwan, and the USA. Part IV calls for the creation of an international support structure that systematically addresses these problems. Five measures that may help to tackle existing difficulties will be introduced.[1]

15.2 A Diversifying Regulatory Landscape

In the mid- and late 2000s, the USA and European Union have emerged as international trendsetters for the regulation of stem cells and human tissue products. Drug regulatory authorities in these regions have issued a risk-based tiered approach that classifies stem cells that are more than minimally manipulated as biological products. These are subject to premarket approval of the US Food and Drug Administration (FDA) and the European Medicines Agency (EMA), which involves systematic preclinical safety studies and the mandatory use of phase I–III drug trials and good clinical practice (GCP) standards (Faulkner 2012; Knoepfler 2015). Stem cells that are only minimally manipulated and intended for homologous use can be applied to patients under compliance with national tissue regulations, which does not involve clinical trials. A problem is that—for clinical investigators and small-to-midsize companies—the high costs of premarket approval procedures are difficult to cover. Geron Corporation, for example, which developed the world's first human embryonic stem cell (hESC) product that entered clinical trials, had to invest about 200 million US dollars in its hESC program, before FDA approval for a phase I trial could be obtained. To get to this point, research was carried out for nearly a decade. As stated by Edward Wirth, the former chief scientist of the hESC program at Geron, to test biodistribution, dosing, delivery, toxicity, tumorigenicity, and immune rejection, the company conducted 24 preclinical studies before an investigational new drug (IND) application could be filed at the FDA in March 2008. These studies included in total 1977 rodents. The IND application that the corporation submitted was 21,000 pages long, with more than 90% consisting of data from the preclinical

---

[1]This chapter is based on three papers that I have published in 2014 and 2015. These papers are (1) Rosemann A (2015) Multi-country stem cell trials: the need for an international support structure. Stem Cell Res 14(3):396–400; (2) Rosemann A (2015) Stem cell treatments for neurodegenerative diseases: challenges from a science, business and healthcare perspectives, Neurodegener Dis Manag 5(2):85–87; and (3) Rosemann A (2014) Standardization as situation-specific achievement: regulatory diversity and the production of value in intercontinental collaborations in stem cell medicine. Soc Sci Med 122:72–80. I have rearranged information from these articles and developed them further in the context of this chapter. These articles have been published open access under a CC-BY license. I am permitted to use and reproduce segments of these texts in this chapter.

studies. According to Wirth, this was the longest application the FDA had received at that time (Wirth 2010). After submission the FDA put the IND on a halt 2 times, for 6 and 13 months, respectively, with the request to carry out additional preclinical studies. Unfortunately, Geron had to halt its first hESC trial after five patients for financial and strategic reasons in 2011, only months after obtaining regulatory clearance by the FDA (Brennan 2011). It is important to note, though, that—because Geron's groundbreaking study took place in an early phase of hESC research—the costs were much higher than for the development of subsequent hESC products (Keirstead 2012). The reason is that drug development costs decrease as a new technology advances and as regulatory frameworks are getting more mature. Nevertheless, in high-income countries extensive preclinical requirements and the obligatory conduct of phase I–III trials and subsequent product release costs are expensive. These costs can run up to several hundred millions of US dollars. For academic investigators and small-to-midsize companies, these high costs are difficult to bear and carry the risk of financial unsustainability.

In the light of these costs, it is an important question whether stem cell-based medicinal strategies can be translated into affordable and widely accessible treatments. As I have argued elsewhere (Rosemann 2015), the translation of the therapeutic potential of stem cells into routine, reimbursable healthcare practice is currently not only hindered by unresolved scientific issues but also by high financial risks for corporations and the question whether evolving treatments will be affordable enough to be reimbursed by national healthcare systems. There is a severe risk—especially for technologically more demanding treatments—investments cannot be amortized. It is expected that the fees for stem cell-based treatments for neurodegenerative diseases will be between 30,000 and 100,000 US dollars (Nuffield Council on Bioethics 2013). For individualized and technologically more complex iPS cell treatments, estimates lie around 200,000 US dollars or more (Knoepfler 2012). As pointed out by the stem cell biologists Tabar and Studer (2014), given the high costs that are associated, for instance, with iPS and other pluripotent cell technologies, the integration of these treatments into routine, reimbursable medical practice is highly unlikely. If this is true, the range of potential users will clearly be delimited, to a relatively small group of wealthy patients. In the light of this situation, the risk of financial unsustainability is high. It remains to be seen whether development costs can be amortized and sustainable profits can be generated.

These risks, high costs, and the long duration of premarket approval procedures have in numerous countries resulted in a refusal to accept the standards of high-income countries in Northern America and Europe. New types of standards and forms of regulation have emerged that significantly diverge from US–European models. Even though drug regulatory authorities in the USA, the European Union, and Canada have recently initiated collaborations that aim to converge regulatory procedures for cellular therapy products, globally harmonized regulatory procedures will be difficult to achieve Rosemann, Bortz and Vasen (2016). What can be observed in the stem cell field is a shift to global regulatory diversification rather than international harmonization. For instance, in 2014, Japan introduced an approval path for stem cell treatments that avoids the phase I–III clinical trial system (Lysaght and Sipp 2014). The new fast-track path enables conditional market approval on the basis of safety and efficacy tests on small numbers of patients, complemented by a 5–7-year period of post-marketing surveillance (Cyranoski 2013). South Korea's Food and Drug Administration still holds onto phase I–III clinical trials but has introduced an accelerated system that, on a global level, enables the fastest path to market approval for stem cell-based medical products (Wohn 2012; Cyranoski 2013). In China and India, health authorities have for many years taken a fairly permissive way to the clinical testing and market introduction of stem cell treatments.

These countries have until now issued only provisional regulations and regulatory guidelines with limited legal options to enforce them (Viswanathan et al. 2013; Rosemann 2013; Tiwari and Raman 2014; Sui and Sleeboom-Faulkner 2015). Although Chinese and Indian regulatory agencies have recently endorsed the mandatory use of controlled trials for market approval of stem cell products (Tiwari and Raman 2014; Rosemann and Sleeboom-Faulkner 2016), hundreds of clinics have provided experimental treatments without evidence from systematic clinical studies (Baker 2005; Cyranoski 2009; McMahon 2014). The providers of these experimental therapies have capitalized on the promissory potential of stem cells, to offer new treatment options for previously incurable diseases and to give rise to new therapeutic markets (Salter 2009). These uncontrolled environments of experimental applications of stem cell research stand in sharp contrast with environments in the USA and Europe, where systematic preclinical safety studies and phase I–III trials are the cornerstones of market approval procedures for medical products (Faulkner 2012). Uncontrolled applications have been reported in more stringently regulated countries, such as Germany and the USA (McMahon 2014), when companies have either ignored state regulation or have exploited regulatory gray areas (Cyranoski 2013; Sipp 2014). These kinds of uncontrolled, commerce-driven applications of stem cell research are generally dismissed in top international journals (Cyranoski 2006; Hyun 2010; Kiatpongsan and Sipp 2008), as well as in official statements by the International Society for Stem Cell Research (ISSCR) (ISSCR 2008; Lindvall and Hyun 2009). While such dismissals may from a methodological and ethical perspective in many cases be justified, they reflect also the position of the global elite of stem cell research, a group composed mainly of researchers from high-income countries (Sleeboom-Faulkner et al. 2016).

15.3 Challenges to the Organization of International Stem Cell Trials

Four types of challenges will be highlighted. First is the need to inquire into and interact with regulatory procedures and law in multiple countries. Second, the interaction with medical authorities in multiple countries is resulting in a very high level of organizational complexity. Third is the challenge of time delays and unexpected costs that result from unclear, emerging, or changing regulatory arrangements in different countries. Fourth is that the high level of regulatory variation across countries necessitates far-reaching forms of scientific self-governance, training, and procedural adjustments in participating clinical trial sites.

15.3.1 The Need to Inquire and Interact with Regulatory Procedures and Law in Multiple Countries

A first challenge to the organization of multi-country stem cell trials is the necessity to conduct long-term in-depth research into the regulatory requirements of drug regulatory authorities in multiple countries (OECD 2011). Stem cell therapies, as pointed out by Martell and colleagues, "do not neatly fit into current regulatory categories," and the barriers of translating stem cell-based approaches in functioning therapies lie "in both technical and regulatory constraints" (Martell et al. 2010, p. 451).

Regulations for the clinical use of stem cells are in many countries emerging only gradually, and far-reaching regulatory differences exist. For clinical investigators and industry, this diversified and rapidly changing situation is confusing and poses significant organizational difficulties (Rosemann 2014a). What is required is a long-term, reflective engagement with the review and approval procedures that are handled by the drug regulatory authorities in the countries in which a trial is conducted. In order to develop study protocols that are compliant with the demands of multiple regulatory agencies, gaps between jurisdictional frameworks must be identified at an early stage of the clinical translation process. This is a difficult task that takes time and may be complicated by language barriers, insufficiently defined regulatory procedures, cultural differences, and disparities in the enforcement of regulatory protocols (Ravinetto et al. 2013). It is complicated, furthermore, because the regulatory issues that are associated with the development of autologous stem cell therapies (Hourd et al. 2014) do in important respects differ from the characteristics that need to be taken into account in the context of clinical trials with pluripotent stem cells (Andrews et al. 2014).

15.3.2 High Level of Organizational Complexity

A second challenge is that the interaction with medical authorities in multiple countries is resulting in a very high level of organizational complexity (Minisman et al. 2012). To file applications at multiple drug regulatory agencies is a time-, cost-, and labor-intensive process that requires specially trained staff and a well-functioning administrative infrastructure (Rosemann 2014b). While for industry-sponsored trials this is not necessarily a problem, for academic research groups and small-to-midsize biotech companies (which at present are the main sponsors of clinical stem cell trials), these resources are often not available and difficult to acquire (Keirstead 2012).

15.3.3 Time Delays and Unexpected Costs from Unclear, Emerging, or Changing Regulatory Arrangements

A third type of challenge are time delays, increased costs, and uncertainties that arise from nonexistent or still emerging regulatory procedures in some countries. In China, for instance, where effective regulatory procedures for the clinical testing of stem cell-based therapeutic approaches have until 2012 been nonexistent, the China Food and Drug Administration (CFDA) has repeatedly refused to accept incoming investigational new drug (IND) applications for stem cell-based products (Rosemann 2013). Such unresolved regulatory issues can cause long-drawn-out delays and additional costs to the sponsors of clinical stem cell trials and result in the need to apply for regulatory approval in another country where regulatory procedures are clearer and to conduct the trial there (Bhagavati 2015). But unresolved regulatory issues and the potential for sudden regulatory changes exist also in countries with highly developed regulatory frameworks. Noteworthy is, in particular, the ongoing debate on who should regulate autologous stem cell interventions (Zarzeczny et al. 2014). In the USA, for instance, think tanks are using the case of autologous stem cells in order to promote broader deregulation, and several companies and professional societies (most prominently the ICMS) have argued that "autologous cell products should be treated as part of medical practice and thus not subjected to marketing approval" (Bianco and Sipp 2014). These calls have resulted in a bill for the Freedom of Choice Act that was put forward to the US congress in April 2014. According to this bill, investigational stem cell technologies could be sold to terminally ill patients, outside of the control of the US Food and Drug Administration (FDA) (Morgan 2014). Similar developments can also be reported from other highly regulated countries. Australia, for instance, has exempted autologous stem cells from the review procedures of its drug regulatory agency (Tuch and Wall 2014), and in Italy the use of autologous mesenchymal stem cells has been taken out of the jurisdiction of the Italian Medicine's Agency in 2013 (Berger et al. 2014). These developments are likely to influence regulations in other countries (Bianco and Sipp 2014). Most importantly, however, the jurisdictional variation in regulatory frameworks and the prospect of ongoing policy changes make the implementation of multi-country stem cell trials more difficult and increase the risk of organizational complications, unexpected or misplaced investments, and time delays.

15.3.4 The Need for Scientific Self-Governance, Training, and Procedural Adjustments in Participating Clinical Trial Sites

A fourth challenge is that the high level of regulatory variation across countries necessitates far-reaching forms of scientific self-governance, training, and procedural adjustments in participating clinical trial sites (Rosemann 2014b). If, for instance, data from clinical trials that are conducted in one country are to be used for investigational new drug applications in other countries (or alternatively, if there are simultaneous clinical trials conducted in multiple countries), the basic regulatory requirements of these countries' drug regulatory authorities must be met. Intercontinental clinical trials conducted in South Africa, India, Mexico, or China, for example, must be congruent with the methodological standards required for the approval of later-stage (or parallel) trials, by the health authorities, in say the UK, Sweden, or the USA. The balancing out of regulatory disparities between countries requires strategic efforts to navigate through a diverse and internationally non-harmonized regulatory environment; the aim is to create compliance with the divergent requirements of drug regulatory authorities and related processes of peer review in multiple countries (cf. Wahlberg et al. 2013). These efforts comprise project-internal forms of scientific self-governance and capacity building that are aimed at compensating regulatory gaps and creating congruence with the auditing demands of diverging regulatory and political systems (Sariola and Simpson 2013; Sleeboom-Faulkner 2013). The enactment of such transnational forms of scientific self-governance involves the restructuration and standardization of local clinical research environments and practices in the involved clinical trial sites (Rosemann 2014a, b). A central reason for this is that regulatory differences between countries give rise to and enable different types of clinical research practice, methodologies, and ethical standards, at the level of local medical institutions. In many countries, moreover, knowledge on the conduct of systematized controlled stem cell trials is often limited among clinical researchers (Li et al. 2014). These disparities between and also within local hospitals form a clear threat to the scientific integrity of international stem cell trials (OECD 2011). As a result, intensive forms of staff training and adjustments of local clinical cultures and research practices are necessary, so that standardized research protocols can be implemented (Ravinetto et al. 2013). Standardization requires, furthermore, the implementation of reliable monitoring and control infrastructures. For academic investigators and small-to-midsize companies, the performances of these tasks pose a significant organizational and financial burden (Keirstead 2012). Unless sufficient funding for these forms of education and scientific self-governance is acquired, multicenter international stem cell trials cannot be conducted. A closely related challenge is that in the stem cell field, long-standing, established clinical trial infrastructures are mostly still absent. In more established fields of medicine, well-functioning international research platforms have emerged over decades that employ unified methodological standards and ethical review procedures; these allow for effective and rapid forms of clinical testing

(Keating and Cambrosio 2012). In the stem cell field, however, the development of clinical infrastructures is often still in initial stages. New alliances between researchers, hospitals, universities, corporations, and government institutions have to be formed, and unified coordination structures must be established. These processes are complicated by regulatory demands for good manufacturing practice (GMP) labs and the development of specialized surgical and injection procedures, which requires cooperation between experts from a multiplicity of disciplines and backgrounds. The formation of such standardized multicenter clinical trial infrastructures is time and labor intensive and involves significant costs. It includes, moreover, tasks and responsibilities for which medical researchers are not trained, and that can only be learned through experience (Keirstead 2008). A substantial amount of energy is necessary to build such infrastructures, long before actual clinical research can be conducted.

In the next section, I will briefly exemplify the different forms that project-internal processes of scientific self-governance and capacity building can take, in order to achieve standardization across participating clinical trial site. The example comes from the China Spinal Cord Injury Network, a clinical research network that has been active across the contexts of China, Hong Kong, and the USA. Three aspects of change shall be highlighted: selection, restructuring, and the forestalling of regulatory gaps.

15.4 Case Study: Scientific Self-Governance in a Sino-American Clinical Trial Network

The China Spinal Cord Injury Network (China SCI Net) is the first intercontinental clinical trial infrastructure in the stem cell field that has emerged between medical researchers in mainland China, Hong Kong, Taiwan, and the USA. The network involves more than 20 spinal cord injury (SCI) centers and is dedicated to the clinical testing of stem cell-based combination therapies for spinal cord injury. It targets the licensing of (potentially) successful treatments in China, Hong Kong, Taiwan, the USA, and potentially other countries. Until April 2014, the China SCI Net had conducted seven clinical studies. An initial noninterventional observational study was carried out between 2005 and 2008 in 22 hospitals to collect diagnostic and long-term follow-up data from up to 600 acute and chronic SCI patients. This study was followed by five phase I and II trials that have been conducted in chronic SCI patients, in two university hospitals in Hong Kong and one military hospital in China. Two of these studies tested the safety and efficacy of lithium in SCI patients, and three studies an experimental combination therapy of umbilical cord blood (UCB) mononuclear cells, lithium, and methylprednisolone. A Phase III trial incorporating more hospitals (including those in Taiwan) is being planned in the nearby future.

15.4.1 Scientific Self-Governance and Capacity Building in the China SCI Net

> What we are trying to do is to bring the international standards of clinical trials to China. [We] bring in the concept of using all the modern standards on how to run a clinical [stem cell] trial, as it is recognized in the West, in the current time. All the conceptions of leading this network ... evolve around that concept. [...] First of all we had to promote the interest [...] to bring in experts from around mainland China, Hong Kong, Taiwan [...], to provide a platform. And the second level is, we would then bring in the knowledge as to how a clinical trial should be run, in an internationally recognized manner.
>
> Prof Kwok-Fai So, Co-Director of China SCI Net, interview with author, Jan 7 2011.

The realization of this objective is based on three different types of intervention: selecting, restructuring, and the forestalling of regulatory gaps.

15.4.2 Selection

A first and crucial task for the formation of the China SCI Net was the selection of hospitals that would have the capacity and qualifications to function as clinical trial sites. This is a continuous process that gains significance with each new clinical trial that is planned and conducted by the network. Selection of suitable hospitals depends, in essence, on the ability of affiliated centers to provide evidence that the standards and criteria required for participation in internationally recognized (multicenter) clinical trials can be met. The leadership of the China SCI Net in this respect is a combination of external and internal assessment parameters. External assessment parameters refer to outward qualification criteria of associated hospitals. These include the Chinese good clinical practice (GCP) certification (i.e., the recognition of hospitals as certified clinical trial units, following a qualification procedure under the National Health and Family Planning Commission [NHFPC; the former Ministry of Health]). They include, furthermore, the availability of good laboratory practice (GLP)-accredited laboratory facilities. Internal assessment parameters refer to criteria that are imposed on affiliated hospitals by the network itself. These internal qualification criteria can be divided into "performance-based" and "organizational" parameters. Organizational criteria cover aspects such as checks of hospital internal institutional review board (IRB) approval procedures, availability of the necessary technical instruments, adequate specialist staff, sufficient hospital beds, insurance protection for patients, and adherence to other technical and clinical conditions that are contractually defined between the China SCI Net's headquarter and affiliated hospitals. Performance-based assessment criteria have been exerted first in the context of the network's observational clinical study that was conducted in 22 hospitals between 2005 and 2008 but have been applied in all further trials that the organization has conducted since then. Performance-based criteria focus, above all, on the compliance (of each participating hospital) to a clinical trial's protocol, which prescribes the exact clinical, methodological, technical, and organizational procedures of a study. The monitoring of protocol compliance involves the observation of the correct handling of inclusion and exclusion criteria, the conduct of physiological examinations and follow-up investigations, the accurate completion of data sheets, and the informed consent procedures. These monitoring tasks are done from the network's headquarters in Hong Kong. The headquarters—which between 2007 and 2012 employed one full-time and two part-time staff—operates under the supervision of the network's board of directors. The Hong Kong office is the nerve center of the network. All operations of the organization, as well as communication with affiliated hospitals, are coordinated from here. In addition to arranging the logistics of the network's clinical trials, and the monitoring of the activities and performance of participating hospitals, the headquarters also plays a central role in the restructuring of institutional arrangements and practices in associated centers.

15.4.3 *Restructuring*

The formation of a standardized multicenter clinical trial infrastructure that operates according to internationally recognized principles requires significant adjustments of local clinical research practices and conditions in network-affiliated hospitals. These changes were achieved by an intensive training program, and the implementation of performance-based assessment procedures, through which required institutional adjustments, could be monitored and—if necessary—corrected. Training for staff members of the relevant departments in the 25 associated research hospitals began in 2005, with three to four meetings per year until 2009. A first target was the standardization of neurological examination procedures to ensure valid and replicable assessment of the injury grade of spinal cord injury patients on the trial.

> When we first came here, the neurological assessment of spinal cord injury – almost everywhere – was completely haphazard. It ranged from, eh, you know … you take a pin, you put it here, you touch a patient, ask "Can you feel it?" There was no discipline … no common languages, no common neurological assessment of the patients.
> Professor Wise Young, director of the China SCI Net, interview with author, June 24 2010.

Standardization of neurological assessment was the first in a long list of methodological, clinical, and organizational issues that were addressed. Training addressed aspects of clinical trial design, such as protocol development, quality assurance measures, reliable use of outcome measures, long-term follow-up of patients, and ethical and legal issues of clinical trials, as well as requirements by foreign drug regulatory authorities and international journals. In its training program, the China SCI Net did not work with an examination system. Instead, new contents and practices were transmitted through demonstrations and educational materials, and compliance to newly introduced standards, protocols, and standardized procedures was then tested in practice.

A crucial endeavor in this respect was the organization of an initial observational (i.e., noninterventional) trial, a multicenter study that was conducted in 22 hospitals in mainland China, Hong Kong, and Taiwan. The purpose of this study was to collect long-term data from 600 chronic and acute spinal cord injury patients, in accordance with international recruitment and measurement protocols. In addition to the scientific value of this study—which was the first longitudinal observational study of chronic and acute spinal cord injury patients in China—it fulfilled a central function for the network to serve as a test trial of the ability of affiliated centers to recruit patients, to conduct standardized neurological assessments [based on the ASIA scheme, developed by the ISCS], to carry out long-term follow-ups, and to document data and data collection procedures in the prescribed, standardized, fashion. This study helped in identifying various challenges:

> The first trial we held was an observational trial. To show that the hospitals can deliver the data ... Now this study revealed a lot of problems I actually had heard about, but never really encountered, until to this point. The number one problem in China is really to get patients to come back [for follow-up investigations]. ... But we [also] observed data that just could not have been. You know – patient data would be the same, over the whole year period. Suggesting that someone had examined the patients very carefully ... It became very clear to us that we need to have very good controls of the protocol.
> Professor Wise Young, director of the China SCI Net, interview with author, June 24 2010.

Due to these problems, instead of the intended 600 patients, only 386 patient profiles were completed in this first—entirely observational—study. These insights into local conditions and related challenges resulted in the wide-ranging restructuring of the control and monitoring structures through which the network operated, such as the introduction of a supervisor–principal investigator double-signing system. With this system, each doctor or nurse involved in examination of patients has to "sign off" the data collection sheet with his or her supervisor and the principal investigator in the institute. Documentation protocols, moreover, were changed from paper to a computerized web-based system for data entry, in order to enhance data insertion and data analysis and to permit continuous checks by the headquarters in Hong Kong. Identification of challenges in this observational study gave rise, too, to adjustments of training procedures, as well as the decision to work with a contract research organization (CRO) during the forthcoming phase III trial.

15.4.4   The Forestalling of Regulatory Gaps

The selection of suitable hospitals, and adjustments of local clinical research practices and conditions, aims at the consistent implementation of fully standardized clinical research protocols. In contrast to multicenter clinical trials that are conducted in a single country, the project-internal forms of self-regulation, capacity building, and institutional restructuring that have been described constitute a long-term strategic endeavor to create congruence with the auditing demands of widely varying regulatory and legal systems. At the time of writing, the clinical trials of the network had been approved exclusively by the regulatory authorities in Hong Kong and mainland China, but the data from these trials shall be used for investigational new drug applications (INDs) in the USA and maybe other countries. This required an enduring anticipatory engagement with the review and approval criteria of the US Food and Drug Administration (FDA) with respect to the "acceptance of foreign clinical studies not conducted under an investigational new drug application (non-IND foreign clinical studies)" (Fink 2009). This constant need for forms of anticipatory audit processes requires the identification and forestalling of regulatory gaps between national jurisdictions from an early stage of the clinical translation process. A brief example will serve to illustrate this point. At the time of writing, the Health Department of the Army General Logistics Department in China (the regulatory agency that approved the China SCI Net's clinical studies in mainland China) did not mandatorily require that clinical studies should be conducted in compliance with ICH-GCP standards. Nor did it require the clinical trials to be conducted exclusively in hospitals certified by the Chinese MOH, as officially recognized clinical trial units. However, the US FDA's list of requirements for the acceptance of "non-IND foreign trials" (in the context of IND applications at the US FDA) states that "accordance with good clinical practice (GCP), including review and approval by an independent ethics committee (IEC)" is obligatory (Federal Register 2008). In order to preempt any difficulty arising from these discrepancies, the China SCI Net tried to forestall regulatory gaps from the outset and ensured their clinical trial protocols were fully GCP compliant and only MOH-certified hospitals were selected. Moreover, in addition to approval by the Army General Logistics Department in Beijing, ethics committee review was also sought by Western IRB, a for-profit IRB in the USA with close ties to the US FDA.

15.5 The Need for an International Support Structure

The International Society of Stem Cell Research (ISSCR) has in 2010 called for the need to harmonize regulations for the clinical translation and commercialization of stem cell-based products and therapies (Martell et al. 2010). However, in 2014 the global regulatory landscape for clinical stem cell research remains as diverse as before. This situation continues to pose problems to the organization of transnational stem cell trials. I suggest that—in the light of the abovementioned challenges—what is needed is the creation of an international support structure, through which the organizational challenges of multi-country stem cell trials can be systematically addressed. International bodies such as the ISSCR or the International Stem Cell Forum have until now focused primarily on the development of guidelines, best practice standards, and various types of recommendation. These documents have concentrated on crucial aspects of the clinical translation process, including the collection, derivation, storage, and clinical application of stem cells, as well as intellectual property rights, commercialization, industry engagement, and ethical issues of stem cell research (Isasi 2012). However, a support structure that specifically addresses the regulatory and organizational challenges of multi-country stem cell trials has so far not yet been developed. Such a scheme could encompass five elements.

15.5.1 The Development of a Web-Based Databank that Provides Detailed Information on Regulatory Requirements and Procedures for Clinical Stem Cell Research and Marketing Approval in a Large Number of Countries

This repository could provide a detailed overview of responsible government units, key contacts, as well as regulatory documents and websites. Regulatory procedures and manuals on how to apply for and conduct stem cell clinical trials in different countries could be introduced in detail. This databank could work with a computerized system that explains differences between the regulatory requirement of specific countries and regions and that helps to clarify what kind of tasks clinical trial sponsors will have to perform to balance out these regulatory gaps. In order to be valuable, such a database would have to be nuanced for different cell types and manufacturing standards. It would also have to provide information for the regulation of combination therapies (such as cell therapy drug or cell therapy device) where different regulatory pathways are required for each element.

15.5.2 The Establishment of an International Task Force that Identifies the Central Challenges to Multi-Country Stem Cell Trials

This task force could consist of researchers and sponsors with experience in the organization of multi-country trials. It should strive for the identification of the key challenges for the clinical translation of stem cell-based therapies in the context of international projects. Such a task force could aim, furthermore, for the development of solution strategies through which international stem cell trials can be conducted in a more time- and cost-efficient way. These measures could include information packages for sponsors and clinical investigators, as well as tools for staff training, project management, and data collection. Such a task force could be initiated by the ISSCR, the International Stem Cell Forum (ISCF), or another international society such as the International Society of Cellular Therapies (ISCT).

15.5.3   The Creation of an Interactive Online Education and Discussion Platform

The establishment of an interactive education and discussion platform would allow for the sharing of critical information and experiences of clinical trial sponsors and investigators. Scientists or sponsors who plan to conduct international stem cell trials can learn in this way from the experiences of other researchers and make practice-based assessments of the tasks, costs, timeframes, and challenges that may lie ahead of them. Such knowledge could also help to make well-informed budgetary estimations and to gain access to other useful information such as information about insurance schemes, the implementation of project-internal monitoring systems, and ethics committee approval (Ravinetto et al. 2013).

15.5.4   Raise Awareness of the Challenges of Multi-Country Stem Cell Trials Among Public, Private, and Charitable Funding Bodies

To facilitate international collaborations in the stem cell field, it will be important to create an awareness of the challenges of multi-country stem cell trials among public, private, and charitable funding bodies. To develop an understanding of these problems will be crucial to prevent unrealistic expectations and to obtain additional money that is required to tackle the challenges associated with international stem cell trials. A first step into this direction has been made by researchers at the University of Alberta in Canada, together with colleagues from McGill University, the University of British Columbia, and the London Regenerative Medicine Network (Bubela et al. 2012). This group has founded the interactive online forum "Enabling Advanced Cell Technologies (EnACT)" [http://enactforum.org] that offers an interactive, moderated discussion platform that aims to develop "solutions to key nonscience barriers" to the clinical translation of cell and stem cell-based treatment pathways (Bubela et al. 2012). The EnACT website features 12 thematic areas where barriers to translational stem cell research emerge.

However, the challenges for the organization of multi-country clinical trial collaborations are not discussed on the forum. Moreover, online forums may not be the best way to identify and/or raise awareness of the challenges of international stem cell trials. The organization of a series of workshops and publications that could be organized by the ISSCR, the ISCF, the ISCT, or another professional organization promises to be a more efficient method. Ideally, such workshops would involve representatives from public as well as charitable funding bodies, the industry, and drug regulatory agencies from multiple countries.

15.5.5   *Promote Forms of Regulatory Harmonization and Lobby for Better Communication Between Drug Regulatory Authorities*

A final field of activity would be the promotion of forms or regulatory harmonization or, at least, to lobby for better communication between drug regulatory authorities, so that some of the challenges regarding multi-country stem cell trials can be prevented or reduced. Harmonization, as recently pointed out in a position paper of the US FDA's Office of Cellular, Tissue and Gene Therapies, does not necessarily imply the production of internationally shared consensus guidelines (as in case of the ICH-GCP standards). Harmonization can refer too to the partial convergence of regulatory perspectives—but based on the independent development of national regulations and guidelines (Arcidiacono et al. 2012). In light of the current level of regulatory divergence in the clinical stem cell field, it is questionable whether such a convergence perspective could really be achieved. Be this as it may, considering the existing challenges for the performance of international stem cell trials, the move toward a more coherent and predictable international regulatory landscape would clearly be advantageous.

15.6   Discussion

There is little doubt that the introduction of such an international support structure for multi-country stem cell trials would be of great value; its feasibility must be viewed from the perspective of possible sponsors. Considering the high costs of clinical translation, it is to be expected that the greater part of phase III trials will involve commercial sponsors. A newly devised support structure, therefore, must be of use to both academic investigators and corporate sponsors. Ideally, representatives of both groups will be involved in the design of these measures. It is not unlikely though that commercial sponsors may prefer to undertake their own work into the regulatory barriers of international stem cell trials, for instance, by external regulatory affair consultants that help companies to navigate and identify existing challenges. For fear of competition, these corporations may not be willing or able to share this information through online forums, publications, or other means. At present, however, the majority of clinical trials in the stem cell field are either investigator-initiated or trials that are organized by small-to-midsize biotech companies, often start-ups that operate under high risks. The financial means of both of these groups are usually limited. Moreover the time and organizational capacities—especially of academic investigators—are highly restricted. For these groups a support structure in which many of the question and practical challenges that emerge in the context of multi-country stem cell trials are discussed and anticipated will allow to save costs and time and facilitate realistic assessments and planning. A problem is, of course, that such initiatives are likely to be expensive. The organization of such a support structure should best lie in the hands of an international professional society such as the International Society for Stem Cell Research or the International Society of Cellular Therapies, which are large-scale organizations that represent large numbers of re- searchers in many countries and that also cater the interests of the industry. Alternatively, the International Stem Cell Forum—which brings together some of the main public funding bodies for stem cell research around the world—would be a suitable umbrella organization for such an initiative. These organizations are firmly grounded into the international scientific community and have a high level of credibility. Most importantly, they are most likely to reach a large number of interested stakeholders, and it also is in the interest of these institutions to provide up-to-date information, to stimulate participation, and to disseminate findings from such an initiative. These large international organizations are also in the best position to attract the funding for such a transnational support structure and to provide an apposite organizational platform. Considering the high expenses and financial risks of multi-country stem cell trials and the potential for regulatory mis-assessments and long-drawn-out delays, the money for such an international support structure seems well invested.

15.7 Conclusion

With a rising number of stem cell therapies entering the clinical development phase, the performance of international stem cell trials is becoming of increasing importance. This chapter has shown, however, that the high level of regulatory diversity in the stem cell field provides important obstacles to the organization of multi-country stem cell clinical trials. Emerging and unclear regulations in some countries, and the absence of internationally harmonized regulatory frameworks, confront investigators with unexpected costs, time delays, or even the need to relocate a trial to another country or region. The high level of regulatory heterogeneity in the stem cell field gives rise to a high level of administrative complexity and requires far-reaching forms of scientific self-governance, training, and the creation of effective coordination and monitoring structures. These forms of self-governance and capacity building constitute a fundamental precondition to successfully navigate through a diverse and internationally non-harmonized regulatory environment. It is important to note, in this respect, that the implementation of standardized clinical research protocol is more difficult to achieve in the field of regenerative stem cell medicine, than in other—more established—areas of medicine research. The key reason for this—aside from the issue of regulatory heterogeneity—is the lack of well-established international clinical research platforms. In oncology research, for instance, long-standing international clinical research infrastructures have evolved over the course of several decades. These transnational platforms have developed their own centralized institutions that are responsible for the coordination of all successive steps of the clinical translation process, including controls of processes of data collection, recording, and analysis. In the stem cell field, however, such infrastructures are only gradually emerging. While international projects such as the China SCI Net show that multi-country clinical research platforms are evolving, this article has illustrated that these processes are complicated by the absence of internationally harmonized regulatory frameworks. The existence of strongly diverging regulatory, institutional, and clinical research cultures across countries and regions makes the performance of standardized multi-country trials to a challenging and risky organizational enterprise.

Acknowledgments This article has benefited from research support provided by the European Research Council (ERC: 283219) and the Economic and Social Science Research Council ESRC: ES/I018107/1). Due to ethical concerns, supporting data cannot be made openly available.

References

Andrews PW, Cavagnaro J, Deans R, Feigal E, Horowitz E, Keating A, Rao M, Turner M, Wilmut I, Yamanaka S (2014) Harmonizing standards for producing clinical-grade therapies from pluripotent stem cells. Nat Biotechnol 32(8):724–726

Arcidiacono JA, Blair JW, Benton KA (2012) US Food and Drug Administration international collaborations for cellular therapy product regulation. Stem Cell Res Ther 3(5):1–5

Baker M (2005) Stem cell therapy or snake oil? Nat Biotechnol 23(12):1467–1469

Berger AC, Beachy SH, Olson S (2014) Stem cell therapies: opportunities for ensuring the quality and safety of clinical offerings: summary of a joint workshop. National Academies Press, Washington, DC

Bhagavati S (2015) Stem cell therapy: challenges ahead. Indian J Pediatr 82(3):286–291

Bianco P, Sipp D (2014) Sell help not hope. Nature 510(7505):336–337

Blasimme A (2013) Translating stem cells to the clinic: scientific societies and the making of regenerative medicine. Quaderni 2:29–44

Brennan P (2011) Geron Corp. shuts down world's first stem cell trial. OC Sci. http://science-dude.ocregister.com/2011/11/16/geron-corp-shuts-down-worlds-first-stem-cell-trial/159883/. Retrieved 15 Dec 2015

Bubela T, Reshef A, Li MD, Atkins H, Caulfield T, Culme-Seymour E, Gold ER, Illes J, Isasi R, McCabe C, Ogbogu U, Piret J, Mason C (2012) Enabling advanced cell therapies (EnACT): invitation to an online forum on resolving barriers to clinical translation. Regen Med 7(6):735–740

Bubela T, Mishra A, Mathews D (2014) Policies and practices to enhance multi-sectorial collaborations and commercialization of regenerative medicine. In: Hogle L (ed) Regenerative medicine ethics. Springer, New York, NY, pp 67–87

Cyranoski D (2006) Patients warned about unproven spinal surgery. Nature 440(7086):850–851

Cyranoski D (2009) Stem cell therapy faces more scrutiny in China: but regulations remain unclear for companies that supply treatments. Nature 459(7244):146–147

Cyranoski D (2013) Japan to offer fast-track approval path for stem cell therapies. Nat Med 19(5):510–610

Faulkner A (2012) Law's performativities: shaping the emergence of regenerative medicine through European Union legislation. Soc Stud Sci 42(5):753–774

Federal Register (2008) Human subject protection; foreign clinical studies not conducted under an investigational new drug application. Fed Regist 73(82):22800–22816

Fink DW (2009) FDA regulation of stem cell-based products. Science 324(5935):1662

Hourd P, Chandra A, Medcalf N, Williams G (2014) Regulatory challenges for the manufacture and scale-out of autologous cell therapies. Stembook; Internet, http://www.ncbi.nlm.nih.gov/books/NBK201975/. Retrieved 15 Dec 2015

Hyun IS (2010) Allowing room for innovative stem cell-based therapies outside clinical trials: ethical and practical challenges. J Law Med Ethics 38(2):277–285

Isasi R (2012) Alliances, collaborations and consortia: the international stem cell forum and its role in shaping global governance and policy. Regen Med 7(6s):84–88

ISSCR (2008) ISSCR guidelines for the clinical translation of stem cells. International Society for Stem Cell Research. Available at http://www.isscr.org/GuidelinesforClinicalTranslation/2480. htm. Retrieved 15 Dec 2015

Keating P, Cambrosio A (2012) Cancer on trial: oncology as a new style of practice. University of Chicago Press, Chicago

Keirstead H (2008) Challenges to the clinical viability of stem cell technology. Spinal Cord Injury—what are the barriers to cure? Cambridge, MA: Bedford Center Workshop 29, p 2008. http://www.spinalcordworkshop.org/portfolio-type/hans-keirstead-phd-challenges-to-the-clinical-viability-of- stem-cell-technologies/. Retrieved 15 Dec 2015

Keirstead H (2012) Translation of stem cell therapies. Spinal Cord Injury—what are the barriers to cure? Bedford Center Spinal Cord Workshop. http://www.spinalcordworkshop.org/portfolio-type/hans-keirstead-phd- translation-of-stem-cell-based-therapies/. Retrieved 15 Dec 2015

Kiatpongsan S, Sipp D (2008) Offshore stem cell treatments. Nat Rep Stem Cells 10:1038

Knoepfler P (2012) Yamanaka on making iPS cells from each patient: 'in reality, we cannot do that. http://www.ipscell.com/2012/11/yamanaka-on-making-ips-cells-from-each-patient-in-reality-we-cannot-do-that/. Retrieved 15 Dec 2015

Knoepfler PS (2015) From bench to FDA to bedside: US regulatory trends for new stem cell therapies. Adv Drug Deliv Rev 82:192–196

Li MD, Atkins H, Bubela T (2014) The global landscape of stem cell clinical trials. Regen Med 9(1):27–39

Lindvall O, Hyun IS (2009) Medical innovation versus stem cell tourism. Science 324:1664–1665

Lysaght T, Sipp D (2014) Dislodging the direct to consumer marketing of stem cell- based interventions from medical tourism. In: Parry B et al (eds) Bodies across borders: the global circulation of body parts, medical tourists and professionals. Ashgate Press, London

Martell K, Trounson A, Baum E (2010) Stem cell therapies in clinical trials: workshop on best practices and the need for harmonization. Cell Stem Cell 7(4):451–454

McMahon D (2014) The global industry for unproven stem cell interventions and stem cell tourism. Tissue Eng Regen Med 11(1):1–9

Minisman G, Bhanushali M, Conwit R, Wolfe GI, Aban I, Kaminski HJ, Cutter G (2012) Implementing clinical trials on an international platform: challenges and perspectives. J Neurol Sci 313(1):1–6

Morgan GH (2014) H.R.4475—Compassionate Freedom of Choice Act of 2014. CONGRESS. GOV. https://www.congress.gov/bill/113th-congress/house-bill/4475. Retrieved 15 Dec 2015

Nuffield Council on Bioethics (2013) Novel neurotechnologies: intervening in the brain. http://nuffieldbioethics.org/wp-content/uploads/2013/06/Novel_neurotechnologies_report_PDF_web_0.pdf. Retrieved 15 Dec 2015

OECD (2011). Facilitating international cooperation in non-commercial clinical trials. Global Science Forum. http://www.oecd.org/science/scienceandtechnologypolicy/49344626.pdf. Retrieved 15 Dec 2015

Ravinetto RM, Talisuna A, Crop M, Loen H, Menten J, Overmeir C (2013) Challenges of non-commercial multicentre North–South collaborative clinical trials. Tropical Med Int Health 18(2):237–241

Rosemann A (2013) Medical innovation and national experimental pluralism: insights from clinical stem cell research and applications in China. BioSocieties 8(1):58–74

Rosemann A (2014a) Why regenerative stem cell medicine progresses slower than expected. J Cell Biochem 115(12):2073–2076

Rosemann A (2014b) Standardization as situation-specific achievement: regulatory diversity and the production of value in intercontinental collaborations in stem cell medicine. Soc Sci Med 122:72–80

Rosemann A (2015) Stem cell treatments for neurodegenerative diseases: challenges from a science, business and healthcare perspective. Neurodegener Dis Manag 5(2):85–87

Rosemann A, Bortz G, Vasen F (2016) Global regulatory developments for clinical stem cell research: diversification and challenges to collaborations. Regen Med 11(7):647–657

Rosemann A, Sleeboom-Faulkner M (2016) New regulation for clinical stem cell research in China—expected impact and challenges for implementation. Regen Med 11(1):5–9

Salter B (2009) State strategies and the geopolitics of the global knowledge economy: China, India and the case of regenerative medicine. Geopolitics 14(1):47–78

Sariola S, Simpson B (2013) Precarious ethics: toxicology research among self-poisoning hospital admissions in Sri Lanka. BioSocieties 8(1):41–57

Sipp D (2014) The domestication of stem cell tourism. In: Joly Y, Knoppers BM (eds) Routledge handbook of medical law and ethics. Routledge, London and New York

Sleeboom-Faulkner M (2013) Latent science collaboration: Strategies of bioethical capacity building in Mainland China's stem cell world. Biosocieties 8(1):7–24

Sleeboom-Faulkner M et al (2016) Comparing national home-keeping and the regulation of translational stem cell applications: an international perspective. Soc Sci Med 153:40–249

Sui S, Sleeboom-Faulkner M (2015) Governance of stem cell research and its clinical translation in China: an example of profit-oriented bionetworking. East Asian Sci Technol Soc 9(4):397–412

Tabar V, Studer L (2014) Pluripotent stem cells in regenerative medicine: challenges and recent progress. Nat Rev Genet 15:82–92

Tiwari SS, Raman S (2014) Governing stem cell therapy in India: regulatory vacuum or jurisdictional ambiguity? New Genet Soc 33(4):413–433

Tuch B, Wall D (2014) Self-regulation of autologous cell therapies. Med J Austin 200:196

Viswanathan S, Rao M, Keating A, Srivastava A (2013) Overcoming challenges to initiating cell therapy clinical trials in rapidly developing countries: India as a model. Stem Cells Transl Med 2:607–613

Wahlberg A, Rehmann-Sutter C, Sleeboom-Faulkner M, Lu G, Döring O, Cong Y et al (2013) From global bioethics to ethical governance of biomedical research collaborations. Soc Sci Med 98:293–300

Wirth E (2010) Preclinical and phase 1 development of oligodendrocyte progenitor cells derived from human embryonic stem cells for the treatment of spinal cord injury. In Conference presentation, the international conference of stem cells and regenerative medicine for neurodegenerative diseases, Hualien, Taiwan, 24 Apr 2012

Wohn DY (2012) Korea okays stem cell therapies despite limited peer-reviewed data. Nat Med 18(3):329

Zarzeczny A, Caulfield T, Ogbogu U, Bell P, Crooks VA, Kamenova K, Master Z (2014) Professional regulation: a potentially valuable tool in responding to "stem cell tourism". Stem Cell Rep 3(3):379–384

Index

© Springer International Publishing AG 2017
P.V. Pham, A. Rosemann (eds.), *Safety, Ethics and Regulations*, Stem Cells in
Clinical Applications, DOI 10.1007/978-3-319-59165-0

CPSIA information can be obtained
at www.ICGtesting.com
Printed in the USA
LVOW05*1702300817

546982LV00001B/30/P